FV

Illustrated Guidebook
to Electronic
Devices and Circuits

Illustrated Guidebook to Electronic Devices and Circuits

FREDRICK W. HUGHES
Lindsey Hopkins Technical Education Center

Prentice-Hall, Inc., Englewood Cliffs, N.J. 07632

Library of Congress Cataloging in Publication Data

Hughes, Fredrick W., 1936–
    Illustrated guidebook to electronic devices and
circuits.

    1. Electronics—Handbooks, manuals, etc.   I. Title.
TK7825.H79   621.381   81-23544
ISBN 0-13-451328-2   AACR2

Editorial/Production Supervision by Barbara Bernstein
Manufacturing Buyer: Gordon Osbourne
Cover Design: Mario Piazza

Printed in the United States of America

10 9 8 7 6 5 4 3 2

ISBN   0-13-451328-2

Prentice-Hall International, Inc., *London*
Prentice-Hall of Australia Pty. Limited, *Sydney*
Prentice-Hall Canada, Inc., *Toronto*
Prentice-Hall of India Private Limited, *New Delhi*
Prentice-Hall of Japan, Inc., *Tokyo*
Prentice-Hall of Southeast Asia Pte. Ltd., *Singapore*
Whitehall Books Limited, *Wellington, New Zealand*

*This book is dedicated to my Father, who during
trying times always looked to hope and
kept a sense of humor.*

Contents

ELECTRONIC COMPONENTS AND BASIC CONCEPTS   1

SOLID-STATE DEVICES   60

OPTOELECTRONIC DEVICES   101

ELECTRONIC VACUUM TUBES   124

POWER SUPPLIES   138

AMPLIFIERS   **154**

OSCILLATORS AND MULTIVIBRATORS   **176**

AUDIO CIRCUITS   **195**

BASIC RADIO SYSTEMS AND CIRCUITS   **219**

10 BASIC TELEVISION SYSTEMS AND CIRCUITS   251

11 INTEGRATED CIRCUITS   275

12 OPERATIONAL AMPLIFIERS (OP AMPS)   307

DIGITAL CIRCUITS 336

MICROCOMPUTERS 399

Preface

This book belongs in the technical library of anyone involved in the field of electronics. It provides the basic theory, components, devices, circuits, and systems of electronics. Not only does this book provide quick and accessible information, but it contains many practical suggestions together with basic testing and troubleshooting procedures. Emphasis is placed on how a particular device or circuit "works." Many of the circuits show component values that, when constructed, produce efficient results for various applications. The purpose of this book is to provide as many important illustrations as needed to help explain the function of a device or circuit while maintaining the amount of reading to a minimum.

This book is a must for the electronics student, instructor, technician, engineer, and experimenter/hobbyist. It is not intended to replace any existing textbooks, but to serve as a supplemental reference book which will aid the instructor in teaching and the student in learning.

For the student, this book gives a wealth of important information that may not be included in a regular electronics course. It allows the student to develop a fuller understanding of the various applications of components, devices, and basic circuits. In many cases, the straightforward approach will enable the student to "break through the learning barrier." Chapter 1 shows how to read the multimeter (voltmeter, current meter, and ohmmeter), the oscilliscope, and how to use a basic signal generator. Each chapter contains a component testing or circuit troubleshooting section which enables the beginner to build quickly the technical skills required in the electronics industry. The student will find the book an excellent source for reviewing for tests and final exams, as well as those employment entrance exams that most companies give before hiring a person.

For the electronics instructor, this book provides an extensive source of illustrations to be used in lectures and from which overhead transparencies can be made. Because time does not permit covering many of the areas in electronics, the instructor can refer the student to specific sections of the book for familiarization of hardware, tools, fundamental relationships, general testing procedures (particularly Section 1-23, Procedure for Testing a Discrete Circuit), and related subjects. Very often, students forget even basic principles learned in previous courses, and specific references to the book can be made which will facilitate and/or enhance their comprehension of a current device or subject during lecture periods. The key point is that the book is easy to locate devices, circuits, and subjects which will encourage the student to perform research and establish learning behavior for future success in courses and when working in industry.

For the technician, this book can be used to update knowledge and skills in the newer areas of electronics, such as field-effect transistors, opto-electronics, integrated circuits, op amps, and digital circuits. A general knowledge can be gained of systems such as radio, television, and micro-computers. Easy reference can be made to devices and circuits that have been forgotten or never encountered. This book should accompany the technician on the job as an aid to understanding the troubleshooting devices and circuits.

For the engineer, this book can provide a general reference to various aspects of electronic devices and circuits. Since many positions in electronic engineering are highly specialized, a general knowledge can be gained from sections such as optoelectronics, integrated circuits, op amps, radio, televi-sion, digital circuits, and microcomputers. Very often, the professional person becomes too specialized and desires general knowledge of other areas in his or her chosen profession.

For the experimenter/hobbyist, this book is a "gold mine" of devices and circuits. Emphasis is placed on "how things work" and the formulas given are for practical applications. By using the various sections as reference, nearly any type of circuit or system can be developed. For example, digital circuits using low voltages ($\approx$+5 V) can be interfaced with higher-voltage ($\approx$12 V, 15 V, or greater) analog circuits employing opto-electronic devices. Perhaps it is desired to construct a circuit from a magazine article, but more information on a particular device is needed to understand better how the circuit operates. Simply turn to the section in this book that illustrates the device to gain the needed information.

Although this book describes many devices and circuits, and presents a general knowledge of electronics, it can by no means provide a complete picture of the vastness of the electronic world. Nevertheless, I hope that it will serve as encouragement for continued research and study within our chosen profession.

Fredrick W. Hughes

Acknowledgments

The following figures are being reproduced with the permission of Prentice-Hall, Inc. (Figure numbers refer to the *Illustrated Guidebook to Electronic Devices and Circuits*.)

Angerbauer, George J. *Electronics for Modern Communications*, 1974.
   Figures 1-32, 2-12, 2-41, 2-42, 2-44, 2-45, 2-46, 4-2, 4-6, 9-1, 9-5, 9-16, 9-18, 9-21, 9-23, 9-26, 9-27, 9-28, 10-7, 10-14, 10-18.

Boyce, Jefferson C. *Microprocessor and Microcomputer Basics*, 1976.
   Figures 14-12, 14-14, 14-15, 14-19.

Coughlin, Robert F., and Frederick F. Dirscoll. *Operational Amplifiers and Linear Integrated Circuits*, 1977.
   Figures 11-19 through 11-25.

Coughlin, Robert F., and Frederick F. Dirscoll. *Semiconductor Fundamentals*, 1976.
   Figures 2-19, 2-20, 2-26, 2-27, 2-30, 2-47, 2-51, 5-4, 6-17.

Gothmann, William H. *Electronics: A Contemporary Approach*, 1980.
   Figures 1-24, 4-3, 4-8, 4-10, 4-12.

Green Clarence R., and Robert M. Bourque. *The Theory and Servicing of AM, FM, and FM Stereo Receivers*, 1980.
   Figures 2-14, 2-52, 4-5, 4-14, 4-16, 4-20, 5-2, 5-3, 5-13, 6-18, 6-19, 6-23, 7-2, 8-12, 8-15, 8-16, 8-17, 8-27, 9-2, 9-3, 9-4, 9-15, 9-17, 9-20, 9-29, 9-36 through 9-40.

Heiserman, David L. *Handbook of Digital IC Applications*, 1980.
   Figures 13-10 and 13-19.

Hughes, Fredrick W. *Op Amp Handbook*, 1981.
   Figures 5-18, 5-19, 8-18, 8-19, 8-20, 12-1 through 12-33.

Hughes, Fredrick W. *Practical Guide to Digital Electronic Circuits*, 1977.
   Figures 2-64, 2-65, 2-66, 3-17, 3-18, 3-21 through 3-24, 3-30, 11-6, 11-36, 11-37,

13-4 through 13-19, 13-21 through 13-29, 13-31 through 13-46, 13-48, 13-50, 13-52, 13-53, 13-55 through 13-82, 13-85, 13-86, 13-88, 13-89, 13-90.

Hughes, Fredrick W. *Workbench Guide to Practical Solid State Electronics*, 1979. Figures 1-33, 1-34, 2-21 through 2-25, 2-28, 2-29, 2-31, 2-32, 2-38, 2-39, 2-40, 2-49, 2-50, 2-54 through 2-60, 2-67 through 2-75, 3-2, 3-4 through 3-8, 3-10 through 3-13, 3-15, 3-16, 3-25, 3-26, 3-28, 3-29, 6-2, 6-8, 6-9, 6-10, 6-14, 6-15, 6-16, 6-21, 6-22, 11-17, 11-26, 11-27, 11-28.

Larson, Boyd. *Transistor Fundamentals and Servicing*, 1974. Figures 2-3, 2-4, 13-3.

Mandl, Matthew. *Fundamentals of Electronics*, 3rd ed., 1973. Figures 1-21, 2-2, 4-1, 4-7, 8-1, 8-2, 8-5, 8-6, 8-14, 8-24, 8-26, 9-6 through 9-9, 9-22, 9-30, 9-31, 9-34, 9-35.

Mandl, Matthew. *Modern Television Systems Theory and Servicing*, 1974. Figures 10-8, 10-10, 10-11, 10-15, 10-22.

Marcus, Abraham. *Electronics for Technicians*, 1969. Figures 4-4, 4-13, 5-5, 5-8 through 5-12, 5-14, 6-6, 6-7, 7-11 through 7-14, 7-16 through 7-19, 8-7, 9-12, 9-13, 9-14, 9-25.

Miller, Gary M. *Linear Circuits for Eelctronics Technology*, 1974. Figures 2-6, 2-8, 2-10, 2-11, 2-48, 6-20, 7-3, 7-4, 7-5, 7-7, 7-8, 7-9, 8-29.

Miller, Gary M. *Modern Electronic Communication*, 1978. Figure 2-13, 9-11, 9-24, 9-32, 9-33, 10-1, 10-3 through 10-6, 10-9, 10-12, 10-13, 10-16, 10-17, 10-19, 10-21, 11-18, 11-29.

Nunz, Gregory J. *Electronics in Our World: A Survey*, 1972. Figures 1-15, 1-16, 1-17, 1-19, 1-28, 1-29, 1-30, 2-5, 2-43, 4-9, 4-11, 4-15, 4-17, 4-19, 8-3, 8-4, 8-21, 8-22, 8-23, 10-2, 10-20, 11-7.

Ryder, J. D., and Charles M. Thomson. *Electronic Circuits and Systems*, 1976. Figures 5-6, 5-7, 6-11, 6-12, 6-13, 7-10, 8-13, 8-28, 9-19.

Tocci, Ronald J. *Digital Systems: Principles and Applications*, 1977. Figures 13-84 and 13-85.

1

Electronic Components and Basic Concepts

1-1 WIRE: TYPES, SIZES, AND DESCRIPTION

Most wire used in electronic circuits is made of soft copper and solder-tinned copper wire. *Solid wire* is easy to handle, but cannot withstand much flexing. Bare solid wire is used for buses, a common connection line for components and other wires. Solid wire with varnish or paint insulation is used for windings of coils, transformers, generators, and motors. Insulated (rubber or plastic) solid wire has advantages in some circuits. Small [American Wire Gauge (AWG) No. 30] insulated solid wire is best to use for integrated circuits (ICs) and connections that are very close.

Flexible wire consists of several strands of twisted wire covered by insulation (rubber, paper, plastic, Teflon, etc.) and is referred to as *stranded wire*. Stranded wire is used extensively in electronic circuits. Two stranded wires may be joined together to form a two-conductor wire, such as that used in electrical appliances.

Shielded wire consists of a stranded inner conductor covered with insulation and then a braid of woven segments of stranded wire. This shield is used as a conductor and also to block electrical interference from the inner conductor. *Coaxial cable* is similar to shielded wire, except that an insulated covering is used around the shield.

Several wires can be put into a larger plastic tube interconnecting cable. Some of these cables may contain shielded or coaxial cable and other special arrangements. The newer type of flat "ribbon" cable has stranded wire sizes of AWG Nos. 22 to 30, which are connected side by side, with up to 64 conductors in a single cable.

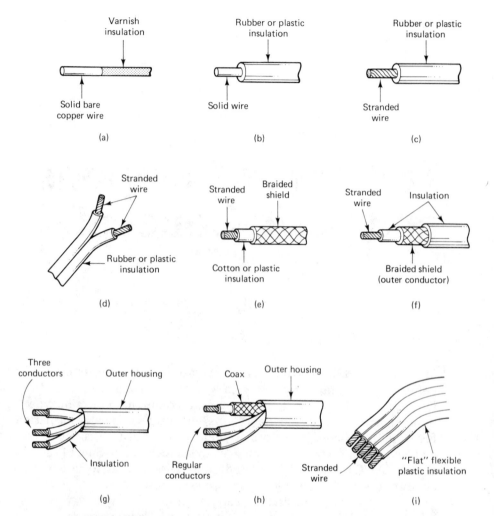

Figure 1-1 Types of wire: (a) solid strand with/without insulation; (b) insulated solid strand; (c) insulated stranded wire; (d) two-conductor wire; (3) shielded wire; (f) coaxial cable; (g) interconnecting (three-wire) cable; (h) special-purpose interconnecting cable; (i) "flat" ribbon cable.

The amount of electrical current that a wire can safely carry depends on its cross-sectional area. The larger the area, the more current it is capable of handling. All conductors exhibit some *resistance*, which can be found by the formula

$$R = \rho \, \frac{l}{d^2}$$

where R is the resistance of the wire in ohms, ρ (rho) the resistance in ohms per cir mil-ft of wire material; l the length of the wire in feet, and d the diameter of the wire in mils. The resistance of the wire increases directly proportional to its length, but varies inversely as the square of its diameter. If the diameter of a wire is doubled, its resistance will decrease by four times. It is important to use the proper wire size to prevent excessive heat, which can melt the insulation of the wire and possibly cause a fire. With the standard American Wire Gauge, the smaller the gauge number, the larger the wire. Table 1-1 compares various gauge wires for diameter, resistance per 1000 ft, and current capabilities.

TABLE 1-1 COMPARISON OF A FEW WIRE GAUGES USING COPPER WIRE AT 20°C

Gauge	Diameter (mils)	Resistance (Ω) per 1000 ft	Ampacity (A)
0000	460.2	0.04901	225
0	324.9	0.09827	125
10	101.9	0.09989	25
12	80.81	1.588	20
14	64.08	2.525	15
18	40.30	6.385	3
22	25.35	16.14	Less than 1
26	15.94	40.81	Less than 1
28	12.64	64.90	Less than 1
30	10.03	103.3	Less than 1

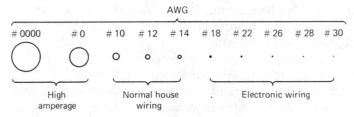

Figure 1-2 Approximate cross-sectional area comparison of several wire sizes.

1-2 PLUGS, JACKS, AND CABLE CONNECTORS

Plugs, jacks, and cable connectors facilitate wire connections between various electronic circuits and equipment. *Plugs* are usually attached to the wire, while their receptacle or *jack* is rigidly mounted to a chassis or some equipment. *Phone plugs* are two or more wire connectors that are used in tele-

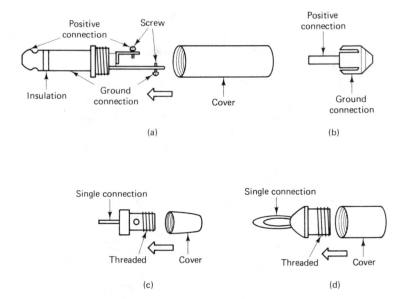

Figure 1-3 Plugs: (a) phone; (b) phono; (c) pin; (d) banana.

phone central offices, radio-television broadcasting, and professional enter-tainment (musical instrument amplifiers and microphones). The smaller *phono plug* is a two-wire connector used in audio systems (i.e., from turn-table to amplifier, from tape deck to amplifier, etc.). Two-wire conductors and shielded and coaxial cables are used with phone and phono plugs. A *pin plug* is a single-wire connector used where circuit changes are sometimes required. The *banana plug* has the same function as the pin plug except that it is larger and the spring connectors on the tip can be adjusted to form a better electrical connection.

The *phone jack* is a connector of two or more wires that matches the phone plug. It may also have contacts that open or close circuits when the plug is inserted. A *phono jack* is a two-wire connector that accepts the phono plug. The pin and banana jacks are single-wire connectors that match their respective plug types and usually are mounted through a panel or chassis.

Coaxial cable connectors are used on high-frequency equipment and test equipment. There are both *screw-type* and *bayonet-type* connectors, which provide good electrical connections. *Pin-contact connectors* require a *male* half and a *female* half. A keyway and key is used on these connectors to properly align the pins and socket holes. *Rack-and-panel connectors* are used with subassemblies or subchassis that fit into a larger mounting rack. This connector serves as a quick-disconnect for rapid test or replacement of subassemblies.

Printed-circuit-board edge connectors provide wire cable or harnessing

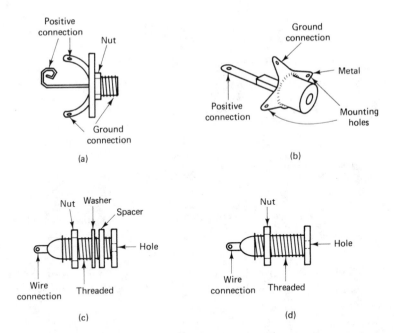

Figure 1-4 Jacks: (a) phone; (b) phono; (c) pin; (d) banana.

Figure 1-5 Cable connectors: (a) coaxial connector (screw type); (b) coaxial connector (bayonet type); (c) pin-contact-type connector; (d) rack-and-panel connector.

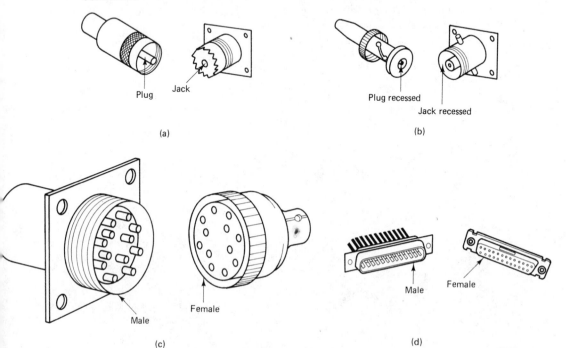

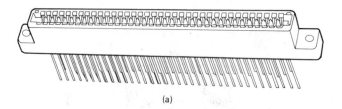

(a)

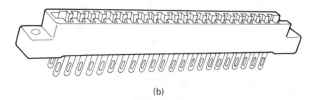

(b)

Figure 1-6 Printed-circuit-board
edge connectors: (a) wire-wrap
type; (b) solder-lug type.

between printed circuit boards in a system. They are of two types, wire-wrap
and solder-lug, and also provide quick-disconnect features.

1-3 TERMINALS AND OTHER HARDWARE

Terminal strips are used as tie points for wire connections. The terminals are
mounted on insulating material, such as Bakelite, phenolic, or porcelain.
A *solder terminal strip* may use one terminal for chassis or ground connections. The *barrier terminal strip* uses screw connections and the wires have
solderless terminals.

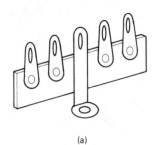

(a)

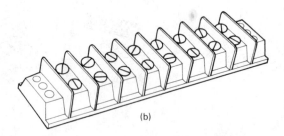

(b)

Figure 1-7 Terminal strips: (a)
solder terminal; (b) barrier (screw-
type) terminal.

Some circuits use *transistor and IC sockets*, which facilitate testing and replacement.   Transistor sockets have three or four terminals, whereas IC sockets may have 8, 14, 16, 18, 20, 22, 24, 28, 36, or 40 terminals.

Heat sinks are used with power transistors to draw the heat away from the transistor, which improves circuit operation and stability, and increases the life span of the transistor.

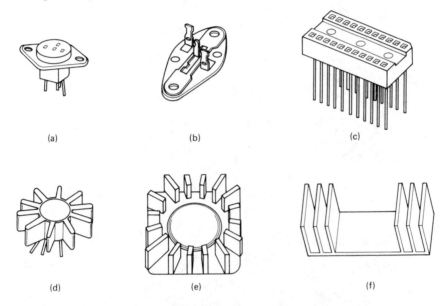

(a)   (b)   (c)

(d)   (e)   (f)

Figure 1-8 Sockets and heat sinks: (a) small-signal transistor socket; (b) power transistor socket; (c) IC socket; (d) heat sink pushes over transistor body; (e) transistor bolts to heat sink; (f) larger fins radiate more heat.

A *printed circuit* (PC) *board* connects discrete components together by means of thin ribbon conductors, which are permanantly bonded to a board made of laminated plastic.   The boards are mass-produced by means of a photographic and etching process.   *Perforated (perfboard) or vector boards* are made of phenolic or epoxy glass and contain numerous holes.   The boards come with various hole sizes and "on center" dimensions.   Circuits are built on these boards and they are often used for engineering prototype and initial testing.

Miscellaneous hardware is usually associated with fasteners and devices for constructing rigid and dependable circuits.   *Metal screws and nuts* are used for bolting.   *Sheet-metal screws* are used to self-tap for securing chassis parts.   *Serrated lock washers* are used to "dig into" the chassis to lock a screw and sometimes to provide a good electrical connection.   *Split-lock washers* are used to lock a screw where a good electrical connection is not essential.   *Flat washers* are used for spacing and also not to cause marring

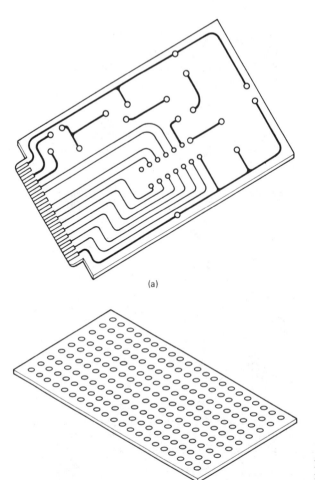

Figure 1-9 Circuit boards: (a) printed circuit board; (b) perforated board ("perfboard") or vector board.

of a chassis or panel. *Fiber washers* are used for insulating purposes. *Solder lugs with a serrated washer feature* are used to provide a chassis or "ground" connection for wires. A *cable clamp* fits around a cable or harness to secure it to the chassis. *Rubber grommets* are used in holes in the chassis where wires must pass through. This prevents the insulation of the wires from becoming cut and causing "shorts" or circuit problems. *Solderless terminals* are used on the ends of wire that must use screw-type terminals. *Alligator clips* are useful as clip leads for testing, troubleshooting, and experimenting with circuits. They may also be used as lead heat sinks, when soldering solid-state components. *Component mounting terminals for PC and vector boards* are used to provide circuit ruggedness and dependability. *Control knobs* of various sizes are needed for switches, potentiometers, and tuning capacitors associated with circuit operation. *Spacers or stand-offs* are used

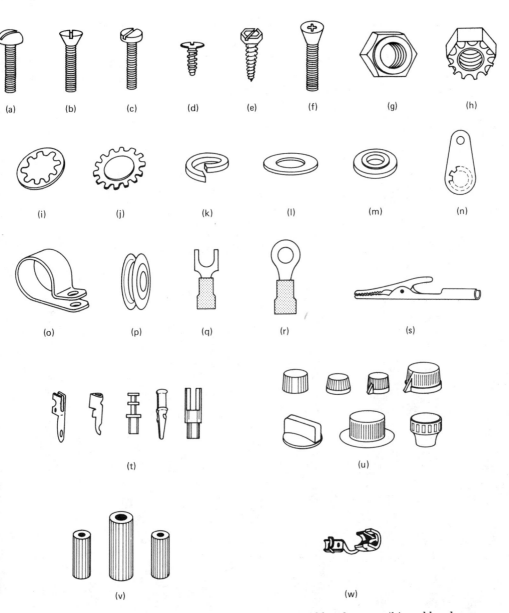

Figure 1-10 Miscellaneous hardware: (a) round-head screw; (b) oval-head screw; (c) binder-head screw; (d) sheet-metal screw; (e) sheet-metal hex-head screw; (f) Phillips-head screw; (g) hex nut; (h) hex nut with lock washer; (i) internal-tooth lock washer; (j) external-tooth lock washer; (k) split-ring lock washer; (l) flat washer; (m) fiber washer with shoulder; (n) solder lug; (o) cable clamp; (p) rubber grommet; (q) solderless terminal—spade tongue; (r) solderless terminal-ring tongue; (s) alligator clip; (t) PC and/or "vector" board component mounting terminals; (u) control knobs; (v) spacers or stand-offs; (w) strain relief.

9

for special component applications. They may be made of metal, phenolic, or ceramic. A *strain relief* has a similar purpose as a rubber grommet and also provides a dependable connection for the line cord to a piece of equipment. It fits into the hole of a chassis, while applying pressure to the line cord that is passing through the hole.

1-4 BASIC HAND TOOLS

Basic electronic hand tools are a necessity for proper and efficient maintenance and construction of electronic circuits. A tool kit should include several sizes of *standard blade-type* and *Phillips-type screwdrivers. Nutdrivers (hex*

Figure 1-11 Basic hand tools: (a) blade screwdriver; (b) Phillips screwdriver; (c) nut driver; (d) wire wrapping/unwrapping tool; (e) long-nose pliers; (f) diagonal cutters; (g) wire strippers; (h) crimper (solderless terminals).

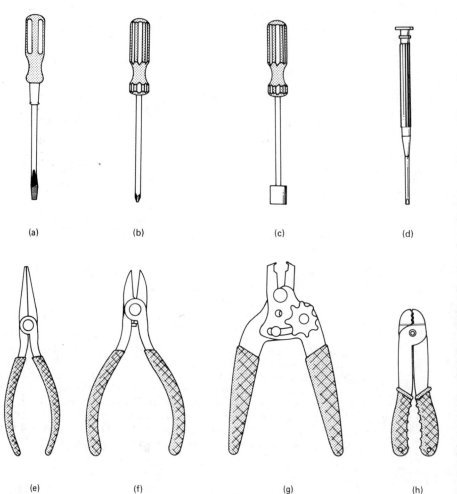

(a)   (b)   (c)   (d)

(e)   (f)   (g)   (h)

type) are similar to screwdrivers and "speed up" assembly and disassembly operations.   Many wire connectors in use require a *wire wrapping and unwrapping tool*.   *Long-nose pliers* are used to form proper bends in component leads, to hold components while soldering, and to reach hard-to-get-to places in a circuit.   *Diagonal cutters* are used for cutting component leads and wires.   They can be used for stripping insulation from wire once expertise is developed so that wires are not nicked or cut.   *Wire strippers* are, of course, best to use in removing insulation from wire, since the exact diameter can be set.   A *crimper* tool is used to attach solderless terminals to the ends of wire.

Proper soldering tools are also essential for working on electronic circuits.   A *soldering iron* (less than 50 W) is plugged in and left on when many connections must be soldered at one time.   A *soldering iron holder* is very

Figure 1-12 Soldering tools; (a) soldering iron; (b) iron holder and cleaning sponge; (c) soldering aids; (d) soldering gun; (e) solder flux and rosin core solder; (f) desoldering tool.

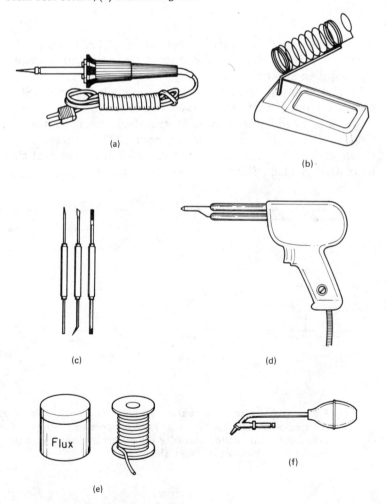

(a)

(b)

(c)

(d)

Flux

(f)

(e)

useful for the protection of materials and the safety of the user. A moistened sponge is used to keep the tip of the iron clean, which results in fast and good solder joints. *Soldering aids* are used to remove "solder shorts" between connections and to clean up solder joints. A *soldering gun* is helpful for making repairs and requires very little warm-up time. *Solder flux* is placed on a solder connection before soldering so that contaminants are burned away during the soldering process, which also speeds up the time needed to solder and makes a better solder joint. A *desoldering tool* is used to remove solder from a solder connection or connections to facilitate the removal of a component. This is particularly important for removing ICs. The connection is first heated and the rubber vacuum bulb is squeezed as the nozzle of the tool is placed on the connection. The bulb is then released and the solder is drawn (or sucked) up into the tool for easy removal. With less solder at the connection, the component is easier to remove from the circuit.

1-5 SWITCHES

Switches are used to open or close an electrical circuit. Their function is to enable or disable a particular circuit operation or transfer operation from one point to another point.

The simplest switch is a *single-pole single-throw* (SPST), which closes a circuit from a common contact (C) to a normally open (NO) contact. A *single-pole double-throw switch* transfers the C contact from a normally closed (NC) contact to a normally open (NO) contact. A *double-pole single-throw switch* operates similarly to a SPST switch except that the two C contacts are "ganged" (tied) together and operate simultaneously. The *double-*

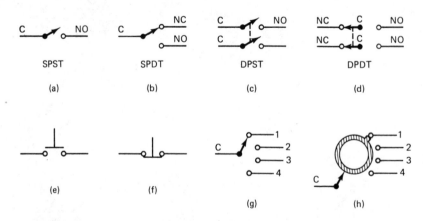

Figure 1-13 Switch schematic symbols: (a) single-pole single-throw; (b) single-pole double-throw; (c) double-pole single-throw; (d) double-pole double-throw; (e) normally open pushbutton; (f) normally closed pushbutton; (g) multiposition; (h) rotary.

pole double-throw switch has the C contacts "ganged" together and operates two NC and NO contacts simultaneously.

A *normally open pushbutton switch* closes a circuit when the button is depressed, whereas a *normally closed switch* opens the circuit. The switch may keep the circuit latched in the desired condition when the button is released, or it may be spring loaded and return the contacts to the original condition. A pushbutton switch may operate several NO or NC contacts or

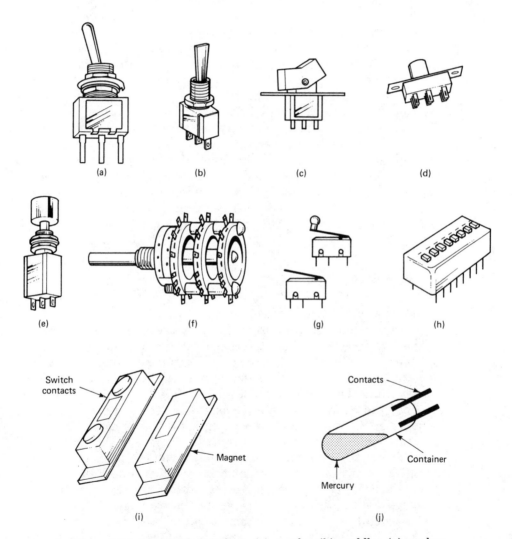

Figure 1-14 Types of switches: (a) toggle; (b) paddle; (c) rocker; (d) slide; (e) pushbutton; (f) multiwafer rotary; (g) subminiature (micro); (h) PC board eight-position; (i) proximity (magnetic); (j) mercury.

a combination of both at a time. In a multiposition switch, the C contact can be set to many position contacts and is referred to as a *rotary switch*.

There are many types and sizes of switches. The distinguishing feature among switches is the type of C contact actuator used. A *toggle-type switch* has a small round lever that moves back and forth. The *paddle-type switch* is like the toggle except that the lever is wider. The *rocker-type switch* rocks back and forth. The *slide-type switch* moves laterally from one side to the other. The *pushbutton-type switch* moves down and up (in or out). A *rotary-type switch* moves in a circular motion. *Subminiature (or micro) switches* are tiny switches that operate with a small plunger or actuator arm, which may have a small roller. *Switch units used on PC boards* are about the size of an IC and fit into a dual-in-line package (DIP) mounting arrangement. They usually have eight separate slide-type switches. Proximity switches have thin metal reed contacts which operate when an external magnet is brought close to a contact. This type of switch is often used in intrusion (burglar) alarm systems. A mercury switch is a small container (often, glass) with a portion of liquid mercury sealed inside. The contacts at one end protrude through the glass for external connections. When the container is tilted, the mercury flows around the contacts, which completes the electrical circuit. This type of switch is used to close or open circuits resulting from physical movement of mechanical parts, such as an interlock switch on a lid or cover leading to a dangerous piece of equipment.

1-6 FUSES AND CIRCUIT BREAKERS

A *fuse* is a safety device placed in series with the power source and a circuit or piece of equipment. Fuses are rated by the amount of current they allow to pass, which can be from fractional parts of an ampere to over hundreds of amperes. A fuse is selected to pass sufficient current for proper circuit operation. When a problem develops in a circuit, so that it draws more current than the rating of the fuse, the fusible link in the fuse melts and "opens" the circuit from the power source, thereby protecting other components in the circuit and preventing a fire. Fast-acting fuses are used in critical circuits where a slight increase in current might damage something. Slow-blow fuses are used in circuits where voltage surges will not necessarily harm the circuit and the operation can continue without the fuse "blowing." The cartridge-type and house-type (screw-in) are two basic types of fuses. Associated with fuses are various types of fuse holders.

A *circuit breaker* has the same function as a fuse, but can be reset after the circuit problem has been cleared. Some circuit breakers operate on the thermal principle, using a bimetallic strip. The bimetallic strip consists of two dissimilar metals joined together. When the strip is heated, the metals

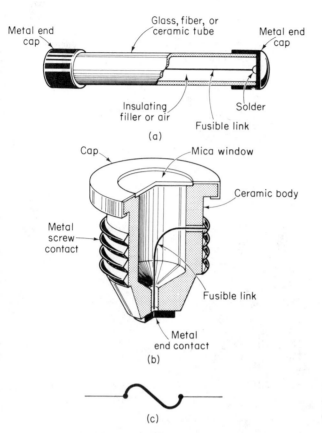

Metal end cap

Glass, fiber, or ceramic tube

Metal end cap

Insulating filler or air

Solder

Fusible link

(a)

Cap

Mica window

Ceramic body

Metal screw contact

Fusible link

Metal end contact

(b)

(c)

Figure 1-15 Fuses: (a) cartridge type; (b) house type (screw-in); (c) electrical symbol.

expand at different rates, causing bending, which opens a set of normally closed contacts, thereby "breaking" the current path to the faulty circuit. Once the bimetallic strip has cooled down and returned to its normal position, an operating lever can again close the contacts. Some circuit breakers have an automatic reset feature, which returns the circuit to operation, such as a thermostat.

1-7 RELAYS

A *relay* is a control device that allows a small current in one circuit to control a large current in another circuit, while providing electrical isolation between the two circuits. A coil of wire is wrapped around a soft iron pole piece. A movable steel armature is pivoted so that it will be attracted to the pole piece when there is current in the coil. When there is no current in the coil, a spring holds the armature away from the pole piece. Electrical contacts are mounted on the relay, which the armature operates when the relay is

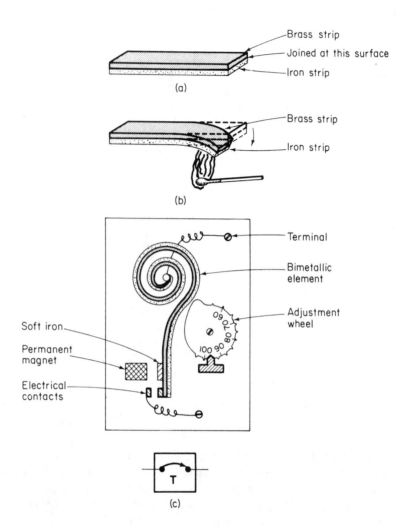

Figure 1-16 Thermostat: (a) bimetallic strip; (b) effect of heat on bimetallic strip; (c) typical wall thermostat and electrical symbol.

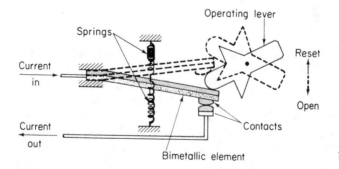

Figure 1-17 Thermal-type circuit breaker.

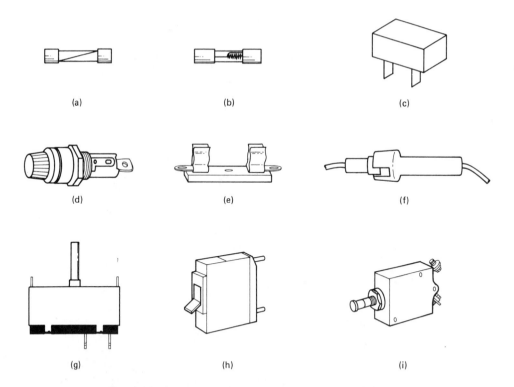

Figure 1-18 Circuit interrupt components; (a) regular "fast-blow" fuse; (b) slow-blow fuse; (c) bel fuse; (d) through-chassis fuse holder; (e) top-mount fuse holder; (f) in-line fuse holder (car radio); (g) television circuit breaker; (h) toggle-switch-type circuit breaker; (i) pushbutton-type circuit breaker.

energized. The common contact is so positioned between other contacts that the motion of the armature will either break contact, make contact, or both. There can be several sets of normally closed (NC) and normally open (NO) contacts on a relay device.

There are many types and styles of relays for various applications. Relays operate with dc or ac voltage. The energizing voltage may be low dc, which controls high-ac voltage through its contacts, or vice versa. A latching relay will remain mechanically "locked up" even after the energizing current has been removed. Another pulse of current is then used to release the latch. There are low-voltage reed relays with operation similar to a reed switch. The thin metal reeds mounted with contacts are fitted inside an air-core coil. Current through the coil induces an electromagnetic action that opens or closes the contacts. Some relays are the size of ICs and have standard DIP configurations for mounting directly on PC boards. Special relays have their own types of sockets and/or special mounting arrangements.

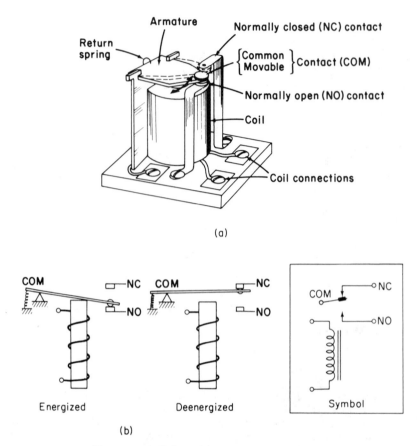

(a)

(b)

Figure 1-19 Relay: (a) construction; (b) operation.

1-8 ELECTRICAL LAMPS AND PILOT LIGHTS

Incandescent miniature and subminiature electrical lamps are most commonly used in numerous applications for panel indicators, flashlights, and pilot lights for consumer electronic products (radios, TV sets, stereosets, etc.). They are used in the fields of computers, medical electronic equipment, aircraft, aerospace, and industrial applications. The major considerations for using these lamps is that the rated voltage of the lamp not be exceeded, and when replacing a specific lamp with a substitute to be aware of a possible high current draw, which could activate the fuse in a piece of equipment. It is wise to know the voltage and current ratings of any lamps being used in circuits.

There are three standard bases for incandescent lamps: screw-type, bayonet-type, and flange-type. Each type of base has to match a corresponding type of socket for proper use.

The *neon* lamp is a cold-cathode type of device, with two electrodes, that contains an inert gas. At a given voltage, the gas ionizes and current flows, causing the electrodes to glow. If used with dc voltage, one electrode glows, whereas both electrodes glow with ac use. This lamp requires a series resistor to limit the current through it.

The *light-emitting diode* (LED) is rugged, requires little power, comes in various colors (red, yellow, green), and is relatively inexpensive. In operation, the LED requires a series current-limiting resistor. Its main disadvantage is that it may not produce enough light for some specific applications (see Section 3-10).

Lamp sockets and indicators are available with different-colored lens and various package styles. Some indicators come complete with lamp, limiting resistor (if needed), and even a switch.

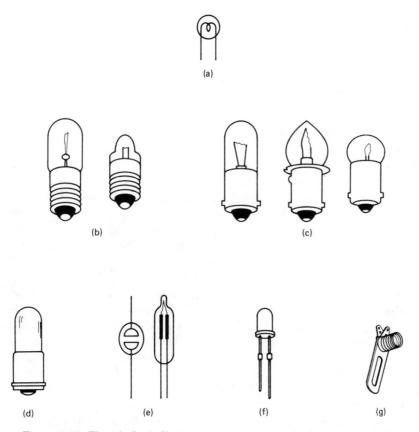

Figure 1-20 Electrical indicator components: (a) lamp electrical symbol; (b) screw-type base lamps; (c) bayonet-type base lamps; (d) flange-base-type lamp; (e) neon lamp; (f) LED; (g) screw-type socket; (h) bayonet-type socket; (i) socket with colored lens; (j) through-panel pilot light indicators.

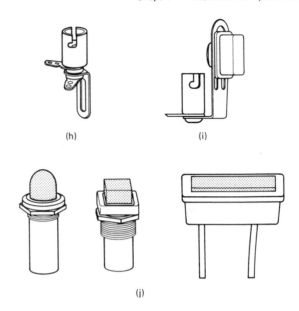

(h)   (i)

(j)

Figure 1-20 Continued.

1-9 PRIMARY CELL BATTERIES

A basic *battery* consists of two dissimilar metal electrodes separated by a chemical composition known as an *electrolyte*. The electrolyte reacts on the metal electrodes to produce a dc *electromotive force* (EMF) or voltage, with one metal being the positive terminal and the other metal being the negative terminal. A cell in use then converts chemical energy into electrical energy. *Primary cell batteries*, often referred to as *dry cell batteries*, are not rechargeable to any great degree. When an external circuit is connected to the battery terminals, a certain amount of the metal is dissolved or used up. Even when batteries are not in use there is still some chemical action taking place, which deteriorates the metals. Batteries left standing too long will lose their original terminal voltage and will not be effective in a circuit. Batteries have an internal resistance (which cannot be measured with an ohmmeter) which affects the current to be supplied to a circuit. As a battery ages, the resistance increases and the resulting internal voltage drop when it is in use will decrease the terminal voltage. This is why a battery voltage measured without a load may indicate that the battery is good, but fails to operate a circuit under load. The best way to check a battery is to measure the terminal voltage when it is being used in a circuit.

A typical flashlight battery using carbon and zinc produces 1.5 V. Increasing the surface area of the metals does not increase the voltage but will increase the amount of current the battery can supply to a circuit.

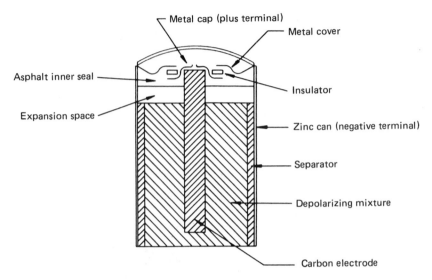

Figure 1-21 Cross-sectional view of battery cell.

Therefore, a group of batteries may produce the same voltage, but the larger ones can supply more current to a circuit.

Primary cells can be placed in series (aiding) to increase the total voltage required by a circuit. A typical 9-V battery contains six 1.5-V cells in series.

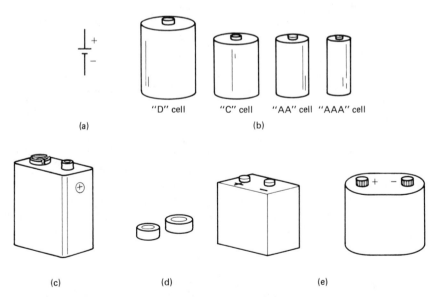

Figure 1-22 Battery types: (a) electrical symbol; (b) 1.5-V flashlight types; (c) 9-V rectangular; (d) 1.35-V miniature types; (e) 6- or 12-V portable lantern types.

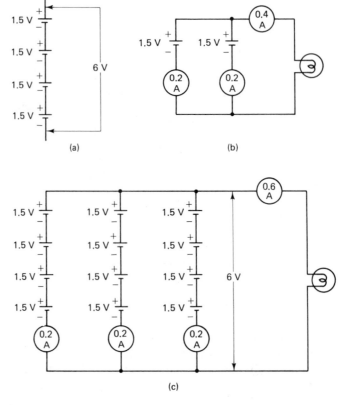

Figure 1-23 Voltage cell connections to increase capabilities: (a) in series to increase voltage; (b) in parallel to increase current; (c) in series-parallel to increase voltage and current.

TABLE 1-2 COMPARISON OF THREE TYPES OF PRIMARY CELL BATTERIES

Battery type	Basic electrode material	Basic electrolyte composition	Terminal voltage (V)	Relative useful life span	Relative cost
Standard	Carbon and ainc	Sal ammoniac with zinc chloride	1.5	Short	Low
Alkaline	Zinc and manganese dioxide	Sodium hydroxide (caustic soda)	1.5	Longer	Medium
Mercury	Mercuric oxide and amalgamated zinc	Potassium hydroxide with potassium zincate	1.4	Longest	Highest

Primary cells can be placed in parallel to increase the current capabilities required by a circuit.   A combination of cells in series and parallel will increase the voltage and current capabilities to a circuit.

Three types of primary cell batteries are compared in Table 1-2.

1-10 SECONDARY CELL BATTERIES

A *secondary cell battery*, also referred to as a *storage battery*, is rechargeable and finds wide application in automative use. The storage battery used in automobiles consists of a negative electrode (or plates) or spongy lead, a positive electrode (or plates) of lead peroxide, and an electrolyte of sulfuric acid diluted with distilled water. A single secondary cell will produce 2.1 V; therefore, six of these cells must be connected in series to produce the standard 12-V battery. The area of the plates determine the current capabilities of the battery. This type of battery is rated in ampere-hours, which indicates how many hours the battery can deliver a given amount of current. For example, a 100 A-h battery could continuously supply 12.5 A for 8 h (12.5 × 8 = 100 A-h), or perhaps 10 A for 10 h (10 × 10 = 100 A-h). However, automotive starters require an initial greater amount of current to turn over the engine, and this *cranking power* must be considered when purchasing a battery.

The condition of this type of battery can be checked with a voltmeter at its terminals when a vehicle is being started. The voltage should not decrease (fall) too far below 10 V. Also, the state of charge can be measured

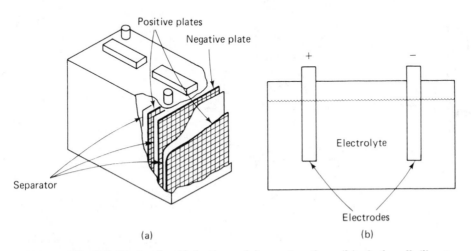

(a)   (b)

Figure 1-24 Lead-acid battery: (a) construction; (b) single-cell illustration.

with a hydrometer, a glass tube containing a calibrated float. Since sulfuric acid is heavier than water, when the battery is fully charged, there is a change in the density of the electrolyte in relation to its discharged condition, which can be measured by the specific gravity (S.G.) indications on the float. Each cell is tested by drawing the electrolyte solution up into the hydrometer with the rubber bulb attached to one end. The float will settle into the solution and the level of the electrolyte on the calibrated scale will give the S.G. indication. A reading of 1280 would indicate a fully charged battery, whereas a reading of 1100 would indicate a fully discharged battery. The battery may be charged by connecting a battery charger to its terminals. Remember to connect the positive terminal of the charger to the positive terminal of the battery (and the same for the negative connections). Making S.G. measurements with the hydrometer during the charging period will determine if the cells are accepting the charge. Newer-type automotive batteries are sealed and are "maintenance free."

There are a couple of other popular rechargeable batteries. The nickel-cadmium (NiCad), with a voltage of 1.25 V per cell, is interchangeable with

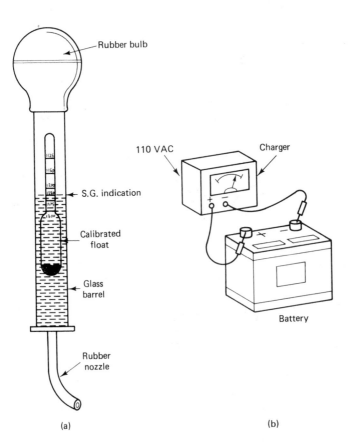

(a)   (b)

Figure 1-25 Lead-acid battery charging: (a) hydrometer; (b) connections for charging battery.

standard batteries used in consumer goods (radios, calculations, flashlights, and other devices). The lithium rechargeable battery has a voltage of 2.1 to 2.4 V per cell and is claimed to have an indefinite life span, will not discharge while stored, and has a higher energy density than other battery types.

1-11 GENERATORS

Generators convert mechanical energy to electrical energy and dc voltage to pulsating dc or ac voltage. The latter type are known as *signal generators*. A basic *ac generator*, also called an *alternator*, consists of a coil of wire which rotates in a permanent magnet field. Slip rings attached to the coil provide current flow to external loads via carbon brushes that "ride" against the slip rings. As the coil rotates, it cuts the magnetic lines of force which induce an electric current in the coil wires. When this current flows through an external load, a voltage is developed that resembles a sinusoidal waveform (sine wave).

A basic *dc generator* is similar to an ac alternator, except that the slip rings are replaced with a split commutator. As the coil rotates in the magnetic field, current flows through the commutator segments and to the load. When 180° is reached, the commutator segments have changed position and

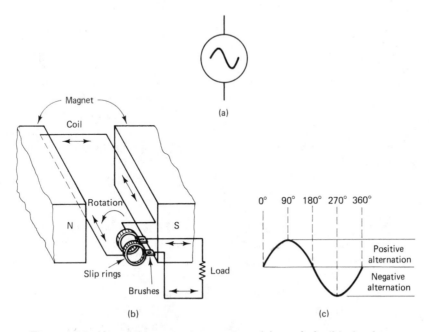

Figure 1-26 Alternating-current generator: (a) symbol; (b) simple ac generator; (c) output voltage waveform.

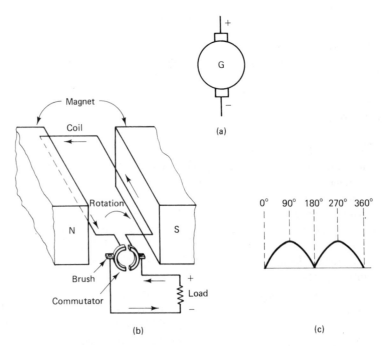

Figure 1-27 Direct-current generator: (a) symbol; (b) simple dc generator; (c) output voltage waveform.

the current continues to flow in the same direction, which produces a pulsating dc voltage waveform across the load.

Practical generators are, of course, much more complicated. There can be several coils at various angles to each other contained in the rotating unit, known as the armature. The output will contain multiphase voltages. Since strong permanent magnets are large, heavy, and bulky, electromagnetic parts are used which receive their current from that which is produced by the generator.

Generators incur losses that decrease their efficiency, some of which are listed below:

Copper losses: The resistance of the copper wire used in the windings limits the current. The induced voltage must overcome this resistance and contribute nothing to the external voltage, producing only heat in the windings.

Eddy currents: Circulating currents set up in the metal core of the windings due to varying magnetic flux, which produces heat and minimizes the effectiveness of the magnetic field. This effect is reduced by laminating the core.

Hysteresis loss: The magnetic resistance of a ferrous metal to a change in magnetic polarity (sometimes called molecular friction). This loss produces heat, which reduces the generator efficiency.

1-12 MOTORS

Motors convert electrical energy to mechanical energy. Essentially, a motor is similar to a generator, except that voltage is applied to the output terminals and the rotating windings, called a *rotor*, revolve. The stationary windings, called a *stator* in a dc motor, set up magnetic fields, which pull the rotor around in one direction or the other, depending on the polarity of the input voltage. The stator windings, also called *field windings*, can be wired various ways with the rotor (or armature) connections to produce certain desired features from a motor. A series-connected motor has a high starting torque under load, but may be destroyed when not connected to a load. A parallel-connected motor produces less torque but maintains a relatively constant speed under varying loads. A compound-connected motor combines the desirable characteristics of the other two types with some trade-off factors between the two: less torque and some speed variation with changing loads.

To a limited extent, a dc motor can operate on ac and is referred to as a *universal motor*. Since both the field and armature windings reverse polarity periodically, the dc motor continues to operate. The series-connected motor

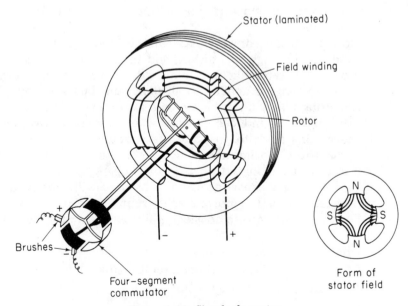

Figure 1-28   Simple dc motor.

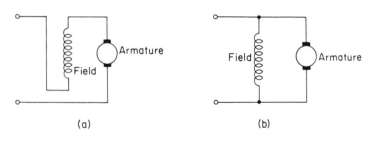

(a)   (b)

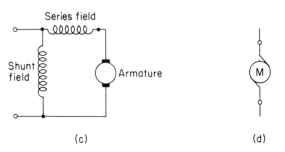

(c)   (d)

Figure 1-29 Types of dc motors: (a) series; (b) shunt; (c) compound; (d) electrical symbol.

is usually preferred and this type of motor finds use for small fans, hand drills, and grinders.

Most ac motors, however, are of the induction type. They have no commutator, slip rings, or brushes, and there is no electrical connection to the rotor. The stator has multiple field windings, which form pole pairs. The same ac voltage is applied to all pole pairs, but is 90° out of phase with its neighboring pole pairs, which create a rotating magnetic field. The rotor inside the stator has induced currents which develop its own fields and react with the stator's fields to push the rotor around. A starting winding is used to get the rotor moving and when the motor is up to running speed, a centrifugal switch mounted on the rotor opens and drops out the starting winding. A starting capacitor can also be used with this type of motor and is disconnected when the motor reaches running speed.

Some specifications to consider when using a motor include:

Operating voltage (ac or dc)
Operating current (for proper circuit fusing)
Revolutions per minute (r/min)
Horsepower
Torque

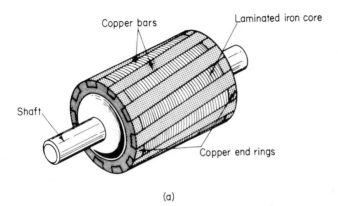

Copper bars   Laminated iron core

Shaft

Copper end rings

(a)

(b)

Figure 1-30 Induction-type ac motor: (a) squirrel cage rotor; (b) rotating field for a four-pole stator.

1-13 RESISTORS

A *resistor* is an electrical component that resists the flow of electrical current. The amount of current (I) flowing in a circuit is directly proportional to the voltage (E) across it and inversely proportional to the resistance (R) of the circuit. This is *Ohm's law* and can be expressed as a formula: $I = E/R$. The resistor is a linear device and its characteristics form a straight line when plotted on a graph.

Resistors are used to limit current flowing to a device, thereby preventing it from burning out, as voltage dividers to reduce voltage for other circuits, as transistor biasing circuits, and to serve as circuit loads.

Generally, resistors consist of carbon composition, wire-wound, and metal film. The size of resistors depend on power ratings. Carbon-type resistors usually come in 1/8-, 1/4-, 1/2-, 1-, and 2-W sizes. Larger sizes are referred to as power resistors. Variable resistors are adjustable power types: rheostats, potentiometers, and trimmer pots. Precision resistors have a tolerance of 1% or less.

Carbon resistors are the most commonly used and have a color code to identify their value. To read a resistor, place the color bands nearest the end of the resistor to the left. Now reading from left to right, the first color indicates the first significant figure; the second color, the second significant

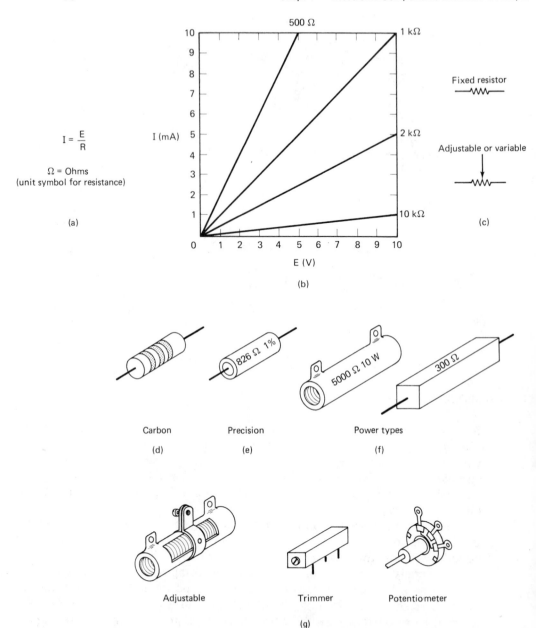

$$I = \frac{E}{R}$$

Ω = Ohms
(unit symbol for resistance)

(a)

Figure 1-31 Resistors: (a) Ohm's law formula; (b) linear characteristic curve; (c) symbols; (d) carbon; (e) precision; (f) power; (g) variable.

figure; and the third color, the multiplier or number of zeros to be added. A fourth color may be present which indicates the tolerance of the resistance. Two examples follow.

1st band = yellow = 4
2nd band = violet = 7
3rd band = red = 2
4th band = no color = 0

$\left.\begin{array}{c} \\ \\ \\ \end{array}\right\}$ 4700 Ω
or at 20%
4.7 kΩ

1st band = green = 5
2nd band = blue = 6
3rd band = gold = $\div$ 10
4th band = gold = 5%

$\left.\begin{array}{c} \\ \\ \\ \end{array}\right\}$ 5.6 Ω
at
5%

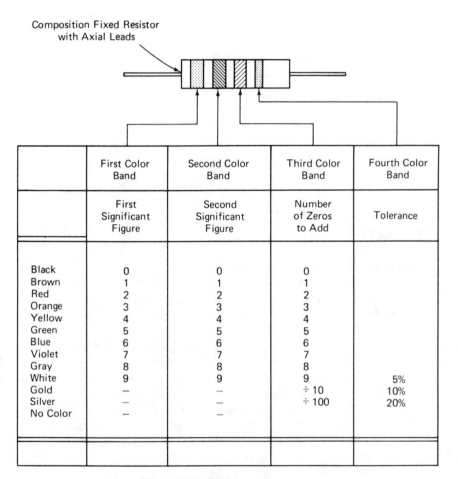

	First Color Band	Second Color Band	Third Color Band	Fourth Color Band
	First Significant Figure	Second Significant Figure	Number of Zeros to Add	Tolerance
Black	0	0	0	
Brown	1	1	1	
Red	2	2	2	
Orange	3	3	3	
Yellow	4	4	4	
Green	5	5	5	
Blue	6	6	6	
Violet	7	7	7	
Gray	8	8	8	
White	9	9	9	5%
Gold	—	—	$\div$ 10	10%
Silver	—	—	$\div$ 100	20%
No Color	—	—		

Figure 1-32 Resistor color code.

1-13.1 Resistor Formulas

The total resistance of resistors placed in series is directly additive. The total resistance of resistors placed in parallel will be less than that of the smallest resistor.

Some resistor formulas are as follows (where R_N is the number of resistors and R is the value of the single resistor):

Resistors in series:

$$R_T = R_1 + R_2 + R_3 + \cdots + R_N$$

Same-value resistors in parallel:

$$R_T = \frac{R}{R_N}$$

Two unequal resistors in parallel:

$$R_T = \frac{R_1 \times R_2}{R_1 + R_2}$$

Unequal resistors in parallel:

$$R_T = \frac{1}{\dfrac{1}{R_1} + \dfrac{1}{R_2} + \dfrac{1}{R_3} + \cdots + \dfrac{1}{R_1}}$$

1-14 NONLINEAR RESISTORS

Standard-type resistors usually maintain their value regardless of external conditions, such as voltage, temperature, and light. These types of resistors are referred to as *linear* resistors. There are other types of resistors referred to as *nonlinear*, whose resistance varies with temperature (termistor), voltage (varistor), and light (see Section 3-8).

The *thermistor* is made from metal oxides, such as manganese, nickel, copper, or iron. Usually, a thermistor has a negative temperature coefficient, where an increase in temperature causes a decrease in its resistance. The typical resistance change is about $-5\%/°C$ with a range of from 1 Ω to more than 50 MΩ. A thermistor might be used to control the heat applied to a liquid. Very often it is used to control the stability of a transistor by being part of the biasing network. The thermistor is mounted close to the transistor and when the temperature increases, its resistance decreases. This results in less forward bias voltage from emitter to base, the current through the transistor decreases, and the circuit becomes more stable. When the temperature decreases, the thermistor resumes its initial value and the normal bias voltage is again present.

Varistors are similar in appearance to thermistors, but their resistance decreases with an increase in voltage. The current that flows in a varistor varies exponentially (V^n) with the applied voltage and may increase as much as 64 times for a given varistor. Most often, varistors are used as protection devices for other circuits, such as being placed in parallel across switch contacts to prevent sparking and in inductive circuits to prevent voltage surges.

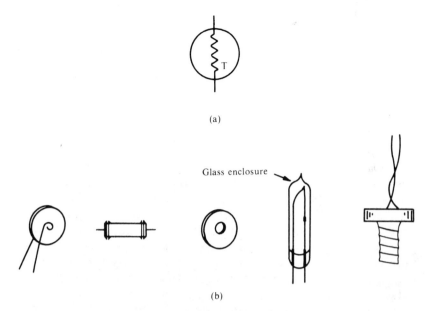

(a)

Glass enclosure

(b)

Figure 1-33 Thermistor: (a) schematic symbol; (b) some physical packages.

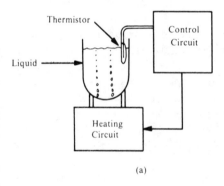

Thermistor

Liquid

Control
Circuit

Heating
Circuit

(a)

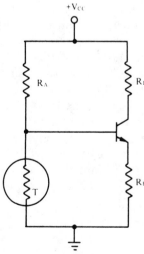

+V_{CC}

R_A

R_1

R_E

T

Figure 1-34 Thermistor applications: (a) temperature control; (b) transistor stability.

(b)

Their schematic symbol is similar to the thermistor symbol except that the T is replaced with a V.

1-15 CAPACITORS

A *capacitor* is an electrical device that can temporarily store a voltage charge. Basically, a capacitor consists of two conductors (metal plates) separated by a dielectric insulating material, which increases the ability to store a charge. The types of dielectrics used are air, wax- or oil-impregnated paper, aluminum oxide, ceramic, tantalum, mica, glass, and plastics, which include polyester (Mylar), polycarbonate, polystyrene, polypropylene, and polysulfane. The amount of capacitance of a capacitor is directly proportional to the area of the plates and inversely proportional to the square of the distance between them, including the type of dielectric used.

A capacitor tends to oppose a change in voltage and the voltage that builds up on it resembles an exponential curve, when its characteristics are plotted on a graph. A time constant is the time in seconds for a dc charge to build up to 63.2% of the total voltage applied to the capacitor. It is also the time a capacitor will discharge to 36.8% of its full charge. This time constant is related to the capacitance and resistance in a circuit and is given by the formula $TC = R \times C$. It takes five time constants to fully charge or discharge a capacitor.

A capacitor will block dc current, but appears to pass ac current by charging and discharging. It develops an ac resistance, known as *capacitive reactance*, which is affected by the capacitance and ac frequency. The formula for capacitive reactance is $X_c = 1/2\pi f_c$, with units of ohms. Capacitors are used for filtering, bypassing signals, for timing circuits, and for radio-frequency (RF) tuning circuits.

Capacitors are available in various shapes and sizes: tubular, disk, cans, tuning, and trimmer types. Usually, the value of capacitance and the working dc voltage are marked on them, but some types use a color code similar to resistors. The first band or dot represents the temperature (tem), followed by the first digit (D_1), the second digit (D_2), the multiplier (M), and the tolerance (Tol). Electrolytic capacitors using aluminum oxide as a dielectric are large-value capacitors that usually have polarity markings to indicate proper circuit connections. Small-value capacitors of mica and ceramic dielectrics are indicated in picofarads (10^{-12}), but only the significant digits are shown on the package. Tuning capacitors (such as used in radio) use air as a dielectric, with one set of plates, which can be rotated in and out of a set of stationary plates. Trimmer capacitors are used for fine adjustment with a screw, and have air, mica, and ceramic as dielectrics.

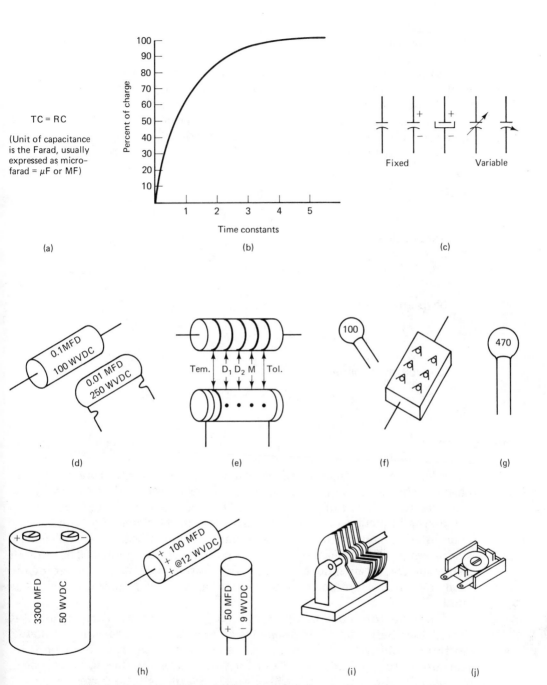

Figure 1-35 Capacitors: (a) time-constant formula; (b) voltage-charge characteristic curve; (c) symbols; (d) tubular (paper); (e) color coded; (f) mica; (g) disk; (h) electrolytic; (i) tuning (air); (j) trimmer (air).

35

1-15.1 Capacitor Formulas

The total capacitance of capacitors placed in series will be less than the smallest capacitor. The total capacitance of capacitors placed in parallel is directly additive.

Some capacitor formulas are as follows (where C_N is the number of capacitors and C is the value of a single capacitor):

Capacitors in series:

$$C_T = \frac{1}{\dfrac{1}{C_1} + \dfrac{1}{C_2} + \dfrac{1}{C_3} + \cdots + \dfrac{1}{C_N}}$$

Same-value capacitors in series:

$$C_T = \frac{C}{C_N}$$

Two unequal capacitors in series:

$$C_T = \frac{C_1 \times C_2}{C_1 + C_2}$$

Capacitors in parallel:

$$C_T = C_1 + C_2 + C_3 + \cdots + C_N$$

1-16 INDUCTORS

An *inductor* is an electrical device which can temporarily store an electromagnetic charge in the field about it as long as current is flowing through it. The inductor is a coil of wire that may have an air core or an iron core, to increase its inductance. A powered iron core in the shape of a cylinder may be adjusted in and out of the core to vary the inductance. Taps on specific windings may also be used to vary the inductance to a circuit. The amount of inductance of an inductor is directly proportional to the square of the number of turns, the cross-sectional area, and inversely proportional to its length.

An inductor tends to oppose a change in electrical current, and the current that builds up on it resembles an exponential curve when its characteristics are plotted on a graph. A time constant is the time in seconds for a dc current to build up to 63.2% of the total current applied to the inductor. It is also the time the current will fall to 36.8% of the total current when the circuit is interrupted. This time constant is related to the inductance and resistance (including the ohmic resistance of the coil's wire) in a circuit and is given by the formula TC = L/R.

An inductor saturates with dc current but has an ac resistance to ac current, known as *inductive reactance*. This inductive reactance is affected by inductance and the ac frequency and is given by the formula $X_L = 2\pi F L$, with units of ohms. Inductors are used for filtering current, increasing the

Figure 1-36 Inductors: (a) time-constant formula; (b) current-rise characteristic curve; (c) symbols; (d) air core; (e) iron core; (f) toroidal; (g) tubular; (h) RF choke; (i) tunable RF coil.

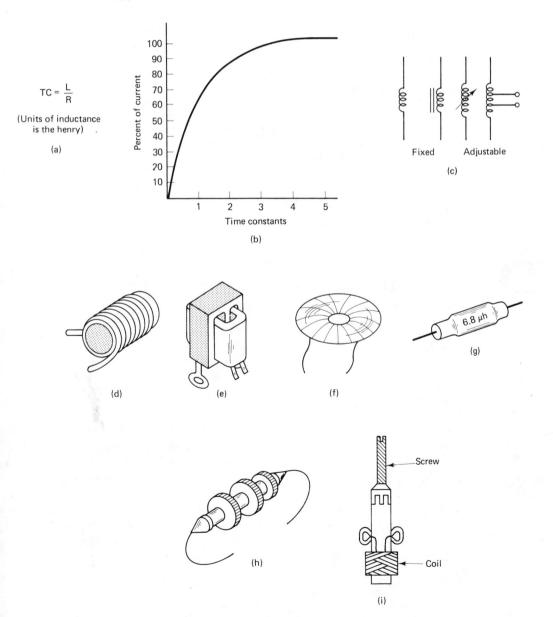

output of the RF amplifier (peaking coils), and to block RF signals and RF tuning circuits.

Inductors are available in a variety of shapes: air core, iron core (which may look like a transformer, but has only two leads), toroidal (doughnut-shaped), small tubular with epoxy, RF choke with separate coils on a cylinder, and tunable RF coil with a screwdriver adjustment.

1-16.1 Inductor Formulas

The total inductance of inductors placed in series is directly additive (not including any mutual inductance). The total inductance of inductors placed in parallel will be less than the smallest inductor (again, not including any mutual inductance).

Some inductor formulas are as follows (where L_N is the number of inductors and L is the value of a single inductor):

Inductors in series:

$$L_T = L_1 + L_2 + L_3 + \cdots + L_N$$

Same-value inductors in parallel:

$$L_T = \frac{L}{L_N}$$

Two unequal inductors in parallel:

$$L_T = \frac{L_1 \times L_2}{L_1 + L_2}$$

Unequal inductors in parallel:

$$L_T = \frac{1}{\dfrac{1}{L_1} + \dfrac{1}{L_2} + \dfrac{1}{L_3} + \cdots + \dfrac{1}{L_N}}$$

1-17 TRANSFORMERS

Transformers are related to the properties of inductors. If an inductor with an ac voltage across it is placed in parallel to a second inductor, the electro-magnetic field of the first inductor will induce an ac voltage into the second inductor. When the two coils referred to as windings are placed on a core, they become a transformer. The input voltage is to the primary winding and the induced voltage is taken off the secondary winding. If the number of turns on the secondary (N_s) is greater than the number of turns on the primary (N_p), the secondary voltage (V_s) will be greater than the voltage on the primary (V_p), referred to as a *step-up transformer*. If the number of

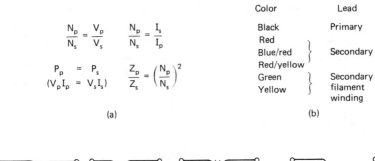

$$\frac{N_p}{N_s} = \frac{V_p}{V_s} \qquad \frac{N_p}{N_s} = \frac{I_s}{I_p}$$

$$P_p = P_s \qquad \frac{Z_p}{Z_s} = \left(\frac{N_p}{N_s}\right)^2$$

$$(V_p I_p = V_s I_s)$$

(a)

Color	Lead
Black	Primary
Red	
Blue/red	Secondary
Red/yellow	
Green	Secondary
Yellow	filament winding

(b)

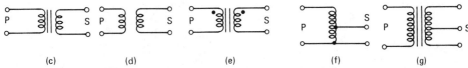

(c)   (d)   (e)   (f)   (g)

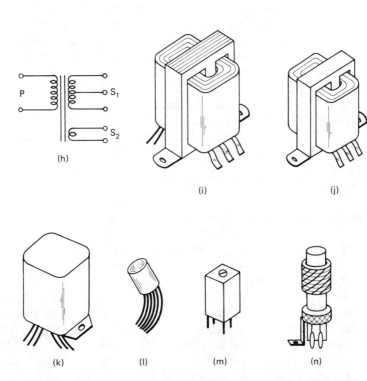

(h)

(i)   (j)

(k)   (l)   (m)   (n)

Figure 1-37 Transformers: (a) formulas; (b) lead color code; (c) iron core; (d) air core; (e) phasing symbols; (f) autotransformer; (g) center-tapped secondary; (h) multiple-secondary windings; (i) power with center tap; (j) autotransformer; (k) can-type; (l) miniature; (m) IF; (n) RF.

turns on the secondary is less, the voltage is less, referred to as a *step-down transformer.*

The power in the primary (P_p) will equal the power in the secondary (P_s); therefore, the product of the voltage and current of the primary must equal the product of the voltage and current in the secondary $(V_p I_p = V_s I_s)$. Also the number of turns and current in the windings are related thus:

$$\frac{N_p}{N_s} = \frac{I_s}{I_p}$$

and the impedance is related by the formula

$$\frac{Z_p}{Z_s} = \left(\frac{N_p}{N_s}\right)^2$$

There are many configurations to a transformer. There are air core, iron core (to increase the mutual inductance between windings), and phased windings, since the secondary voltage is $180°$ out of phase with the primary voltage. An autotransformer has only three leads; one is a common connection. Some transformers have center-tapped secondaries and multiple secondaries to develop various voltages. They are also available in various wattage ratings, sizes, and shapes.

Transformers are used to step-up or step-down voltage for power supplies, for electrical isolation, for impedance matching, to couple signals between circuits, and for tuning intermediate-frequency (IF) and RF circuits.

Normally, dc is not useful on a transformer unless it is varying and it can saturate the windings. Therefore, a transformer is usually considered an ac device.

1-18 UNDERSTANDING MULTIMETERS

A *multimeter* is a general-purpose meter capable of measuring dc and ac voltage, current, resistance, and in some cases, decibels. There are two types of meters: *analog,* using a standard meter movement with a needle, and *digital,* with an electronic numerical display. Both types of meters have a positive (+) jack and a common jack (−) for the test leads; a function switch to select dc voltage, ac voltage, dc current, ac current, or ohms; and a range switch for accurate readings. The meters may also have other jacks to measure extended ranges of voltage (1 to 5 kV) and current (up to 10 A). There are some variations to the functions used for specific meters.

The analog meter usually includes the function and range switches in a single switch. It may also have a polarity switch to facilitate reversing the test leads. The needle will have a screw for mechanical adjust to set it to zero and also a zero adjust control to compensate for weakening batteries when measuring resistance. An analog meter can read positive and negative

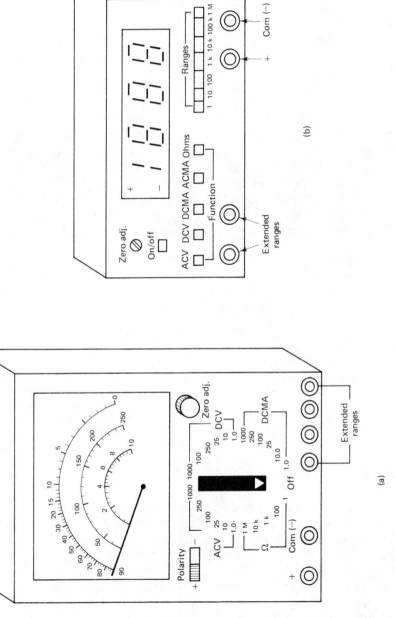

Figure 1-38 Multimeters: (a) analog VOM; (b) digital VOM (DVM).

voltage by simply reversing the test leads or moving the polarity switch. A digital meter usually has an automatic indicator for polarity on its display.

A meter of reasonable quality will have an input resistance of 20 kΩ per volt or greater to prevent loading down a circuit, which causes an error in the reading. For example, if a dc voltmeter was set on the 10-V scale, its input resistance would be 200 kΩ. If it were placed across a 200 kΩ resistor in a circuit, the total effective resistance at that point would be 100 kΩ and would certainly cause an erroneous reading.

Meters must be properly connected to a circuit to ensure a correct reading. A voltmeter is always placed across (in parallel) the circuit or component to be measured. When measuring current, the circuit must be opened and the meter inserted in series with the circuit or component to be measured. When measuring the resistance of a component in a circuit, the voltage to the circuit must be removed and one end of the component opened from the circuit (to prevent any parallel paths from affecting the reading) and the meter placed in parallel with the component.

Special probes are used with meters for specific circuits. These include shielded cable, high-voltage, and capacitance types and RF detectors.

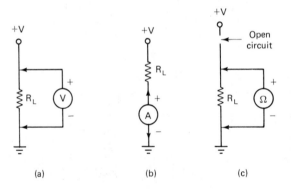

Figure 1-39 Proper meter connections: (a) measuring voltage (parallel); (b) measuring current (series); (c) measuring ohms (open circuit).

1-19 READING MULTIMETERS

On a standard analog meter there is a scale for ohms, dc, and ac. When the function switch is set on 250ACV, a full-scale needle deflection would indicate that the meter was measuring 250 V ac. If the needle was at 150 the meter would be measuring 150 V ac. The various ranges of a specific function would use the same scale; therefore, the individual gradient values have to be determined. For example, if the 250ACV scale is used, there are 10 gradients between the numbers and the value between numbers is 50 V. The value of each gradient can be found by dividing 50 by 10 (50 ÷ 10 = 5) which results in 5 V per gradient. The same scale would be used for 25ACV except that the number 250 = 25, 200 = 20, 150 = 15, and so on. Now

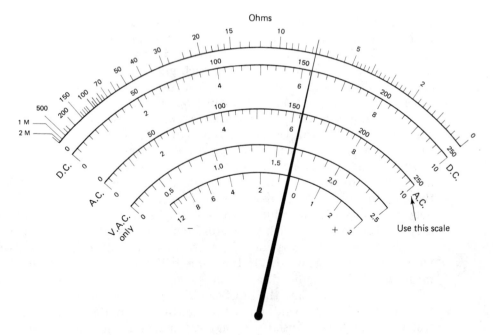

Figure 1-40 Reading ac voltage (range switch set at 250ACV, meter reads 155 V).

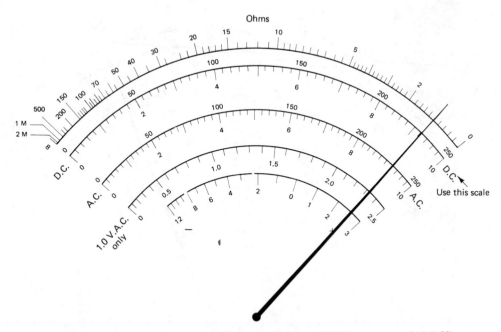

Figure 1-41 Reading dc voltage (range switch set at 10DCV, meter reads 9.2 V).

there is a 5-V difference between numbers, so each gradient is worth 0.5 V. If the range switch is set to 10ACV, the same scale is used and each gradient is worth 0.2 V. When the range switch is set to 100ACV, the number 10 = 100, 8 = 80, 6 = 60, and so on, and each gradient is worth 2 V. All voltage and current scales are used the same way, remembering that the ac voltage is the effective or root-mean-square (rms) value.

The ohm scale is a nonlinear scale that may be indicated in reverse to the other scales. The resistance function is used as a multiplier indicator. The function switch is placed to the desired range and the test leads are shorted together. The zero adj. (ohms adj.) control is then used to set the needle to zero on the scale. The leads are then opened and placed across the desired resistor to be read. If the function switch is set at 1 kΩ and the needle goes to 10, the value of the resistor being read is 10 kΩ. The meter may need to be zeroed each time a different range is selected.

With a digital meter all values of dc, ac, and ohms measured will fall within the range selected. If the value being measured is greater than the range selected, an indication will be given, such as the display going blank, blinking, or perhaps only the most significant digit will light.

A user should spend some time getting orientated to meters and any test equipment being used. Equipment manuals will give detailed instructions as to their proper use.

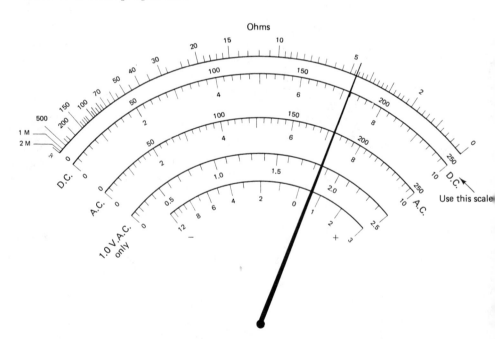

Figure 1-42 Reading dc current (range switch set at 25DCMA, meter reads 18 mA).

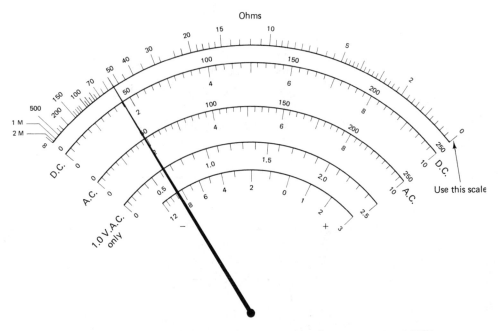

Figure 1-43 Reading ohms (range switch set at 1KΩ, meter reads 55 KΩ).

1-20 UNDERSTANDING THE OSCILLOSCOPE

The oscilloscope presents an accurate electronic picture of changing voltages within a circuit. An electron beam is created, focused, accelerated, and properly deflected to display the voltage waveforms on the face of a cathode-ray tube (CRT). The basic circuits and controls of an oscilloscope are:

Power supply: provides high dc voltage (up to a few thousand volts) for the CRT and lower dc voltages for other circuits.

Intensity control: adjusts brightness of display.

Focus control: adjusts sharpness of display.

Time-base generator: provides the basic sawtooth voltage, which moves the trace on the face of the CRT from left to right horizontally.

Time/CM selector: adjusts the frequency of the time-base generator.

Horizontal amplifier circuits: amplifies the output of the time-base generator and applies it to the horizontal deflection plates.

Horizontal gain control: adjusts full trace horizontally on face of CRT.

Horizontal positioning control: centers trace horizontally on face of CRT.

Vertical input: accepts voltage to be measured, either dc or ac.

Vertical attenuator: reduces input voltage amplitude so as not to over-drive trace on face of CRT.

V/CM selector: selects desired input voltage attenuation.

Vertical amplifier circuits: amplifies input voltage and applies it to the vertical deflection plates.

Vertical gain control: manually adjusts amplitude of input voltage displayed on face of CRT.

Figure 1-44 Block diagram of basic oscilloscope.

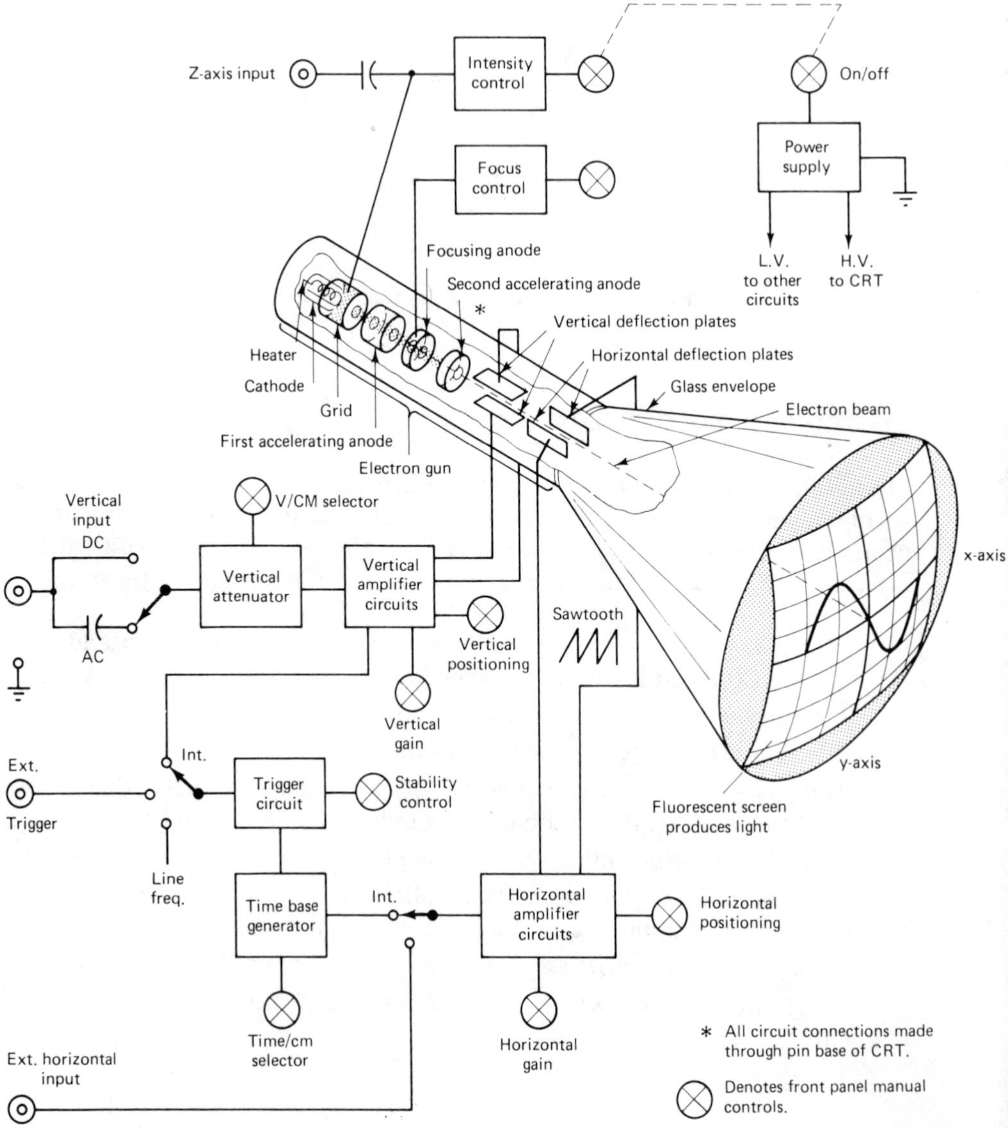

Control name	Alternate names
Intensity/On-off	Brightness/On-off
Focus	Usually none
Horizontal position	Hor. or X: positioning, centering
Horizontal gain	Hor. or X: gain, vernier
Horizontal selector	Hor. or X: select, sense, input control
Vertical position	Ver. or Y: positioning, centering
Vertical gain	Ver. or Y: gain, vernier
Volts/cm	Volts/div, Ver. or Y: range, attenuator, sensitivity, input control
Time/cm	Time/div, time base, sweep, sweep range, sweep frequency, course frequency, frequency range, range
Stability	Trigger level, sync: adjust, lock, signal
Trigger selector	Sync: select, input control, function

Types of probes

1. × 10 attenuation
2. Low capacitance–high impedance
3. Demodulator or detector
4. Ac (inductive coupling)
5. High voltage

Preferred 'scope qualities

A. Sensitivity: 10 mv/div
B. Bandwidth: to 20 MHz
C. Automatic triggering
D. Dual trace

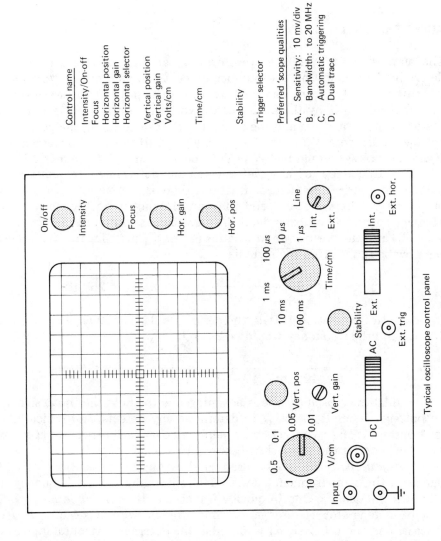

Typical oscilloscope control panel

Figure 1-45 Oscilloscope controls.

Vertical positioning control: centers trace vertically on face of CRT.

Trigger circuit: synchronizes time-base generator with input frequency another external frequency, or 60-Hz line frequency. Enables trace to be stopped for accurate measurements.

Stability control: manual control for "locking in" display.

External horizontal input: synchronizes horizontal trace for special measurements and displays as Lissajous patterns.

Z input: used for intensity modulation of electron beam, perhaps for frequency measurements.

1-21 READING THE OSCILLOSCOPE

The amplitude of a voltage waveform on an oscilloscope screen can be determined by counting the number of centimeters (cm), and/or fractions thereof, vertically, from one peak to the other peak of the waveform and then multiplying it by the setting of the volts/cm control. As an example, if the amplitude was 4 cm and the control was set on 1 V/cm, the peak-to-peak voltage would be 4 V (4 cm $\times$ 1 V/cm = 4 V). If the control was set on 0.5 V/cm, the voltage would be 2 V peak to peak (4 cm $\times$ 0.5 V/cm = 2 V).

The frequency of a waveform can be determined by counting the number of centimeters, and/or fractions thereof, horizontally, in one cycle or period of the waveform and then multiplying it by the setting of the time/cm control. For example, if the waveform is 4 cm long and the control is set at 1 ms, the period would be 4 ms (4 cm $\times$ 1 ms = 4 ms). The frequency can now be found from the formula

$$f = \frac{1}{p} = \frac{1}{4 \text{ ms}} = \frac{1}{4 \times 10^{-3} \text{ s}} = 0.25 \times 10^3 = 250 \text{ Hz}$$

If the control was set on 100 μs, the period would be 400 μs (4 cm $\times$ 100 μs = 400 μs) and the frequency would be 2.5 kHz:

$$f = \frac{1}{p} = \frac{1}{400 \text{ ms}} = \frac{1}{4 \times 10^{-4}} = 0.25 \times 10^4 = 2500 \text{ Hz}$$

A dual-trace oscilloscope is advantageous to show the input signal and output signal simultaneously, to determine any defects, and indicate phase relationships. The two traces may be placed over each other (superimposed) to indicate better the phase shift between two signals.

Lissajous patterns can be used to show the phase relationship of two signals of the same frequency and to determine an unknown frequency from a known frequency. One frequency is placed at the vertical input (f_v). The time-base generator is disengaged when the horizontal selector is set to external and the other signal is placed at the external horizontal input. This

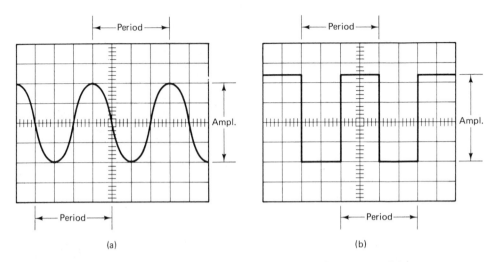

Figure 1-46 Oscilloscope voltage waveforms: (a) sine wave; (b) square wave.

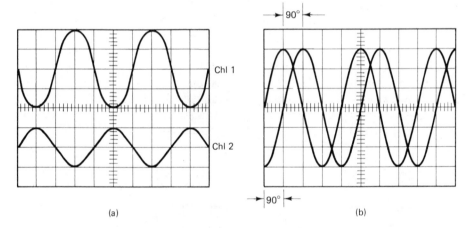

Figure 1-47 Dual-trace oscilloscope voltage waveforms: (a) input/output signals of an amplifier; (b) both channels superimposed to show phase shift of two signals.

signal (f_H) now drives the horizontal sweep section. If both signals are the same frequency, a circle will appear on the face of the oscilloscope. If f_v is twice f_H, a "bow-tie" type of pattern will appear on the screen. The two peaks at the top (or bottom) and the single peak (side) indicate a ratio of 2:1. If the known frequency was at the horizontal input, say f_H = 1 kHz, then the frequency at the vertical input, f_v, would be twice that at the horizontal input, or 2 kHz. If the frequencies are reversed at the inputs, the "bowtie" will turn on its side and indicate a 1:2 ratio. Other frequency ratios are also possible.

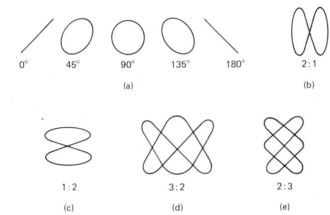

Figure 1-48 Lissajous patterns: (a) two signals of the same frequency varying in phase; (b) $f_v = 2f_H$ (2:1); (c) $f_H = 2f_v$ (1:2); (d) $f_v = 1.5 f_H$; (e) $f_H = 1.5 f_v$.

1-22 USING THE BASIC SIGNAL GENERATOR

A signal generator converts dc to ac or varying dc in the form of sine waves, square waves, triangle waves, or other types of voltage waveforms. The signal generator is used to inject a signal into a circuit or piece of equipment for troubleshooting or for calibration. Some generators may be used for audio, RF, or higher frequencies, whereas others have overlapping frequency ranges. A standard function generator usually has three types of waveforms. All generators will have a frequency range switch, a fine adjustment control for selecting a specific frequency, an amplitude control for varying the peak-to-peak output voltage, and output terminals.

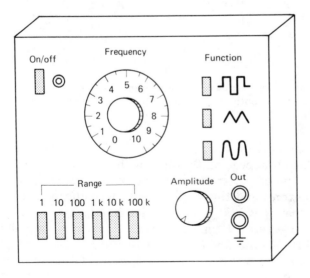

Figure 1-49 Basic signal generator.

To select a sine wave of say 5 kHz, the user would set the function switch to the sine wave, set the range switch to 1 k, and then adjust the frequency fine adjust control to 5. The amplitude control would then be adjusted to establish the desired peak-to-peak voltage output.

Some generators may have a dc component at the output terminals that could upset the circuit it is connected to. In this case, a capacitor connected in series with the positive output terminal and the circuit will block the dc component.

In some cases a very small signal is required from the generator, but the noise at the output terminals may be too objectionable or the signal too large when the amplitude control is turned way down. To remedy this, the user can place a large-value resistor (100 kΩ to 1 MΩ) in series with the positive output terminal and the circuit. Sufficient voltage can be developed at the output terminals to overcome the problems mentioned, while the resistor drops some of the voltage, which permits the correct signal amplitude to be placed on the circuit.

1-23 PROCEDURE FOR TESTING A DISCRETE CIRCUIT

It may be necessary to disconnect a PC board from a system to check it separately. The following procedure can be used as a guide for setting up the equipment to check a PC board or for experimenting with a new circuit on a breadboard.

1. Have the proper circuit schematic in front of you.
2. Have the proper equipment, parts, and test leads in front of you.
3. Construct the circuit if it is an experiment.
4. Connect the power supply to the circuit, which may be positive and ground, negative and ground, or both positive–negative and ground.
5. Connect all equipment grounds to the common circuit ground (as indicated by the dashed lines).
6. Turn on the power supply and set the proper dc voltages. Measure with a voltmeter.
7. Use a voltmeter to check dc voltages on the circuit.
8. Connect the output device to the circuit.
9. Connect the input device to the circuit.
10. Set the desired input signal to the circuit.
11. Use the oscilloscope or voltmeter to check the input signal at point A.
12. Use the oscilloscope or voltmeter to check the output signals at point B.
13. Observe the output device for the correct indication.

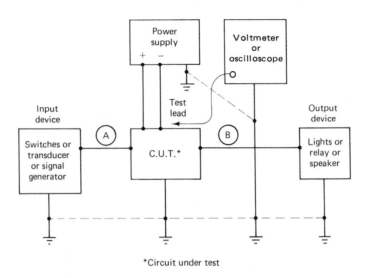

*Circuit under test

Figure 1-50 General setup for testing a discrete circuit.

1.24 BASIC DC CIRCUIT FORMULAS

The current in a circuit is directly proportional to the voltage across the circuit and inversely proportional to the resistance of the circuit. Power (the rate of doing work) is equal to voltage times current. The following formulas indicate how one unknown variable can be found when the other two variables are known.

$$I = \frac{E}{R} \qquad P = EI = I^2 R = \frac{E^2}{R}$$

$$E = IR$$

$$E = \frac{P}{I} = \sqrt{PR}$$

$$R = \frac{E}{I}$$

$$I = \frac{P}{E} = \sqrt{\frac{P}{R}}$$

$$R = \frac{P}{I^2} = \frac{E^2}{P}$$

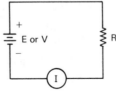

Figure 1-51 Relationship of E, R, and I.

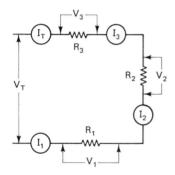

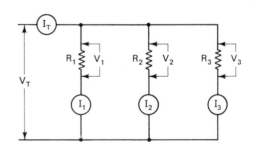

Figure 1-52 Voltage drops and current of a series circuit.

Figure 1-53 Voltage drops and current of a parallel circuit.

The voltage drops in a series circuit will add up to the total voltage applied to the circuit. The current in a series circuit is the same throughout the circuit. The following formulas indicate these conditions:

$$V_T = V_1 + V_2 + V_3 \qquad I_T = I_1 = I_2 = I_3$$
$$V_1 = V_T - (V_2 + V_3)$$
$$V_2 = V_T - (V_1 + V_3)$$
$$V_3 = V_T - (V_1 + V_2)$$

The voltage is the same across each branch in a parallel circuit. The current in each branch will add up to the total current of a parallel circuit.

$$V_T = V_1 = V_2 = V_3 \qquad I_T = I_1 + I_2 + I_3$$
$$I_1 = I_T - (I_2 + I_3)$$
$$I_2 = I_T - (I_1 + I_3)$$
$$I_3 = I_T - (I_1 + I_2)$$

Figure 1-54 Relationship of a rotating vector ($360°$) to a sine wave.

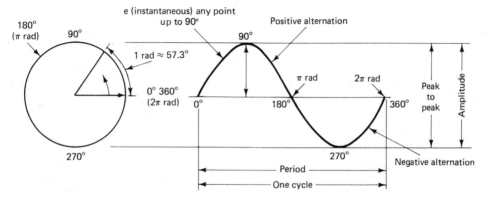

Alternating current (ac) is a flow of current that reaches maximum in one direction, decreases to zero, then reverses itself and reaches maximum in the other direction. This cycle is repeated continuously and the number of cycles per second (hertz) is the frequency. The following formulas relate to ac.

$$e \text{ (instantaneous voltage)} = V_{peak} \times \sin \theta$$

$$A_V \text{ (average voltage)} = 0.637 \times peak = 0.9 \times rms$$

$$rms \text{ (effective voltage)} = 0.707 \times peak = 1.11 \times A_V$$

$$peak = 1.414 \times rms = 1.57 \times A_V$$

$$peak \text{ to } peak = 2.82 \times rms = 3.14 \times A_V$$

$$\text{cycle} = \text{period} = \frac{1}{\text{frequency}} \qquad I = \frac{V}{Z}, \ Z = \frac{E}{I}, \ E = IZ$$

$$\text{frequency} = \frac{1}{\text{period (cycle)}} \qquad P = IE \cos \theta, \text{ where } \theta = \text{phase angle}$$

$$\omega \text{ (angular velocity)} = 2\pi f \qquad \text{power factor (PF)} = \frac{\text{true power}}{\text{apparent power}}$$

To convert degrees to radians:

$$\text{multiply degrees by } \frac{\pi}{180}$$

$$= \frac{EI \cos \theta}{EI}$$

$$\theta = 0°, PF = 1$$

To convert radians to degrees:

$$\theta = 90°, PF = 0$$

$$\text{multiply radians by } \frac{180}{\pi}$$

(a) $Z = R$, $\theta = 0°$
(b) $Z = X_c$, $\theta = -90°$
(c) $Z = X_L$, $\theta = +90°$
(d) $Z = \sqrt{R^2 + X_c^2}$, $\theta = \text{arc tan} \dfrac{X_c}{R}$

(e) $Z = \sqrt{R^2 + X_L^2}$, $\theta = \text{arc tan} \dfrac{X_L}{R}$

(f) $Z = X_L - X_c$, $\theta = -90°$ when $X_L < X_c$, $\theta = 0°$ when $X_L = X_c$ (resonance), $\theta = +90°$ when $X_L > X_c$

(g) $Z = \sqrt{R^2 + (X_L - X_c)^2}$, $\theta = \text{arc tan} \dfrac{X_L - X_c}{R}$

Figure 1-55 Reactance to RCL components: (a) R only; (b) C only; (c) L only; (d) series RC; (e) series RL; (f) series LC; (g) series LCR; (h) parallel RC; (i) parallel RL; (j) parallel LC; (k) parallel LCR; (l) series-parallel RC-RL.

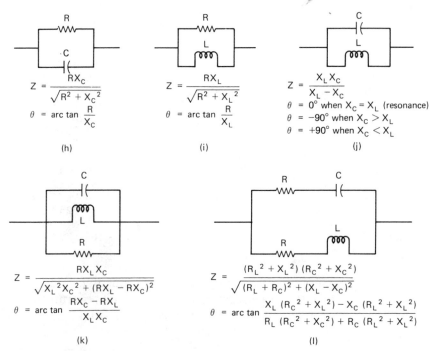

$$Z = \frac{RX_C}{\sqrt{R^2 + X_C{}^2}}$$

$$\theta = \text{arc tan } \frac{R}{X_C}$$

(h)

$$Z = \frac{RX_L}{\sqrt{R^2 + X_L{}^2}}$$

$$\theta = \text{arc tan } \frac{R}{X_L}$$

(i)

$$Z = \frac{X_L X_C}{X_L - X_C}$$

$\theta = 0°$ when $X_C = X_L$ (resonance)

$\theta = -90°$ when $X_C > X_L$

$\theta = +90°$ when $X_C < X_L$

(j)

$$Z = \frac{RX_L X_C}{\sqrt{X_L{}^2 X_C{}^2 + (RX_L - RX_C)^2}}$$

$$\theta = \text{arc tan } \frac{RX_C - RX_L}{X_L X_C}$$

(k)

$$Z = \sqrt{\frac{(R_L{}^2 + X_L{}^2)(R_C{}^2 + X_C{}^2)}{(R_L + R_C)^2 + (X_L - X_C)^2}}$$

$$\theta = \text{arc tan } \frac{X_L (R_C{}^2 + X_L{}^2) - X_C (R_L{}^2 + X_L{}^2)}{R_L (R_C{}^2 + X_C{}^2) + R_C (R_L{}^2 + X_L{}^2)}$$

(l)

Figure 1-55 *Continued.*

The impedance and phase angle of an ac circuit containing resistance and reactive elements such as capacitance and inductance can also be calculated with formulas.

1-26 FREQUENCY SPECTRUM AND ALLOCATIONS

The following tables show the radio-frequency spectrum and specific Federal Communications Commission (FCC) allocations.

Radio-frequency spectrum

VLF (very low frequencies)	3–30 kHz
LF (low frequencies)	30–300 kHz
MF (medium frequencies)	0.3–3 MHz
HF (high frequencies)	3–30 MHz
VHF (very high frequencies)	30–300 MHz
UHF (ultra-high frequencies)	0.3–3 GHz
SHF (super-high frequencies)	3–30 GHz
EHF (extra-high frequencies)	30–300 GHz

General allocations

Power frequencies	Less than 100 Hz
	(U.S. = 60 Hz, Europe = 50 Hz)
Audio frequencies	20-20,000 Hz
Human voice	300-3000 Hz
Maritime communications and navigation	30-535 kHz
Standard AM radio broadcasting	535-1605 kHz
Amateur radio and international shortwave	1.605-30 MHz
Television broadcast channels 2 to 4	54-72 MHz ⎫
Television broadcast channels 5 and 6	76-88 MHz ⎬ VHF
Television broadcast channels 7 to 13	174-216 MHz ⎭
Television broadcast channels 14 to 83	470-890 MHz UHF
FM radio broadcasting	88-108 MHz
Citizen's band radio	26.48-28 MHz,
	220-225 MHz,
	450-470 MHz
Aeronautical navigation	108-122 MHz

Government (police, fire, etc.), experimental, industrial, scientific, medical, aeronautical navigation, amateur, and nongovernment broadcasts are interspersed throughout the radio-frequency spectrum.

1-27 POWERS OF 10 AND DECIMAL PREFIXES

Mathematical units used in the field of electronics involve very large and extremely small values. Therefore, a method called *scientific notation* or engineer's shorthand is used with the aid of powers of 10 for writing these numbers in simple abbreviated form. Powers of 10 is an exponential way of writing large numbers involving many zeros.

$$1 = 10^0 \text{ (any number to the zero power equals 1)}$$
$$10 = 10^1$$
$$100 = 10^2$$
$$1000 = 10^3$$

$10,000 = 10^4$ Examples: $2400 = 2.4 \times 10^3$

$100,000 = 10^5$ $2,200,000 = 2.2 \times 10^6$

$1,000,000 = 10^6$ $0.015 = 15 \times 10^{-3}$

$0.1 = 10^{-1}$ $0.000003 = 3 \times 10^{-6}$

$$0.01 = 10^{-2}$$
$$0.001 = 10^{-3}$$
$$0.000001 = 10^{-6}$$

Decimal prefixes (Table 1-3) also simplify electronic calculations.

TABLE 1-3 VALUES AND SYMBOLS FOR DECIMAL PREFIXES

Values	Submultiples and multiples	Prefix	Symbol
quintillionth	1×10^{-18}	alto	a
quadrillionth	1×10^{-15}	femto	f
trillionth	1×10^{-12}	pico	p
billionth	1×10^{-9}	nano	n
millionth	1×10^{-6}	micro	μ
thousandth	1×10^{-3}	milli	m
hundredth	1×10^{-2}	centi	c
tenth	1×10^{-1}	deci	d
ten	1×10^{1}	deca	da
hundred	1×10^{2}	hecto	h
thousand	1×10^{3}	kilo	k
ten-thousand	1×10^{4}	myria	my
million	1×10^{6}	mega	M
billion	1×10^{9}	giga	G
trillion	1×10^{12}	tera	T

Examples:

$$V = I \cdot R$$
$$= 2 \text{ mA} \times 5 \text{ k}\Omega$$
$$= 2 \times 10^{-3} \text{ A} \times 5 \times 10^{3} \ \Omega$$
$$= 10 \text{ V (milli and kilo cancel each other)}$$

$$T = R \cdot C$$
$$= 100 \text{ k}\Omega \times 3 \ \mu\text{F}$$
$$= 100 \times 10^{3} \ \Omega \times 3 \times 10^{-6} \text{ F}$$
$$= 300 \times 10^{-3} \text{ s}$$
$$= 300 \text{ ms (kilo and micro partially cancel, leaving milli)}$$

1-28 BASIC MATHEMATICAL DATA AND FUNCTIONS

1-28.1 Symbols

+	positive, plus, add	$\gg$	much greater than
−	negative, minus, subtract	$<$	less than
$\times$ or $\cdot$	multiply	$\ll$	much less than
$\div$ or :	divide	$\geqslant$	greater than or equal to
$\pm$	positive or negative, plus or minus	$\leqslant$	less than or equal to
		$\equiv$	identity
$\mp$	negative or positive, minus or plus	$\therefore$	therefore
		$\angle$	angle

= or ::	equal to
≃ or ≈	approximately equal to
≠	not equal to
>	greater than

Δ	increment
⊥	perpendicular to
∥	parallel to
$\lvert n \rvert$	absolute value of n

1-28.2 Transposition of Terms

1. If $A = \dfrac{B}{C}$, then $B = AC$ and $C = \dfrac{B}{A}$.

2. If $\dfrac{A}{B} = \dfrac{C}{D}$, then $AD = BC$ and $A = \dfrac{BC}{D}$, $B = \dfrac{AD}{C}$, $C = \dfrac{AD}{B}$, and $D = \dfrac{BC}{A}$.

3. If $A = \dfrac{1}{B\sqrt{CD}}$, then $A^2 = \dfrac{1}{B^2 CD}$, $B = \dfrac{1}{A\sqrt{CD}}$, $C = \dfrac{1}{A^2 B^2 D}$, and $D = \dfrac{1}{A^2 B^2 C}$.

4. If $A = \sqrt{B^2 + C^2}$, then $A^2 = B^2 + C^2$, $B = \sqrt{A^2 - C^2}$, and $C = \sqrt{A^2 - B^2}$.

1-28.3 Exponents and Radicals

$$a^0 = 1$$
$$a^x + a^y = a^{x+y}$$
$$(ab)^x = a^x b^x$$
$$(a^x)^y = a^{xy}$$
$$\frac{a^x}{a^y} = a^{x-y}$$
$$\left(\frac{a}{b}\right)^x = \frac{a^x}{b^x}$$

$$a^{-x} = \frac{1}{a^x}$$
$$a^{1/x} = \sqrt[x]{a}$$
$$a^{x/y} = \sqrt[y]{a^x}$$
$$\sqrt[x]{ab} = \sqrt[x]{a}\ \sqrt[x]{b}$$
$$\sqrt[x]{\frac{a}{b}} = \frac{\sqrt[x]{a}}{\sqrt[x]{b}}$$
$$\sqrt[x]{\sqrt[y]{a}} = \sqrt[xy]{a}$$

1-28.4 Right-Triangle Relationships

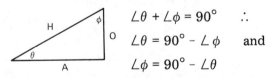

$$\angle\theta + \angle\phi = 90° \qquad \therefore$$
$$\angle\theta = 90° - \angle\phi \qquad \text{and}$$
$$\angle\phi = 90° - \angle\theta$$

H = the hypotenuse
A = the side adjacent to θ and opposite ϕ
O = the side opposite θ and adjacent to ϕ

$$H = \sqrt{A^2 + O^2}$$
$$A = \sqrt{H^2 + O^2}$$

θ = the angle formed by H and A

ϕ = the angle formed by H and O

$$O = \sqrt{H^2 - A^2}$$

$$\text{sine (sin) } \theta = \frac{O}{H} \qquad\qquad \text{secant (sec) } \theta = \frac{H}{A}$$

$$\text{cosine (cos) } \theta = \frac{A}{H} \qquad\qquad \text{cosecant (csc) } \theta = \frac{H}{O}$$

$$\text{tangent (tan) } \theta = \frac{O}{A} \qquad\qquad \text{cotangent (cot) } \theta = \frac{A}{O}$$

$\sin \theta = \cos \phi$

$\cos \theta = \sin \phi$ $\qquad \dfrac{1}{\sin \theta} = \csc \theta \qquad \dfrac{1}{\csc \theta} = \sin \theta$

$\tan \theta = \cot \phi$

$\csc \theta = \sec \phi \qquad \dfrac{1}{\cos \theta} = \sec \theta \qquad \dfrac{1}{\sec \theta} = \cos \theta$

$\sec \theta = \csc \phi$

$\cot \theta = \tan \phi$ $\qquad \dfrac{1}{\tan \theta} = \cot \theta \qquad \dfrac{1}{\cot \theta} = \tan \theta$

Solid-State Devices

2-1 INFORMATION ON SOLID-STATE DEVICES

Three- and four-lead solid-state devices are manufactured in a variety of packages, as shown. Except for diodes, which are two-lead components, any one of the packages may contain a PNP or NPN bipolar transistor, JFET, MOFET, SCR, TRIAC, UJT, or Darlington transistor. Therefore, it is important to have the component identification number and specifications when using solid-state devices. Packages TO-3, TO-66, and TO-220 are larger than the other packages and are designed for higher power capability.

There is no standard lead identification for the devices, but some follow the common pattern as shown. It is always advisable to have the device's specification sheet showing the lead arrangement.

Generally, solid-state devices consist of n-type and p-type semiconductor material made from germanium or silicon. The p-type material has holes as its majority current carriers and electrons as the minority current carriers, whereas n-type material has electrons as its majority current carriers and holes as the minority current carriers.

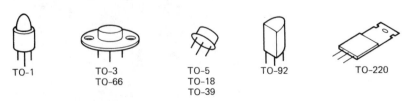

Figure 2-1 Solid-state device packages.

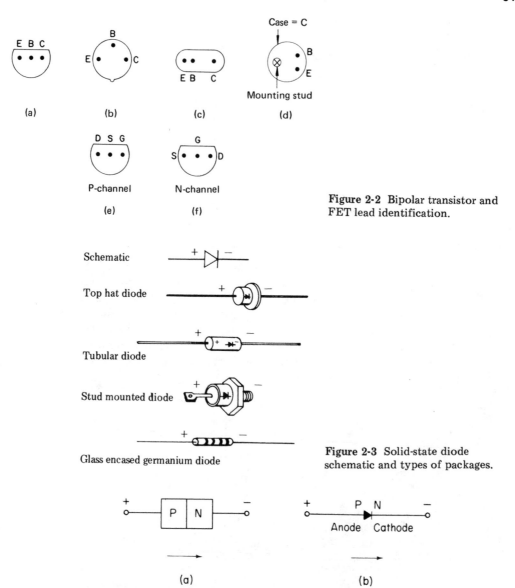

Figure 2-2 Bipolar transistor and FET lead identification.

Figure 2-3 Solid-state diode schematic and types of packages.

Figure 2-4 *Pn* junction and schematic diagram: (a) conventional current flow; (b) low-resistance direction for conventional current flow.

2-2 SEMICONDUCTOR DIODE

A *semiconductor diode* consists of a *pn* junction made of semiconductor material. The *p*-type material is called the *anode*, while the *n*-type material is called the *cathode*. The diode schematic diagram is painted on larger

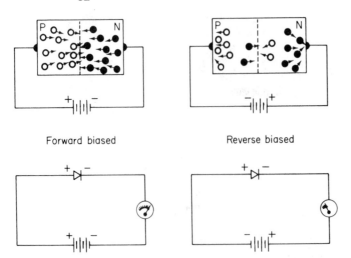

Forward biased

Reverse biased

Figure 2-5 *Pn*-junction biasing.

diodes and shows which leads are the anode and cathode. On smaller diodes, a band around one end indicates the cathode. Glass diodes with several multicolored bands will usually indicate the cathode by a black band.

During the manufacturing process, some holes and electrons combine at the junction to produce an imaginary battery called the *potential battery, potential barrier, space charge region,* or *depletion region.* When the poten-

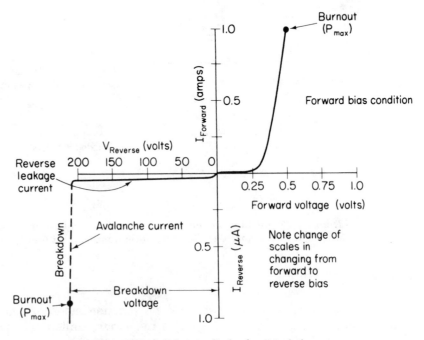

Figure 2-6 Solid-state diode characteristic curve.

tial barrier is overcome in the forward-bias condition, the depletion region is small, and holes and electrons combine easily. In the reverse-bias condition, the depletion region is large and only a few holes and electrons combine to produce what is called *leakage current*.

A diode is *forward biased* when the anode is more positive than the cathode (greater than the turn-on voltage, which is approximately 0.3 V for germanium and 0.7 V for silicon). Actually, the positive end of the battery repels holes toward the junction, while the negative end of the battery repels electrons toward the junction, facilitating combination of holes and electrons. In this condition the internal resistance of the diode is low and a large current will flow through the diode (depending on the external circuit resistance). The diode is *reverse biased* when the anode is less positive than the cathode. In this case the holes are attracted to the negative end of the battery, while the electrons are attracted to the positive end of the battery, thereby making a combination of holes and electrons almost impossible. Now, the internal resistance is extremely high, resulting in very little current flow (depending on the diode's leakage current). Excessive reverse-biased voltage can break down the diode. Too much current through the diode in either direction will destroy it.

2-3 ZENER DIODE

A *zener diode* is a specially doped diode that operates the same as a regular diode in the forward-biased condition (it allows maximum current to flow). However, in the reverse-biased condition, it will not conduct until it reaches the zener voltage (V_Z) at which it is designed. At this point the zener diode conducts current in the reverse direction, while maintaining the zener (or reference) voltage across its terminals. The amount of current that flows through it is determined by two factors, a series (current-limiting) resistor (R_S) and the parallel load resistance (R_L). Resistor R_S is found by the formula $R_S = V_{R_S}/I_Z$, where $V_{R_S} = V_{\text{source}} - V_Z$. With no load, a specific amount of current ($I_Z = I_{R_S}$) flows through the zener diode and R_S. Voltage drops V_{R_S} plus V_Z will equal V_{source} (V_{source} should be at least 1 V greater than V_Z). When a load is connected across the zener diode, the diode current decreases by the amount drawn by the load so that the current through R_S remains constant. ($I_Z = I_{R_S} - I_{R_L}$). The zener diode maintains the voltage across its terminals by varying the current that flows through it.

B   L

Figure 2-7 Zener diode schematic diagrams.

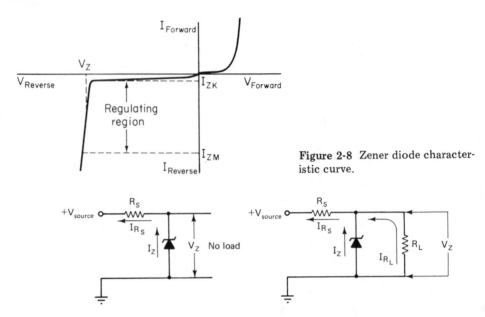

Figure 2-8 Zener diode characteristic curve.

Figure 2-9 Zener diode operation.

2-4 TUNNEL DIODE

The *tunnel diode* is a semiconductor diode in which the p and n materials are heavily doped, resulting in an extremely thin depletion region. Useful operation of the tunnel diode occurs below the turn-on voltage for a regular germanium diode. When forward bias is applied, a significant amount of forward current flows. As the forward bias is increased, amazingly, the current decreases nearly to zero. Although the forward-biased voltage increases, this decrease in current is called the *negative resistance region* (point A to point B).

It is in this region that the tunnel diode is very useful for switching circuits and in high-frequency amplifiers and oscillators. In effect, current has tunneled through the normal potential barrier region of the diode. When the

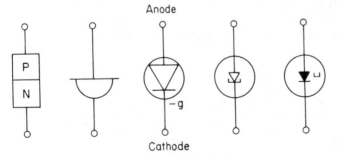

Figure 2-10 Tunnel diode *pn* junction and schematic diagrams.

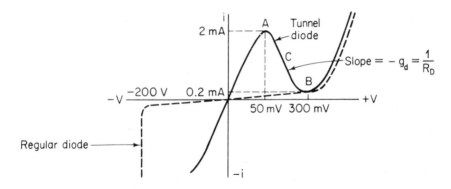

Figure 2-11 Tunnel diode versus regular diode characteristic curve.

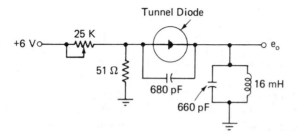

Figure 2-12 Tunnel diode oscillator circuit.

forward bias continues to increase, normal diode action results as the normal potential barrier is overcome. Because of the heavy doping, reverse bias also results in a significant current flow.

In the typical oscillator circuit shown, the 25-kΩ potentiometer and the 51-Ω resistor form a voltage divider to bias the tunnel diode. The tunnel diode acts as a switch to sustain the oscillations of the tank circuit (the 660-pF capacitor and the 16-mH inductor). This circuit operates at approximately 100 kHz.

2-5 VOLTAGE-VARIABLE CAPACITANCE (VVC) DIODE

A reverse-biased diode exhibits the properties of a *voltage-variable capacitor* (VVC). The *p* and *n* regions serve as the plates, and the depletion region acts as the dielectric. The least amount of reverse-biased voltage produces the greatest capacitance; increasing the bias voltage reduces the amount of capacitance. The range of capacitance is in picofarads. If the diode becomes forward biased, the capacitance is destroyed and the diode performs as a normal diode.

VVC diodes are used in electronic tuning circuits. They appear the same as normal diodes and are rugged, reliable, and have a fast response and

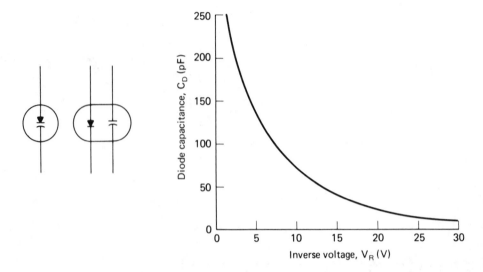

Figure 2-13 VVC diode schematic diagrams and characteristic curve.

a fairly wide frequency range. Since the VVC diode can be tuned electrically rather than mechanically, a good advantage is found in remote tuning.

Several manufacturers produce VVC diodes, under such trade names as Varactor, Varicap, Epicap, and Voltacap.

In the tuning circuit shown, the VVC diode is reverse biased and the amount of dc control voltage changes the capacitance, which is in parallel to the tank circuit. The change in capacitance accomplishes the tuning of the circuit. The series capacitor blocks dc voltage from the tank circuit.

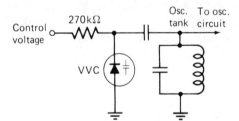

Figure 2-14 VVC diode tuning circuit.

2-6 MISCELLANEOUS DIODES

A *backward diode* is similar to a zener diode, except that it is specially doped and breaks down sooner in the reverse-biased condition than in the forward-biased condition. This diode can be used to rectify small level signals that are difficult to amplify before rectification.

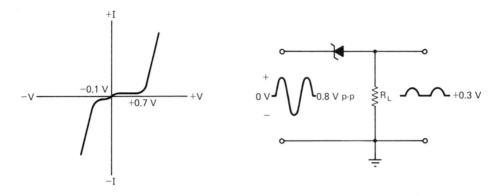

Figure 2-15 Backward diode characteristic curve and circuit application.

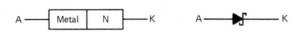

Figure 2-16 Schottky diode junction structure and schematic symbol.

The *Schottky diode* is a unipolar device in which the anode is made of metal and the cathode is made of *n*-type material. The metal has no holes, so there is no depletion layer or potential barrier present, enabling the diode to switch off and on faster than a bipolar diode. A Schottky diode is well suited for high-frequency rectification.

A *step-recovery diode* (sometimes called a snap diode) is especially doped so that the concentration of impurities drops off near the junction. A regular diode rectifies well at low frequencies, but at high frequencies the depletion region does not have time to switch back and forth, resulting in somewhat of a reverse current. This also occurs in a snap diode, except at a certain point in the reverse-biased condition, the diode appears to "snap" open. This diode finds applications in RF amplifiers and frequency multipliers.

The *p-i-n diode* has a thin layer of intrinsic (undoped) silicon between the *p* and *n* regions. With no bias applied, the intrinsic (*i*) region is empty of free charges. In the forward-biased condition electrons from the *n* region flow into the *i* region and lower its ac resistance. The greater the dc current, the lower the ac resistance. This particular phenomenon finds application in RF modulator circuits.

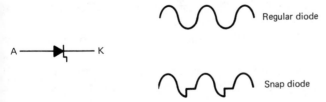

Figure 2-17 Snap diode schematic symbol and comparison of rectification with regular diode.

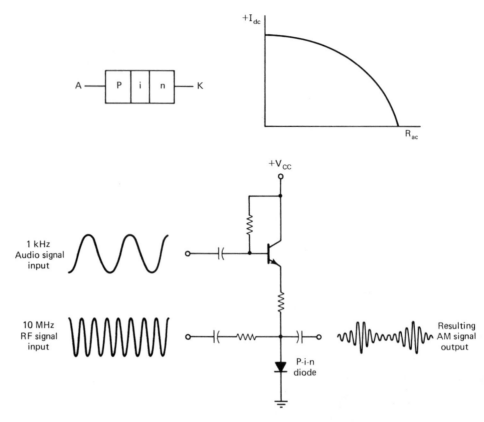

Figure 2-18 *P-i-n* diode junction structure, ac resistance characteristic curve, and circuit application.

2-7 SILICON-CONTROLLED RECTIFIER (SCR)

The *silicon-controlled rectifier* (SCR) belongs to the thyristor family (four-layer PNPN solid-state devices). Its official name is reverse-blocking triode thyristor. The SCR acts as a normal diode except that it is usually triggered into conduction in the forward-biased condition by a positive pulse applied to the gate lead. Once it is triggered into conduction, the gate loses control over the current flow and the SCR acts like a latched switch. To turn off the SCR, the current must be decreased below its minimum holding current. This is usually accomplished by opening the cathode or anode lead or by reverse biasing.

The SCR can be used in ac control; however, it operates only during the positive alternation of the cycle. With the simplest circuit using only a

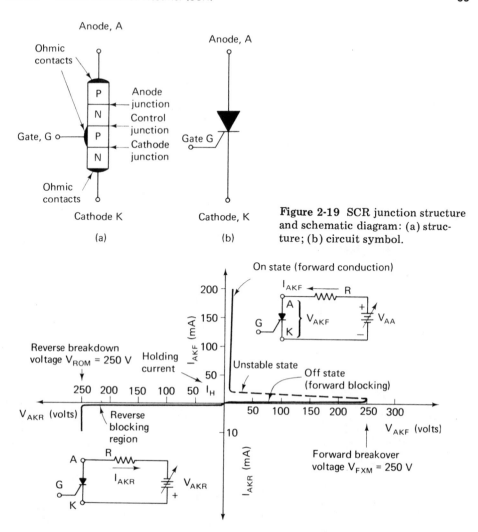

Figure 2-19 SCR junction structure and schematic diagram: (a) structure; (b) circuit symbol.

Figure 2-20 SCR characteristic curve.

potentiometer control, it can be triggered from about 0 to 90°, while the conduction time will be from about 0 to 180°. By adding a capacitor to the control circuit, the SCR trigger time can be extended to 180°.

When the SCR is open (not conducting) the applied voltage will appear across its terminals, while the voltage across the load will be zero, since no current is flowing. After being triggered, the resistance of the SCR is low and the voltage across it is minimal, while the voltage across the load will be the remaining applied voltage.

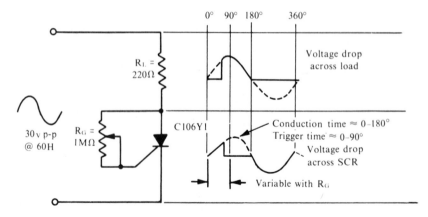

Figure 2-21 Basic ac triggering.

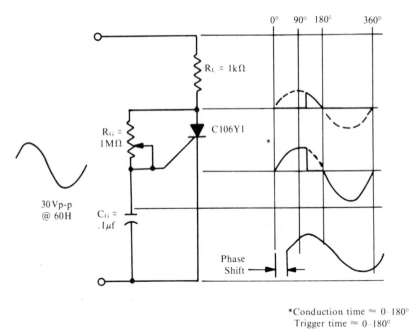

*Conduction time ≈ 0–180°
Trigger time ≈ 0–180°

Figure 2-22 *RC* phase-control triggering.

2-8 BIDIRECTIONAL DIODE THYRISTOR (DIAC)

The *bidirectional diode thyristor* (DIAC) operates similar to two zener diodes back to back. The DIAC will conduct current in either direction, whenever the breakover voltage is exceeded. While in conduction, the voltage across the DIAC will be less than the breakover voltage due to the negative resistance action. When the current through the DIAC decreases below the minimum holding current, the device stops conducting and appears as an open circuit. Like other solid-state devices, the DIAC should have some external series resistance connected to it to limit current and prevent possible burnout. DIACs can be used as protection devices by limiting peak pulses. They are also used quite extensively with other thyristor circuits, since a sharp current pulse is available at the voltage breakover.

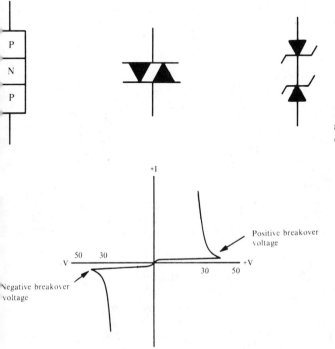

Figure 2-23 Diac structure, schematic symbol, and equivalent circuit.

Figure 2-24 Diac characteristic curve.

Figure 2-25 Ac limiting with a Diac.

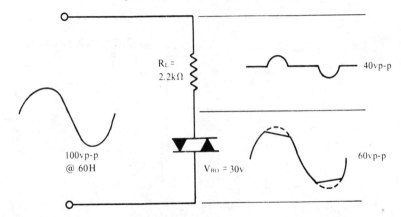

2-9 BIDIRECTIONAL TRIODE THYRISTOR (TRIAC)

The *bidirectional triode thyristor* (TRIAC) is capable of conducting current in either direction and can be triggered by a positive or a negative voltage. This is possible because the metal leads contact a p and an n region simultaneously. The TRIAC is similar in operation to that of two reversed SCRs in parallel and, when triggered, the gate also loses control over current flow. The lead names "anode" and "cathode" are replaced by terminal 1 and terminal 2, since the TRIAC conducts in both directions. The four modes of operation are presented in Table 2-1.

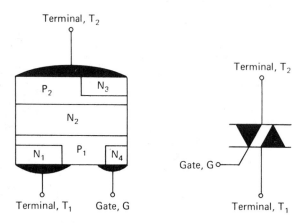

Figure 2-26 Triac structure and schematic symbol.

Figure 2-27 Triac forward and reverse characteristics.

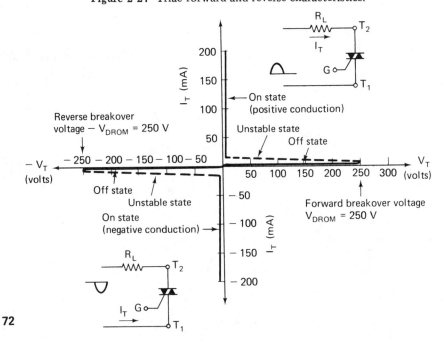

TABLE 2-1 MODES OF OPERATION OF TRIACS

T_1	T_2	Gate trigger
Positive	Negative	Positive
Positive	Negative	Negative
Negative	Positive	Positive
Negative	Positive	Negative

The TRIAC is well suited for ac control, because it can conduct during the positive and negative alternations of a cycle. A simple resistive control will control up to 90° of each alternation. The addition of a capacitor will increase the control for nearly the entire cycle. This type of circuit with the DIAC is used in the popular dimmer switches. When the TRIAC is not conducting, the input voltage will appear across its terminals. During conduction, the voltage across the TRIAC will be nearly zero, while the voltage across the load (R_L) will be the input voltage.

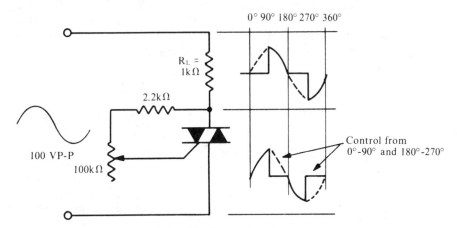

Figure 2-28 Simple ac triggering of a triac.

Figure 2-29 *RC* phase control triggering of a triac.

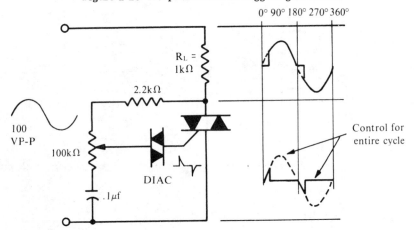

The *Shockley diode*, often called a reverse-blocking diode thyristor, is a four layer device that operates similarly to a DIAC. It is triggered into conduction in the forward-biased condition by exceeding the anode breakover voltage. This diode is used for SCR triggering, timing circuits, and pulse generators.

A *silicon-controlled switch* (SCS), also called a reverse-blocking tetrode thyristor, operates like an SCR, but can be triggered on by either gate lead. A positive pulse on the cathode gate lead will turn it on, while a negative pulse to the anode gate lead will accomplish the same thing. It is primarily used in drivers, alarm circuits, and counters.

The *silicon unilateral switch* (SUS) is triggered on by breakover voltage or a gate signal. A zener junction added to the gate determines the positive voltage trigger level, and it can also be triggered by negative voltage. The SUS is used in low-voltage trigger circuits, timing circuits, and threshold detectors.

A *silicon bilateral switch* (SBS) operates similar to a SUS, but will conduct in either direction. It can be triggered on by exceeding the breakover voltage in either direction or a positive or negative gate voltage. The SBS has applications as a low-voltage TRIAC trigger and threshold detector.

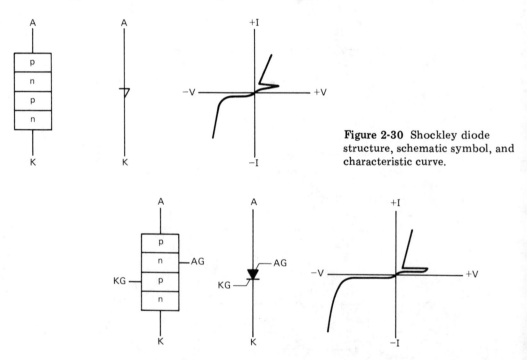

Figure 2-30 Shockley diode structure, schematic symbol, and characteristic curve.

Figure 2-31 SCS structure, schematic symbol, and characteristic curve.

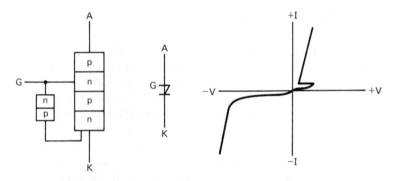

Figure 2-32 SUS structure, schematic symbol, and characteristic curve.

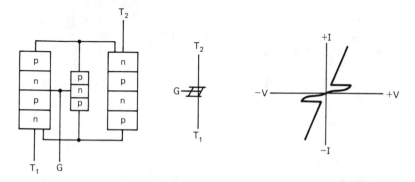

Figure 2-33 SBS structure, schematic symbol, and characteristic curve.

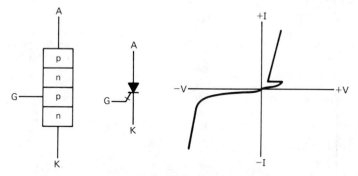

Figure 2-34 GCS structure, schematic symbol, and characteristic curve.

The *gate-controlled switch* (GCS), also called a turnoff thyristor, is similar to an SCR. A positive pulse to the gate will trigger it into conduction, while a negative pulse to the gate will turn it off. The GCS is used in dc switching, inverters, and choppers.

2-11 UNIJUNCTION TRANSISTOR (UJT)

The *unijunction transistor* (UJT), originally called the double-base diode, consists of a bar of n-type silicon and an emitter (E) area of p-type material. In a circuit, base 2 (B_2) is made more positive than base 1 (B_1). A small amount of current will flow from B_1 to B_2, even though the UJT is considered off. A voltage gradient exists along the bar, and when the emitter voltage is equal to the gradient voltage opposite the pn junction (V_G), plus the 0.7-V drop (V_D), the junction is forward biased and the UJT turns on. This firing or peak voltage (V_P) is expressed as $V_P = V_G + V_D$. Upon firing, the resistance between B_1 and E decreases rapidly and the current increases from B_1 to B_2, while there is also a substantial current from E. To turn off the UJT, the voltage at the emitter must be below the valley voltage (V_V). The emitter firing voltage (V_P) is determined by the intrinsic standoff ratio, $\eta = R_{B_1}/R_{BB}$, where R_{B_1} is the resistance from B_1 to E and R_{BB} is the resistance from B_1 to B_2.

In a UJT relaxation oscillator the capacitor charges up to V_P. When the UJT fires, the capacitor discharges back through the B_1 to the E circuit and

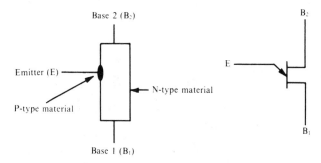

Figure 2-35 UJT structure and schematic symbol.

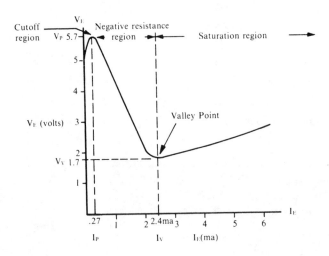

Figure 2-36 UJT emitter characteristic curve.

76

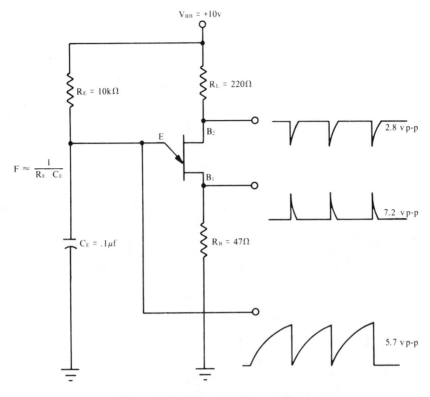

Figure 2-37 UJT relaxation oscillator.

the cycle is then repeated. If V_P is about 63% of V_{BB}, the approximate frequency can be found from the formula $F \approx 1/R_E C_E$.

2-12 PROGRAMMABLE UNIJUNCTION TRANSISTOR (PUT)

The *programmable unijunction transistor* (PUT) belongs to the thyristor family. However, its characteristics allow it to be used in UJT-type circuits. The PUT works as a normal diode except that it is triggered into conduction in the forward-biased condition by a negative voltage applied to the gate lead. Once it is triggered into conduction, the gate loses control over the current flow and the PUT acts like a latched switch. To turn off the PUT, the current must be decreased below its minimum holding current. When the gate and cathode are at a fixed voltage and the voltage at the anode is increased to a higher potential, the PUT will turn on and has the same effect as applying a negative voltage to the gate. In this manner a PUT can perform as a UJT and the voltage at which it turns on is determined (programmed) by the resistive voltage divider at the gate. The turn-on voltage (V_p) is found

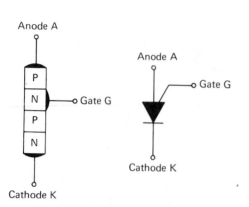

Figure 2-38 PUT structure and schematic symbol.

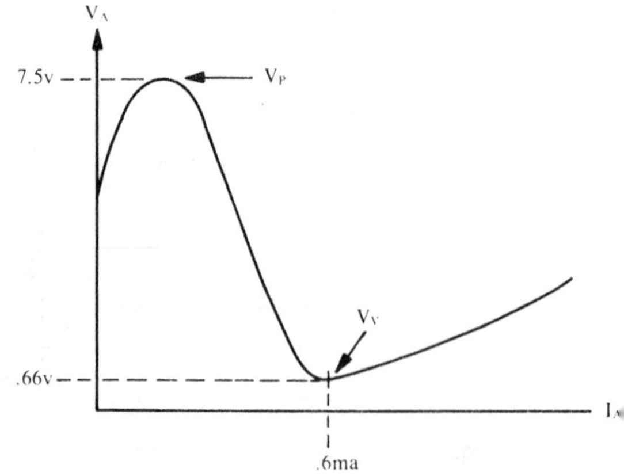

Figure 2-39 PUT characteristic curve.

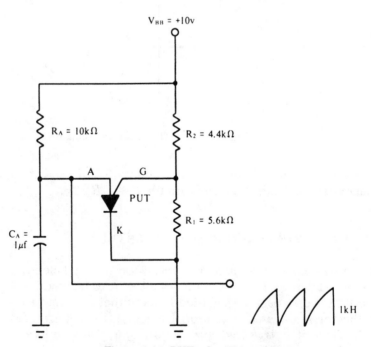

Figure 2-40 PUT relaxation oscillator.

thus: $V_p = V_G + V_D$, where V_G is the voltage at the gate and V_D (0.7 V) is the voltage drop across the forward-biased junction. Therefore,

$$V_G = R_1 \frac{V_{BB}}{R_1 R_2}$$

The *NPN bipolar transistor* consists of an *n*-type emitter (*E*), *p*-type base (*B*), and *n*-type collector (*C*). In a normal circuit, the *EB* junction is forward biased and the *BC* junction is reverse biased. A bipolar transistor is a current-operated device and a small current from *E* to *B* controls a large current from *E*, through *B*, to *C*. The base region is exceptionally thin, allowing the large current to pass through from *E* to *C*. The amount of collector current (I_C) is directly proportional to the amount of base current (I_B). The collector current (I_C) will be less then the emitter current (I_E), since a small base current (I_B) must flow to turn on the transistor. The relationship of the currents is $I_C = I_E - I_B$ or $I_B = I_E - I_C$ or $I_E = I_B + I_C$. The ratio of I_B to I_C is called the current gain of the transistor and indicates its ability to amplify. This current gain is called beta (β) and is expressed as $\beta = \Delta I_C / \Delta I_B$, when the voltage from *C* to *E* (V_{CE}) is held constant.

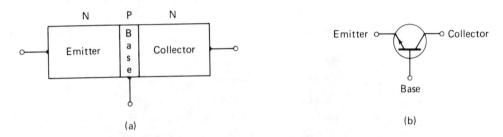

(a)

(b)

Figure 2-41 NPN bipolar transistor structure and schematic symbol.

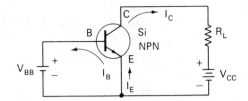

Figure 2-42 Current paths in an NPN transistor.

To turn on an NPN bipolar transistor, the base must be more positive than the emitter (about +0.6 V). When it is turned on hard (in saturation), this voltage is about +0.7 V and the resistance from *C* to *E* is low and may even appear almost as a "short." When the transistor is off, the resistance from *C* to *E* is high and may appear as an "open." A small leakage current (I_{CO}) or (I_{CBO}) from *C* to *B* is always present and may cause stability problems for a transistor circuit.

79

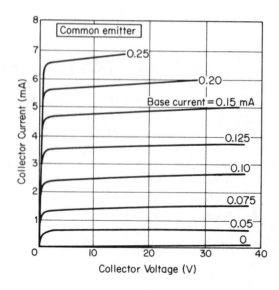

Figure 2-43 NPN transistor output characteristics for a common-emitter circuit.

2-13.1 Some Transistor Specifications

V_C	dc voltage from collector to ground
V_B	dc voltage from base to ground
V_E	dc voltage from emitter to ground
V_{EB} or V_{BE}	dc bias voltage from emitter to base
V_{CE} or V_{EC}	dc voltage from collector to emitter
BV_{CEO}	dc breakdown voltage, collector to base with base open circuited
BV_{CBO}	dc breakdown voltage, collector to base, with emitter open circuited
BV_{EBO}	dc breakdown voltage, emitter to base, with collector open circuited
P_T	total collector power dissipation ($V_C \times I_C$)
h_{fe}	same as beta (β)
f_{ab}	frequency response of the transistor

2-14 PNP BIPOLAR TRANSISTOR

The *PNP bipolar transistor* consists of a p-type emitter (E), n-type base (B) and p-type collector (C). Its operation is the same as an NPN bipolar transistor, except that the supply voltages are reversed (see Section 2-13). Also the current through the transistor consists mainly of hole flow. The PNP bipolar transistor is turned on when B is more negative then E. Notice that

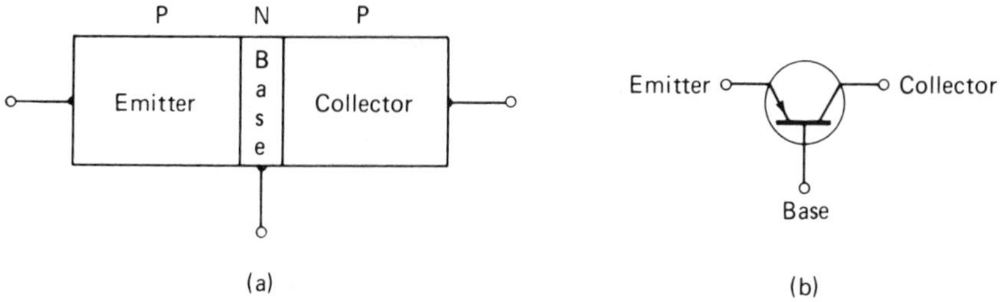

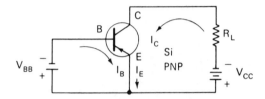

Figure 2-44 PNP bipolar transistor structure and schematic symbol.

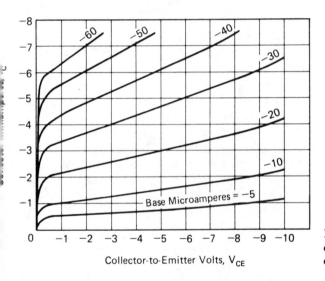

Figure 2-45 Current paths in a PNP transistor.

Figure 2-46 PNP transistor output characteristics for a common-emitter circuit.

the voltage at C (V_C) will be less negative (closer to zero) when the transistor is on, while V_C will be maximum negative when the transistor if off.

2-15 DARLINGTON PAIR TRANSISTORS

Darlington pair transistors are formed by connecting together the collectors of two transistors, while feeding the emitter of one to the base of the other. The advantages of this arrangement are extremely large current gain, less loading effect, high input resistance, and impedance matching.

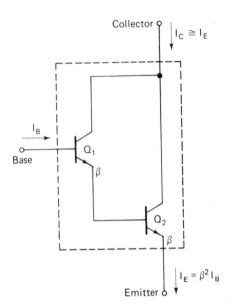

Figure 2-47 Darlington pair NPN transistors.

If the β for Q_1 is 100 and the β for Q_2 is 100, then the total current gain is 10,000, as given by the formula $\beta_T = \beta_1 \times \beta_2$. When 0.1 mA flows in the base of Q_1, the current drawn by the emitter of Q_2 would be 1.0 A, as given by the formula $I_E = \beta_1 \times \beta_2 \times I_B$.

The input resistance can be calculated if the emitter resistor is known, by the formula $R_i = \beta_1 \times \beta_2 \times R_E$. For example, an R_E of 1 KΩ would yield an R_i of 10 MΩ. Any biasing resistors or stages before or after the Darlington pair would be in parallel with this R_i of 10 MΩ. Darlington pairs can be constructed from discrete transistors or are available commercially. A commercial unit will usually have matched betas and the leads will correspond to a normal transistor (i.e., E, B, C).

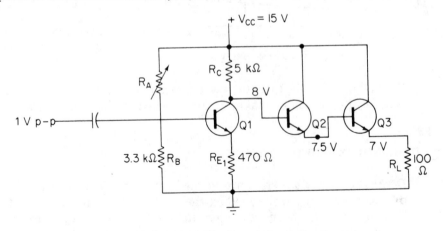

Figure 2-48 Darlington pair transistor application.

2-16 POWER TRANSISTORS

Power transistors are the solid-state workhorses that operate speakers, relays, motors, and devices requiring a specific amount of power. Their characteristics are similar to low-power, small-signal transistors; however, they are manufactured to handle higher values of voltage and current. Three common power transistor packages are the TO-5 with flange, TO-3 (TO-66), and TO-220. The collector is usually internally connected to the case. This case is mounted to the chassis or heat sink to provide heat dissipation. A heat sink is made of metal and may consist of a radial fin which pushes over the case of the transistor, or a simple fin (or multiple fin) to which the transistor is bolted. Since the case is part of the electrical circuit of the transistor, it may have to be insulated from the heat sink or chassis by nylon bushings and special mica insulators.

Other types of solid-state devices are also found in these power device packages; therefore, part identification numbers and device specifications are extremely important in servicing equipment.

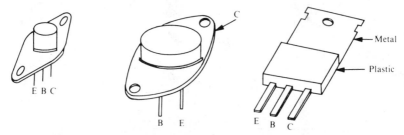

Figure 2-49 Power transistors: TO-5 with flange, TO-3 (TO-66), and TO-220.

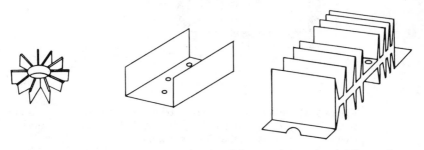

Figure 2-50 Heat sinks: radial fin for TO-5 package, simple fin, and multiple fin.

2-17 JUNCTION FIELD-EFFECT TRANSISTOR (JFET)

An *n*-channel *junction field-effect transistor* (JFET) consists of a bar of *n*-type silicon with a source terminal at one end and a drain terminal at the other end. *P*-type silicon is diffused on each side or completely around the bars with a gate terminal attached. The drain is connected to a positive potential, while the source is connected to a negative potential. When the gate is shorted to the source ($V_{GS} = 0$ V), maximum current (electrons) flows from source to drain, called drain current (I_D). If V_{GS} is made more negative, a depletion region occurs, which reduces the n-channel width, increasing the resistance of the JFET and decreasing I_D. (The depletion region acts similarly to the electrostatic charge on the control grid of a vacuum tube.) It would appear that if the voltage from drain to source (V_{DS}) were varied, I_D would also vary. However, the voltage gradient along the channel varies, which changes the depletion width, allowing more or less electrons to flow so that I_D remains constant. The current through the JFET is not cut off, but rather is "pinched off" for various values of V_{GS}. If V_{GS} is made positive, the gate would draw current due to normal pn-junction forward biasing and upset the stability of the JFET.

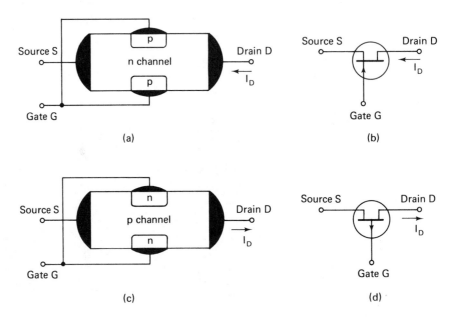

Figure 2-51 Junction field-effect transistor structures and schematic symbols: (a) n channel; (b) circuit symbol for an n channel; (c) p channel; (d) circuit symbol for a p channel.

Two major advantages of the JFET are its high input impedance (because the junction is normally reverse biased) and the fact that it can operate from low- and high-voltage power supplies. The JFET is considered a "normally on" device, since with no bias ($V_{GS} = 0$ V), maximum I_D flows. A p-channel JFET operates the same way, except that all the voltages are reversed and its current flow is in the form of holes.

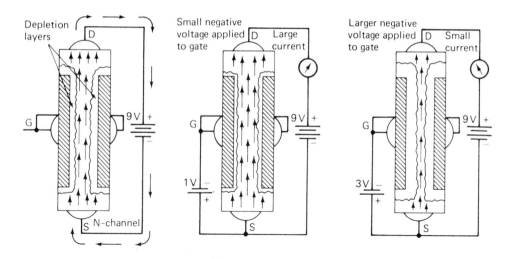

Figure 2-52 *N*-channel JFET operation.

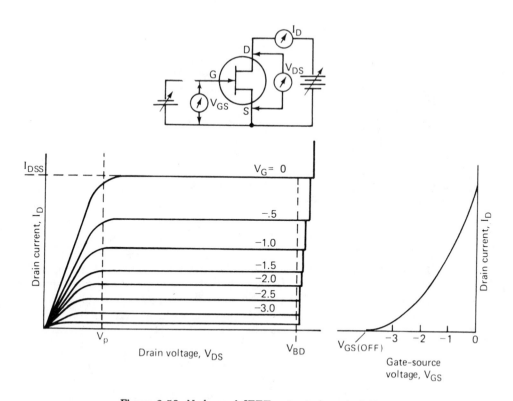

Figure 2-53 *N*-channel JFET output characteristics.

2-17.1 FET Circuit Nomenclature

V_{DD}	power supply voltage applied to the drain
V_{SS}	power supply voltage applied to the source (in many cases this is ground)
V_D	voltage from drain to ground
V_{DS}	voltage across drain and source
V_G	voltage from gate to ground
V_S	voltage from source to ground
V_{GS}	bias voltage across the gate and source
V_P	pinch-off voltage
I_D	drain current (current from source to drain)
I_{DSS}	saturation drain current, when $V_{GS} = 0$ V
I_{GSS}	gate-to-source leakage current with the drain shorted to the source
G_{FS}	forward transconductance or the amplification ability of the FET, usually given in μmhos:

$$G_{FS} = \frac{\Delta I_D}{\Delta V_{GS}}$$

Y_{FS}	forward transadmittance, the same as G_{FS}
BV_{GSS}	gate-to-drain reverse breakdown voltage with drain shorted to source (this is the algebraic difference of V_D and V_G)
BV_{DGS}	drain-to-source breakdown voltage with the source shorted to the gate

2-18 DEPLETION-TYPE METAL-OXIDE SEMICONDUCTOR FIELD-EFFECT TRANSISTOR (MOSFET)

The *depletion-type metal-oxide semiconductor field-effect transistor* (MOSFET) has a source (S) and drain (D) connected by a narrow channel of the same type of semiconductor material. A metal gate (G) is placed above the channel but is insulated from the semiconductor material by a thin layer (approximately 1×10^{-6} m) of silicon dioxide (SiO_2) that has a resistivity of about 1×10^8 MΩ. The SiO_2 layer is so thin that an electrostatic field produced by the bias voltage (V_{GS}) on the gate, still penetrates and controls the conductivity of the channel.

With an n-channel depletion-type MOSFET, the drain is connected to a positive potential and the source to a negative potential or ground. When $V_{GS} = 0$ V, a specific amount of drain current (I_D) flows from S to D. If V_{GS} is made negative, the depletion region extends into the channel area, increasing resistance and decreasing I_D. Since, the gate and channel are

86

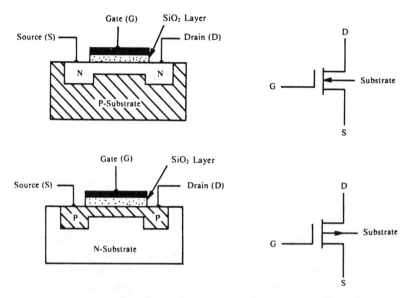

Figure 2-54 Depletion-type MOSFET: *n*-channel and *p*-channel.

Figure 2-55 Operation of a depletion-type *n*-channel MOSFET; (a) V_{GS} is zero; (b) V_{GS} is negative; (c) V_{GS} is more negative (d) V_{GS} is positive.

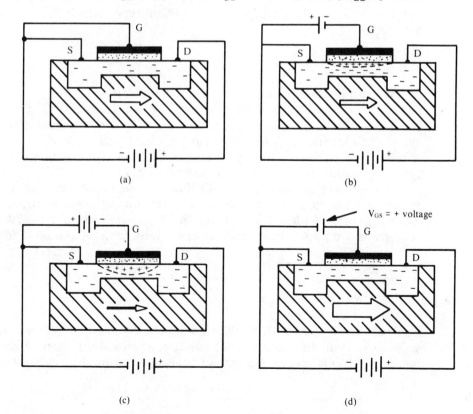

(a)

(b)

(c)

(d)

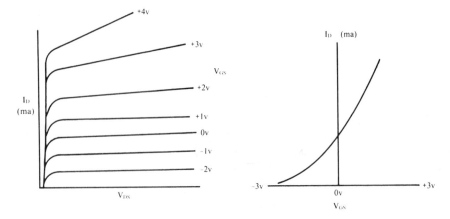

Figure 2-56 Drain characteristic curves and transfer curve for a depletion-type n-channel MOSFET.

insulated (no pn junction), V_{GS} may go positive, which enhances or creates more current flow than when V_{GS} = 0 V.

A p-channel depletion-type MOSFET operates the same way, except that all voltages are reversed and the current through the device is hole flow.

A graph of output characteristics shows that a depletion-type MOSFET operates in the depletion region with reverse bias and in the enhancement region with forward bias. Bias voltage V_{GS} controls drain current I_D. A transfer characteristic curve also shows how V_{GS} can be made positive and negative and its effect on I_D. The depletion-type MOSFET is a "normally on" device.

2-18.1 A Word of Caution

Special care in handling must be observed with MOS devices, since the thin SiO_2 layer can be ruptured by a static charge as little as 100 V. This rupture occurs between gate and channel and usually results in excessive gate leakage current. Soldering tools and a person working with MOS devices must be grounded or discharged of any static charges. Once in the circuit, the MOS device is usually protected by other circuit components.

Some MOS devices have zener diode protection from gate to source, which is fabricated into the same structure during manufacturing. These diodes protect the device against potential damage from in-circuit transient voltages and out-of-circuit handling operations, without the need for external shorting devices.

2-19 ENHANCEMENT-TYPE MOSFET

The *enhancement-type MOSFET* has a source (S) and drain (D) region of the same semiconductor material, but is not connected by a channel region. A metal gate (G) is placed above the nonchannel region, but is insulated by a

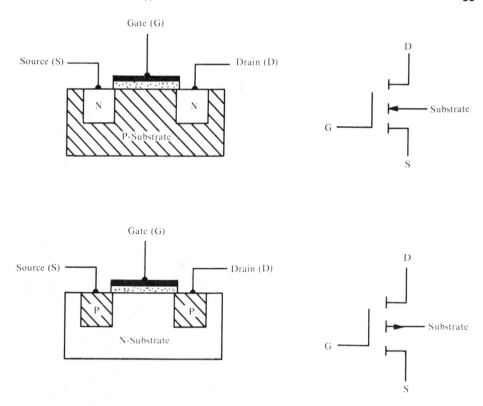

Figure 2-57 Enhancement-type MOSFET; n-channel and p-channel.

thin layer (approximately 1×10^{-6} m) of silicon dioxide (SiO_2), having resistivity of about 1×10^8 MΩ. This SiO_2 layer is so thin that an electrostatic field produced by a forward-biased voltage (V_{GS}) on the gate still penetrates and controls the conductivity of the channel. The enhancement-type MOSFET is a "normally off" device and a channel is created when forward bias is applied to the gate.

With an n-channel enhancement-type MOSFET, the drain is connected to a positive potential, while the source is connected to a negative potential. When V_{GS} = 0 V, there is no current flow through the device, because there is no path from S to D. If a positive voltage is placed on the gate, electrons from the substrate are attracted to the underside of the gate, thereby creating a channel from S to D, which allows current I_D to flow. An increase in V_{GS} attracts more electrons from the substrate, widens the channel, and I_D increases.

A graph of the output characteristics shows that forward bias must be applied to the gate to cause current flow (a similar operation to that of turning on bipolar transistors). The transfer characteristics curve also shows that reverse bias prevents any current flow and keeps the MOSFET off.

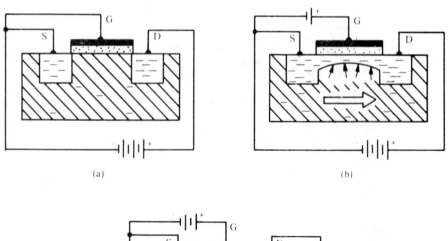

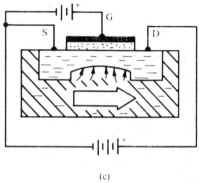

(c)

Figure 2-58 Operation of an enhancement-type n-channel MOSFET:
(a) V_{GS} is zero; (b) V_{GS} is positive; (c) V_{GS} is more positive.

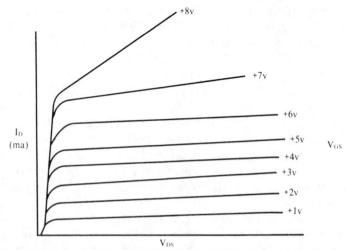

Figure 2-59 Drain characteristics of an enhancement-type n-channel MOSFET.

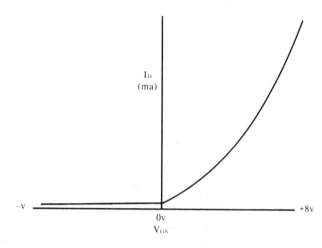

I_D
(ma)

−v

0v
V_{GS}

+8v

Figure 2-60 Transfer characteristic curve for an enhancement-type *n*-channel MOSFET.

A *p*-channel enhancement-type MOSFET operates the same way except that the voltages are reversed and the current flow is holes.

Care must also be used with handling enhancement-type MOSFETs, so as not to rupture the thin SiO_2 layer with static charges. Some of them also have diode input protection.

Because of the SiO_2 layer at the gate, MOSFETs have a high input impedance. MOSFETs can be used for low-frequency applications, but their best use is in high-speed switching circuits, RF circuits, linear ICs, and digital ICs.

2-20 VERTICAL MOSFET (V-MOSFET)

The *vertical MOSFET* (V-MOSFET) has the same schematic symbol and operates similar to the enhancement-type *n*-channel MOSFET except that it is fabricated with the gate in the form of a V. This feature increases the channel cross-sectional area, enabling the MOSFET to handle more current. This increased power capability makes the V-MOSFET more desirable than bipolar transistors in such circuits as switching power supplies, motor controllers, high-efficiency switching amplifiers, and audio amplifiers. MOS technology alleviates the minority-carrier storage delay of bipolar devices and the V-MOSFET has a decreased channel length, both of which increase operating speed. An increased operating speed makes the V-MOSFET particularly advantageous for RF circuit applications. RF circuits using power V-MOSFETs range upward to 4 GHz.

Other advantages of the V-MOSFET are: high voltage–current–power operation, good linearity, high input impedance, low noise, excellent thermal stability, and simple dc biasing, since no dc input power is consumed. The

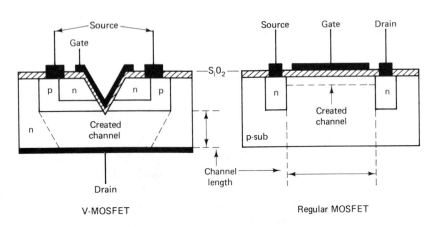

Figure 2-61 V-MOSFET schematic symbol and *pn* structure compared with a regular MOSFET structure.

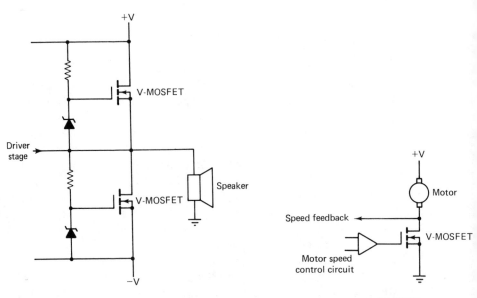

Figure 2-62 V-MOSFET audio output.

Figure 2-63 V-MOSFET motor control.

main disadvantage is high input capacitance, which may make impedance matching difficult in some applications.

2-21 TRANSISTOR-TRANSISTOR LOGIC (TTL OR T²L)

Transistor–Transistor logic (TTL or T²L) refers to a type of solid-state circuit configuration used in digital logic circuits. The transistors are directly coupled where the condition of Q_1 controls the condition of Q_2, which in

Figure 2-64 Basic TTL circuit: NAND gate logic symbol and schematic diagram.

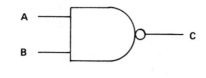

(a)

(b)

turn controls the condition of Q_3 and Q_4. Components R_3, Q_3, D_1, and Q_4 are referred to as a "totem-pole" output. The output transistors Q_3 and Q_4 will never be on at the same time and D_1 is used to make sure of this condition. The inputs to the circuit are the multiple emitters of Q_1.

When all inputs are high (positive), current (conventional flow) flows through R_1 and the BC junction of Q_1 into the base of Q_2, turning it on. This allows current to flow through R_2, Q_2, R_4, and the base of Q_4, turning it on. Q_4, being on, "sinks" or pulls the output toward ground. Q_3 is off at this time.

When any or all inputs are low (at ground), current flows through R_1 and the BE junction of Q_1 to ground. Q_2 is off at this time. Current also flows through R_2, to the base of Q_3, turning it on, which allows current to flow through R_3, Q_3, and D_1 to the output. The output is now "pulled up" toward V_{CC} through D_1, Q_3, and R_3. The 7400 IC logic family uses this type of circuitry, where V_{CC} = +5 V. A high-output condition will equal approximately +3.6 V; a low-output condition will equal about +0.2 V.

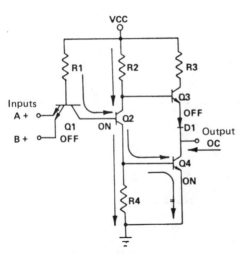

Figure 2-65 When all inputs are high (1).

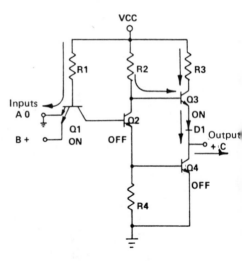

Figure 2-66 When any one or all inputs are low (0).

2-22 COMPLEMENTARY METAL-OXIDE SEMICONDUCTOR (CMOS)

The *complementary metal-oxide semiconductor* (CMOS) is an integrated circuit in which an *n*-channel enhancement-type MOSFET and a *p*-channel enhancement-type MOSFET are fabricated together at the same time to form a complementary circuit. Applications using CMOS are found in digital and linear circuits. A basic CMOS circuit used as a digital inverter has the drains and gates connected together. The source of the *p*-channel MOSFET is connected to a positive voltage (V_{DD}), while the source of the *n*-channel MOSFET is connected to a negative voltage (V_{SS}) or ground.

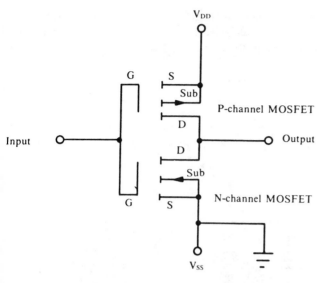

V_{DD}

G   S

Sub

P-channel MOSFET

D

Input   Output

D

Sub

N-channel MOSFET

G   S

V_{SS}

Figure 2-67 Basic CMOS circuit.

When the input at the gates is 0 V, the *p*-channel MOSFET is on and the *n*-channel MOSFET is off. The output at the drains would be pulled up toward V_{DD} or indicate a high (1).

When the input is made positive (high or 1), the *n*-channel MOSFET will be on and the *p*-channel MOSFET will be off. The output is now pulled down toward ground or indicates a low (0).

The advantages of CMOS ICs over other types of ICs include:

Operation from a wider range of power supply voltages (approximately 1.3 to 15 V).

Less power dissipation because of lower operating current.

Easier to manufacture.

MOSFETs require less space than bipolar devices, thereby increasing the packing density (more circuits per given area).

Power supply regulation is less critical.

Extremely high input impedance, resulting in fewer loading problems.

Inherently good noise immunity.

Excellent thermal operating characteristics.

Many CMOS ICs are compatible pin for pin with 7400 TTL ICs.

2-23 TESTING SOLID-STATE DEVICES

Commercial test instruments are always the best to use when testing solid-state devices; however, simple GO/NO GO functional tests can be performed on most devices with an ohmmeter. A forward-biased *pn* junction exhibits low resistance, while a reverse-biased *pn* junction shows high resistance. A shorted *pn* junction will show a low-resistance reading in both directions, whereas an open *pn* junction will show a high-resistance reading in both directions.

2-23.1 Testing Diodes

Place the ohmmeter leads as shown. A low-to-high meter reading of $1:10$ for rectifier diodes is considered good, while a $1:100$ ratio is acceptable for switching diodes.

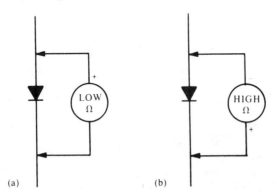

(a)   (b)

Figure 2-68 Testing a diode: (a) forward biased—minimum resistance; (b) reverse biased—maximum resistance.

2-23.2 Testing Zener Diodes

Zener diodes can be checked for opens and shorts with an ohmmeter. The circuit shown can be used to determine if a zener diode is operating properly in the breakdown region. The power supply voltage should be a few volts more positive than the zener voltage (V_Z).

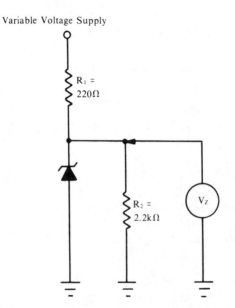

Figure 2-69 Testing a zener diode.

2-23.3 Testing SCRs

Set the ohmmeter on the R × 1 scale and place across the SCR in the for-ward-biased condition. The meter should indicate high resistance. Place a clip lead from the positive lead of the ohmmeter to the gate. The meter should indicate low resistance. When the clip lead is removed, the meter should indicate low resistance. If it does not, the ohmmeter may not be sup-plying enough holding current to keep the SCR on.

2-23.4 Testing PUTs

A PUT is tested in the same way as an SCR except that the clip lead is placed on the negative lead of the ohmmeter. The PUT is then triggered on by applying the clip lead to the gate.

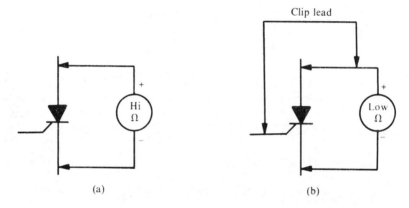

(a)   (b)

Figure 2-70 Testing an SCR: (a) high resistance before firing; (b) low resistance after firing and after clip lead is removed, depending on holding current.

2-23.5 Testing TRIACs

A TRIAC is tested similarly to an SCR except that both directions of conduc-tion must be checked as indicated.

2-23.6 Testing Other Thyristors

Other thyristors using a gate for triggering can be tested in a similar manner as for SCRs and TRIACs. Nontriggered thyristors, such as DIACs, will have to be tested using a power supply and limiting resistors similar to the testing method for a zener diode.

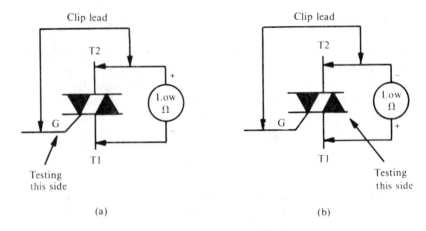

Figure 2-71 Testing a triac: (a) testing for positive triggering; (b) testing for negative triggering.

2-23.7 Testing UJTs

The resistance from base 1 to base 2 of a UJT should be several thousand ohms. When a clip lead is placed from the positive lead of the ohmmeter to the emitter, the resistance should decrease, perhaps to one-half, as indicated. When the clip lead is removed, the resistance will return to its original value.

2-23.8 Testing Bipolar Transistors

The *pn* junctions of NPN and PNP transistors can be tested as indicated.

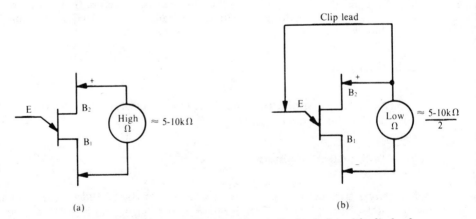

Figure 2-72 Testing a UJT: (a) without clip lead; (b) with clip lead.

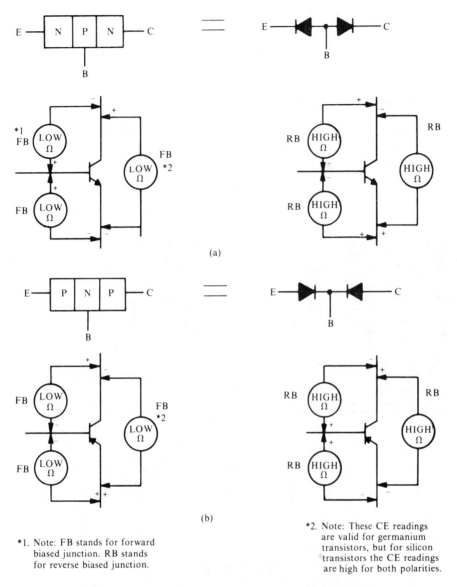

(a)

(b)

*1. Note: FB stands for forward
   biased junction. RB stands
   for reverse biased junction.

*2. Note: These CE readings
   are valid for germanium
   transistors, but for silicon
   transistors the CE readings
   are high for both polarities.

Figure 2-73 Testing bipolar transistors: (a) NPN; (b) PNP.

2-23.9 Testing JFETs

The JFET can be tested for its channel resistance between drain and source.
It can then be checked for *pn*-junction biasing, as indicated.

Excessive gate leakage current can cause instability in a JFET. A microhmmeter and low-voltage power supply can be used to check this as shown. A reading of more than a few microamps would indicate a defective JFET.

2-23.10 Testing MOSFETs

MOSFETs are usually not tested with an ohmmeter, because of the insulated gate. Also, any static charges could damage MOSFETs that did not have diode protection.

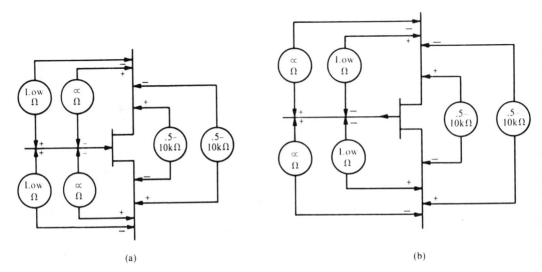

(a)   (b)

Figure 2-74 Testing a JFET: (a) *n*-channel; (b) *p*-channel.

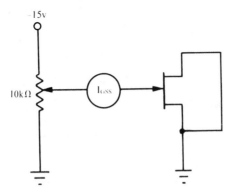

Figure 2-75 Testing JFET gate leakage current.

3

Optoelectronic Devices

3-1 INFORMATION ON OPTOELECTRONICS

Optoelectronic devices combine the technologies of optics and electronics. *Optics* refers to the science of vision and properties of light. *Light* is technically described as electromagnetic radiation of a frequency that can be perceived by the human eye. Also, light is conceived as consisting of minute particles known as *photons*.

Two fundamental actions are used in optoelectronic devices: converting electrical energy into light energy, known as *light emitting* or a *light source*, and converting light energy into electrical energy, known as *light detecting* or *sensing*.

Light emission occurs in semiconductor material when electrons combine with holes. The electrons fall to a lower energy level as they travel from one atom to another and during this process photons are released. Light detection occurs in semiconductor material when light strikes the material and releases valence electrons, creating free electrons and holes.

Germanium (Ge) and silicon (Si) are the most widely used semiconductor materials and are responsive to light. This is the reason standard units of transistors, thyristors, and other devices must have lightproof packages for proper operation. Other semiconductor materials used in optoelectronic devices which are more responsive to light are cadmium sulfide (CdS), gallium phosphide (GaP), cadmium selenide (CdSe), and gallium arsenide (GaAs).

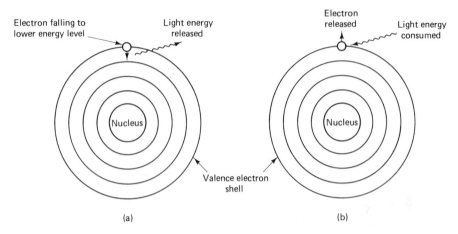

Figure 3-1 Radiant energy versus electrical energy in an atom: (a) electron falling to a lower energy level gives up photon; (b) photon striking valence electron raises the energy level so it can move to another atom.

3-2 PHOTODIODE

Photodiodes are similar to normal diodes except that they are constructed with a small glass window which allows light to strike the *pn* junction. In normal operation, photodiodes are reverse biased, which causes the depletion region to increase. With no light striking the diode very little current flows; this current is referred to as *dark current* (I_D). Practically speaking, the diode represents an open circuit. When light (photon energy) strikes the diode, electron–hole pairs are generated which cross the depletion region and combine to produce a larger current known as *light current* (I_L). The diode is now conducting, but in the reverse direction, and the amount of current flow is proportional to the amount of light striking the diode.

A photodiode can be used in the biasing of an NPN transistor. When no light strikes the diode, there is very little base current (I_B) and the transistor is off. The load, which might be a light bulb, LED, relay, alarm, or other component, will not operate. When light strikes the diode, it conducts, causing I_B to increase, which turns on the transistor and allows current to reach the load.

3-3 PHOTOTRANSISTOR

A *phototransistor* is similar to a normal transistor except that it has a small glass window which allows light to strike the base junction region. It operates similar to a photodiode, but has current gain. The slight current induced in

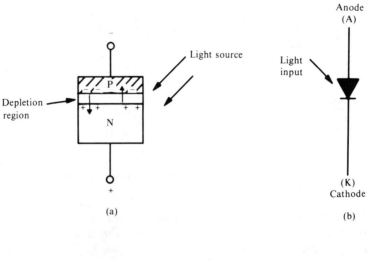

(a)

(b)

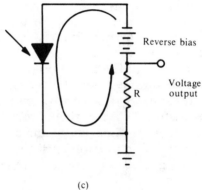

(c)

Figure 3-2 Photodiode: (a) structure; (b) schematic symbol; (c) circuit operation.

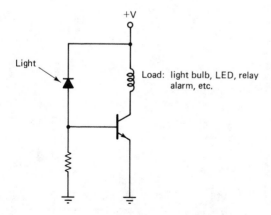

Figure 3-3 Photodiode used to bias a transistor.

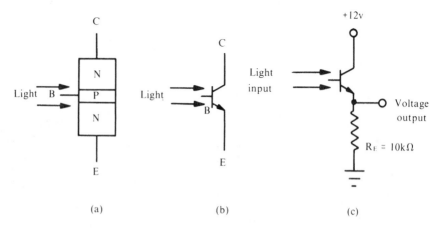

Figure 3-4 Phototransistor: (a) structure; (b) schematic symbol; (c) circuit operation.

the base I_B from the light allows current to flow from the emitter to the collector. In a common-emitter configuration, this photoinduced I_B is amplified by the transistor's beta and can be expressed as $I_C \approx I_E \approx \beta I_B$. Also, the base lead can be left floating or used in a biasing arrangement to provide more circuit control. With no light striking the phototransistor, there is very little current flow (I_D) and essentially the circuit is open. When light strikes the phototransistor, a large current flows (I_L) and the circuit is said to be closed. The amount of current flowing through the circuit is proportional to the intensity of the light. In a simple circuit using only an emitter resistor (R_E), a positive output is present when light reaches the phototransistor. With a circuit using a collector resistor and an emitter resistor, a negative-going output and a positive-going output, respectively, are available. It may also have a resistive biasing arrangement, with one of the resistors being variable to provide more circuit control.

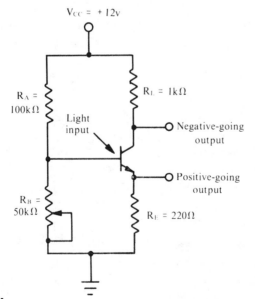

Figure 3-5 Phototransistor circuit with bias control.

3-4 PHOTODARLINGTON TRANSISTOR

A *photodarlington transistor* can provide an increase in load current and serve as isolation between circuits. A phototransistor is combined with a regular transistor to produce the photodarlington transistor. The emitter of Q_1 is directly connected to the base of Q_2 and the collectors are connected together. The package usually has the same appearance as a phototransistor or other photosensing device.

When light is blocked from the unit, the transistors are off and little or no current can flow to the load (R_L). When light strikes the photo-transistor, both transistors turn on, allowing current to flow to the load. The total current is the sum of the two collector currents, $I_T = I_{CQ1} + I_{CQ2}$. Expressed another way, the total collector current is the photoinduced base current of Q_1 multiplied by the beta of each transistor: $I_T = I_{B1}\beta_1\beta_2$.

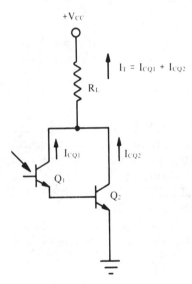

Figure 3-6 Photodarlington transistor.

3-5 PHOTO FIELD-EFFECT TRANSISTOR (PHOTOFET)

The *photo field-effect transistor* (photofet) is similar to a conventional JFET (see Section 2-17), with the exception of a lens for focusing light onto the gate junction. Light falling on the gate area causes valence electrons to be released, which slightly increases nominal gate current I_G. Since I_G flows through gate resistor R_G, a voltage drop is developed, which changes the gate-to-source voltage V_{GS}. This change in the bias voltage is multiplied by the photofet's transconductance (G_{fs}) and a change occurs in the drain current I_D. The change in I_D produces a change in the voltage drop across drain resistor R_D. The drain-to-source voltage changes, thus creating a

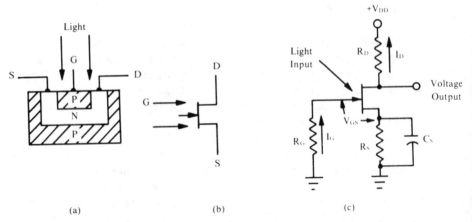

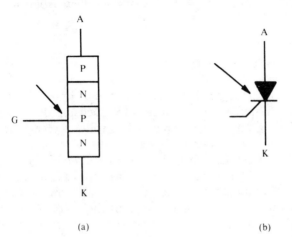

Figure 3-7 Photofet: (a) structure; (b) schematic symbol; (c) circuit operation.

change in the output voltage. The photofet provides a useful combination of a photodetector with high input impedance and low noise in a single device.

3-6 LIGHT-ACTIVATED SCR (LASCR)

The *light-activated SCR* (LASCR) is similar to a conventional SCR, except for a window and lens which focuses light on the gate junction area.

The LASCR operates like a latch. It can be triggered on by a light input to the gate area, but will not turn off after the light source is removed. Similar to a conventional SCR, it can be turned off by reducing the current through it below its minimum holding point.

In a working circuit, gate resistor R_G can be made adjustable to set the trigger threshold point for varying intensities of light. However, too low a

Figure 3-8 Light-activated SCR (LASCR): (a) structure; (b) schematic symbol.

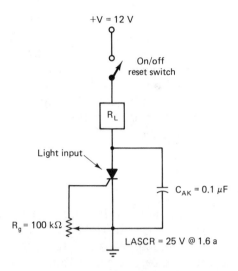

+V = 12 V

On/off
reset switch

R_L

Light input

C_{AK} = 0.1 μF

R_g = 100 kΩ

LASCR = 25 V @ 1.6 a

Figure 3-9 LASCR application for latching up a load.

value of R_G will not allow the LASCR to turn on and too high a value of R_G may cause it to be too sensitive and difficult to adjust. A capacitor from anode to cathode (C_{AK}) prevents false triggering when power is applied to the circuit. On the basis of size, the LASCR is capable of handling larger amounts of current than are photodiodes and phototransistors.

3-7 INFRARED DETECTION SYSTEM

Most standard photosensors will respond to the invisible light of infrared radiation. Infrared detection systems are very well suited in situations where the ambient-light conditions are not easily controlled and may cause unwanted false triggering in a circuit. Infrared LEDs made from gallium arsenide (GaAs) can be used as a photo source for such a system. They respond in the same way as a normal LED (see Section 3-10) except that in

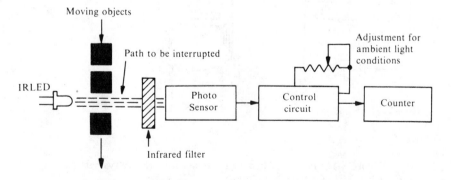

Figure 3-10 Infrared detection system.

the forward-biased condition, there is no visible light. An infrared filter is placed in front of the photodiode, phototransistor, photofet, etc.) to block out most, if not all, of the visible light. The invisible infrared radiation will easily pass through the filter to trigger the photosensor. A solid object moving through the infrared radiation path will stop the infrared radiation from reaching the photosensor and the circuit will respond accordingly. A control circuit may be added to adjust the sensitivity threshold due to any unwanted radiation that might reach the sensor.

3-8 PHOTORESISTOR

Photoresistors are made from cadmium sulfide (CdS) or cadmium selenide (CdSe) semiconductor materials. The CdS or CdSe material is deposited on a ceramic substrate with attached metallic leads and mounted in a case with a window. When light strikes the material, electron–hole carriers are released. In a circuit, the electrons will travel toward the negative source. The current of the circuit has increased; therefore, the resistance of the photoresistor has decreased. The amount of decrease in resistance is proportional to the intensity of light. Light resistance (R_{light}) is the minimum resistance for a given amount of light. Dark resistance (R_{dark}) is the maximum resistance, when there is no light.

Photoresistors can be used in voltage-divider circuits to produce varying outputs with respect to light. If the output is taken across the photoresistor, the output voltage will decrease when light is present. If the output is taken across the normal series resistor R_s, the output voltage will increase when light is present.

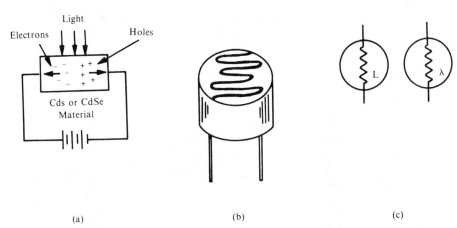

(a)   (b)   (c)

Figure 3-11 Photoresistor: (a) structure; (b) physical package; (c) schematic symbols.

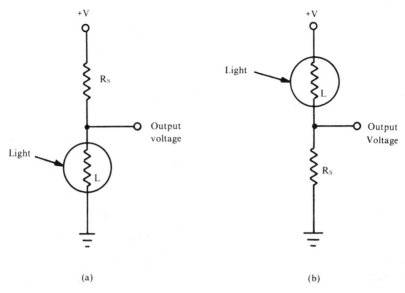

Figure 3-12 Photoresistor voltage dividers: (a) decreasing output; (b) increasing output.

3-9 PHOTOCELL (SOLAR CELL)

The *photocell* or *solar cell* is a photovoltaic semiconductor device that produces a small voltage and current when exposed to light. It is classified as a photodetector. The older-type selenium cell consists of a metallic collector strip above a transparent electrode, followed by a unilateral (one-way) barrier layer, a selenium material, and finally a metallic base plate. When light strikes the cell, electron–hole pair carriers are released. Electrons flow toward the transparent electrode, where they are trapped by the barrier layer and form the negative terminal. The holes travel toward the metallic base plate and form the positive terminal.

The newer-type silicon cell is a specially doped *pn* junction. When light strikes the cell, electron–hole pair carriers are released. Electrons are forced into the *n*-type region as holes are forced into the *p*-type region. The depletion region increases slightly and a small voltage is stored between the positive and negative terminals as long as light is present. A single solar cell may produce from 0.25 to 0.6 V across its terminals and be able to supply up to 50 mA of current.

Like conventional voltaic cells, solar cells can be connected in series to produce a larger voltage and connected in parallel to provide more current. Series–parallel combinations of solar cells produce usable solar batteries.

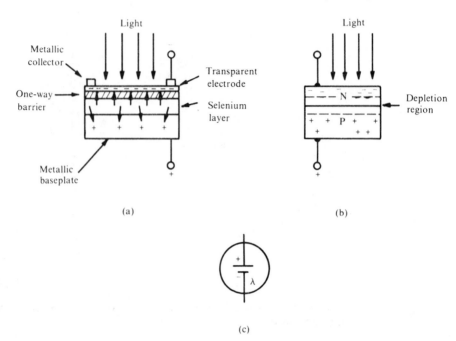

Figure 3-13 Photovoltaic cells: (a) selenium cell; (b) silicon cell; (c) schematic symbol.

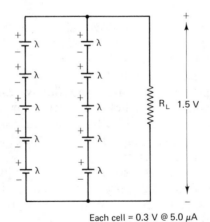

Each cell = 0.3 V @ 5.0 μA

Figure 3-14 Series–parallel combination of photo cells to produce a total battery of 1.5 V at 10.0 μA.

3-10 LIGHT-EMITTING DIODE (LED)

Light-emitting diodes (LEDs) are *pn* junctions made from gallium phosphide (GaP) or gallium arsenide phosphide (GaAsP). Biasing for an LED is the same as for regular diodes. When it is forward biased (anode positive

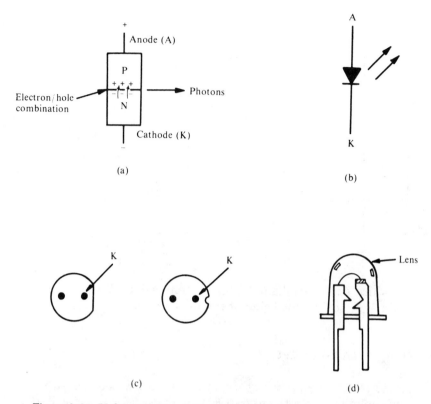

Figure **3-15** Light-emitting diode (LED): (a) structure; (b) schematic symbol; (c) lead identification (bottom view); (d) side view of package.

cathode negative), electrons cross the junction and combine with holes. This action produces light energy. Actually, little appreciable current flow or light is produced until the forward-biased voltage is equal to or greater than the inherent forward voltage drop (V_F) of the LED; the V_F is typically about 2 V. In the reverse-biased condition, little or no current flows, hence no light is produced.

The light emitted by the junction is very small and a glass or epoxy lens is attached above the junction during manufacturing to focus the light. These lenses may be clear or diffused to give a more visible light source. LEDs come in various sizes and have clear and colored lenses (red, yellow, orange, green, and blue).

In most applications the LED must be protected with a series-limiting resistor (R_s). The value of this resistor is easily calculated using Ohm's law. The V_F of the LED is fairly constant; therefore, the voltage across R_s (V_{RS}) is the difference between V_F and the voltage applied across the circuit (V_{cc}): $V_{RS} = V_{cc} - V_F$. A safe current must be chosen for the LED and still

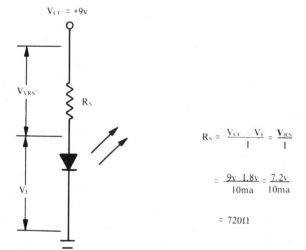

$$R_S = \frac{V_{CC} - V_F}{I} = \frac{V_{RS}}{I}$$

$$= \frac{9v - 1.8v}{10ma} = \frac{7.2v}{10ma}$$

$$= 720\Omega$$

Figure 3-16 Calculating current-limiting resistor (RS).

produce sufficient light. This current, which flows through R_S, is divided into V_{RS} to find the value of the resistors: $R_S = V_{RS}/I$.

3-11 LED 5 × 7 MATRIX DISPLAY

A *LED 5 × 7 matrix display* contains 36 LEDs arranged in seven rows and five columns, with the extra LED being used for a decimal point (D.P.). This display can produce alphanumeric characters (i.e., letters, numbers, and

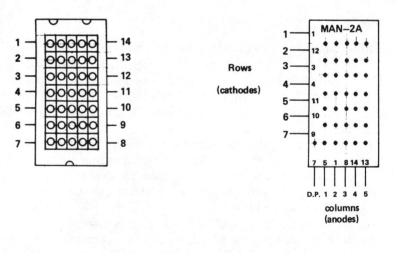

(a)

(b)

Figure 3-17 LED 5 × 7 matrix display: (a) 14-pin DIP; (b) schematic diagram.

special symbols). The row inputs are connected to the cathodes of the LEDs and the column inputs are connected to the anodes. To produce a character, various rows are brought low, while each column is brought high with a scanning operation. For example, to produce the letter F: when column 1 is high, all seven rows are low; when column 2 is high, rows 1 and 4 are low; when column 3 is high, rows 1 and 4 are low; when column 4 is high, rows 1 and 4 are low; and when column 5 is high, only row 1 is low. A pulse sequencer is connected to the columns and is synchronous with the addressing of a RAM or ROM, which provides data to the rows. The scanning rate is from 100 Hz upward, and the LEDs are "refreshed" (relighted) about 100 times a second, so the eye perceives them as being permanently lit. Some LED matrix displays need external series-limiting resistors in their anode circuits, while other types have these resistors (fabricated during manufacturing) included in the package.

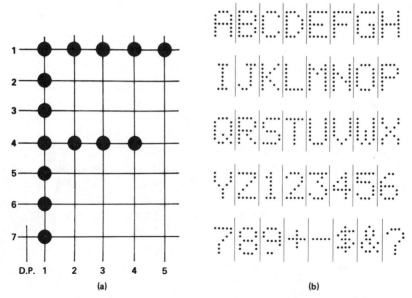

Figure **3-18** Forming symbols on LED 5 × 7 matrix display: (a) letter F; (b) other symbols.

3-12 ALPHANUMERIC DISPLAY

Alphanumeric displays are similar to the seven-segment display (see Section 13-28) but contain more segments which can produce numbers, letters, and special characters. There are two popular versions in use; one has 13 segments and the other has 16 segments. These displays are produced in the form of LED, fluorescent, liquid crystal, and gas discharge (Nixie-tube). Special decoder devices are needed to produce the various characters.

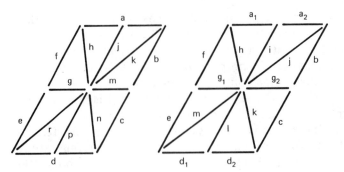

Figure 3-19 Alphanumeric displays.

Alphanumeric displays can be fabricated into multidigit displays which show large arrays of alphanumeric information.

3-13 FLUORESCENT DISPLAY

When bombarded by electrons, fluorescent material gives off a glow (typically green or blue-green in color). The fluorescent material is formed into seven-segment anodes on an insulated supporting structure. Very thin wires or a fine grid forming the cathode is placed in front of the seven-segment pattern anodes. This cathode is heated directly with a low voltage. The unit is then placed in an evacuated glass container. A high positive voltage is used to select the desired anode segments. Electrons flow from the cathode wires or grid to the selected anodes, which in turn give off a glow indicating the desired numeral or character. A typical display, 0.5 inch in height, might require 1.5 V for the cathode filaments, 22 V for the anode segments, and perhaps draw 0.5 mA of current for each segment.

Fluorescent displays are manufactured in single-digit and multidigit units. Many small calculators use this type of display.

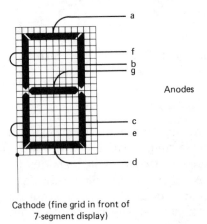

Anodes

Cathode (fine grid in front of
7-segment display)

Figure 3-20 Fluorescent display.

3-14 LIQUID CRYSTAL DISPLAY (LDX, LCD)

The *liquid crystal display* (LDX or LCD) is not a light-generating device, but depends on good ambient light or some other form of light source for its operation. Unlike the light-generating type of displays that tend to "wash out" as the ambient light increases, the LDX becomes increasingly more legible. An older type of dynamic-scatter LDX produced milky white characters, whereas the newer and more often used, field-effect LDX has either black or clear characters. It is constructed of a polarizing filter, followed by a glass back plate that is completely covered by a microscopically thin layer of deposited metal; a nematic liquid (liquid crystal); metallization of seven-segment numerals, colons, and perhaps alphanumeric characters on another panel of glass; and finally another polarizing filter. The metallization of symbols is so thin that the glass appears transparent. The entire unit is sealed and electrical connections are made through edge connectors. In a transmissive system, light passes through the display, whereas a reflective

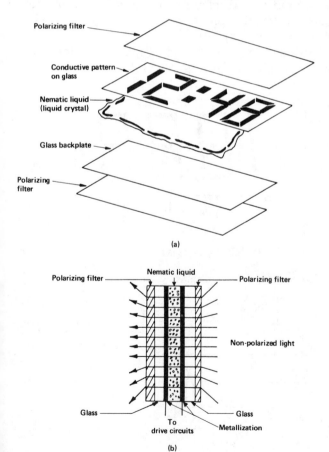

Figure 3-21 Liquid crystal display (LCD): (a) construction; (b) transmissive LCD operation.

115

system has a reflector, which reflects the light back through its front panel. When the selected segments are energized with ac or a train of square-wave pulses, the liquid crystal beneath these segments rotates or "twists" 90° and alters the light passing through. Light is blocked by this twisting action and the desired character will appear either black or clear.

LDX displays draw only a few nanoamperes of current, but are slower acting than other displays.

3-15 MULTIDIGIT DISPLAYS

Multidigit displays consist of two or more seven-segment displays contained in a single package or module. Seven inputs select the segments of each digit. A decimal-point input is usually included. Other inputs are used to select the individual digits, depending on the number of digits in the unit. As an example, the anodes may be the seven segments to each digit and the cathode inputs select each digit. A multidigit display requires a multiplex operation, where the data to the seven segments (anode inputs) are synchronized with the proper digit selected (cathode input).

In a basic counter system using a multidigit display, the clock (CLK) provides the synchronization. The address counter selects the counters via

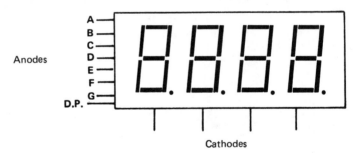

Figure 3-22 Multidigit display.

Figure 3-23 Multidigit display system.

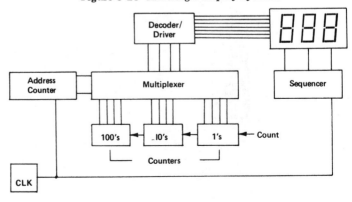

the multiplexer in order (i.e., 1 s, 10 s, 100 s) and then repeats. The data of each counter are sent through the decoder/driver to the seven-segment inputs. At the same time the sequencer selects the proper digit from right to left. The clock frequency is such that the multiplexing rate makes each digit appear to be on continuously. The result is that each digit display shows the contents of its appropriate counter.

3-16 NIXIE-TUBE DISPLAY

The gas-discharge display, often referred to as a *Nixie-tube* (a trademark of the Burroughs Corporation), operates on the principle of the cold-cathode neon glow indicator. In a numerical display each number 0 to 9 is formed from fine wire; each of these is a cathode and they are placed one behind another. A fine-mesh-grid type of anode is placed around the cathodes and placed into a vacuum tube containing neon gas. Decimal points are included in most numerical displays. Approximately +150 to 200 V is required by the anode. A current-limiting resistor (R_L) is needed to prevent the fine-wire cathodes from burning out. When one of the numerical cathodes is

Figure 3-24 Gas-discharge tube (Nixie-tube, trade name of Burroughs Corporation): (a) front view; (b) side view; (c) schematic diagram.

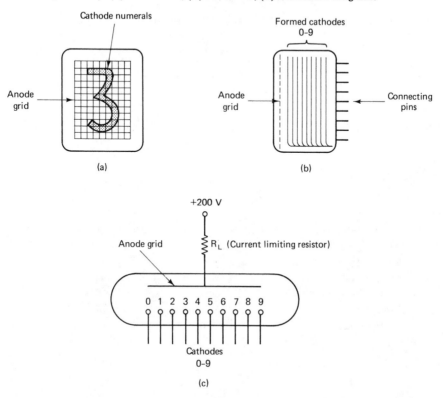

selected by grounding, current flows from that cathode to the anode. A gas discharge (typically a yellow-orange glow) appears near the surface of the cathode, thus displaying the desired number. Transistor drivers or special integrated circuits are used to energize the cathodes.

Gas discharge displays also come with alphabetical characters, special characters, and seven-segment and multidigit panels.

3-17 OPTOISOLATORS

Optoisolators, also called optically coupled isolators, optocouplers, photon couplers, photocouplers, and other names, are used to isolate one circuit from another electrically, while allowing one circuit to influence another. An optoisolator consists of a photo source, such as a LED, and a photo-sensor. These two devices are placed into a light-tight container that provides light coupling only from the photo source to the photosensor. A LED may

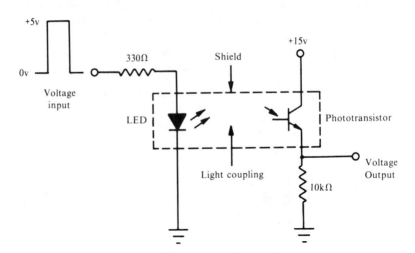

Figure 3-25 Optoelectronic isolation.

be turned on by a +5 V pulse, which controls a phototransistor connected to +15 V, or a higher-power source. These units are commercially available, and a popular version is the six-pin mini-DIP integrated circuit. Some other combinations of opto-isolators are LED/Darlington pair transistors, LED/photoresistors, and LED/LASCRs. These devices not only provide interfacing between circuits, but are not plagued by the contact bounce, electrical noise, and physical wear associated with their electromechanical counterparts.

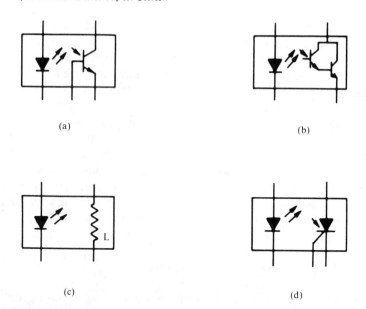

(a)

(b)

(c)

(d)

Figure 3-26 Types of optoisolators: (a) LED/phototransistor; (b) LED/Darlington; (c) LED/photoresistor; (d) LED/LASCR.

3-18 OPTOELECTRONIC INTERRUPTER SWITCH

An *optoelectronic interrupter switch*, sometimes called an emitter-sensor unit or emitter-detector pair, is similar to an optoisolator (see Section 3-17). Interrupter modules have a photosource (emitter) and a photosensor (photo-diode, phototransistor, etc.), which are separated by a narrow channel or slot. The light (or infrared radiation) emitted by the photo source crosses

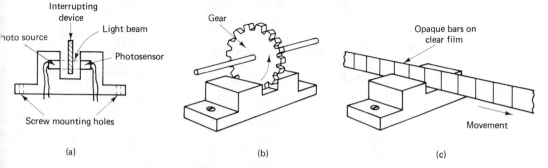

Figure 3-27 Interrupter module: (a) side view of structure; (b) timing a gear; (c) reading data from clear film.

the slot and is detected by the photosensor. When a substance of sufficient density to block the light is placed in the slot, the light will not reach the photosensor and the proper indication can be made by a circuit.

An emitter-detector pair can be used to time a gear shaft. The teeth of the gear pass through the slot in the module which sends the pulse data to control circuits that regulate the speed of the gear shaft motor. Data from clear acetate or Mylar film can be used with an emitter-detector module to control motion or other operations in an industrial application. There are countless applications involving electromechanical switching devices using moving contacts. However, the moving contacts arc, generate noise, and wear out. Substituting optoelectronic devices for their electromechanical counterparts is easy, inexpensive, more reliable, and in many cases provides a direct interface with digital-logic circuits.

3-19 SOLID-STATE AUTOMATIC LIGHT SWITCH

The *solid-state automatic light* switch utilizes three optoelectronic devices; a photoresistor, a LED, and a LASCR. An optoisolator is formed by the LED and LASCR and isolates a +12 V dc circuit from a 120-V ac circuit.

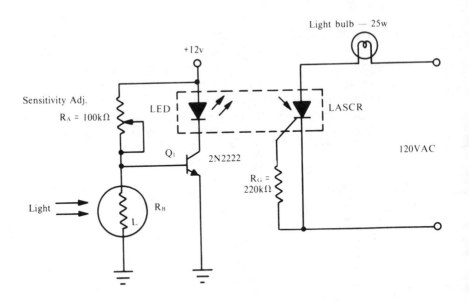

LASCR — 200PRV @ 1 amp or greater
LED — 1.8 to 2.1v @ 20ma or greater
R_B — Clairex CL7032 or equivalent

Figure 3-28 Solid-state automatic light switch.

When light is striking the photoresistor, its resistance is low, the voltage drop is low (in effect, the base is pulled down toward ground), Q_1 is cut off, and no current reaches the LED. Light is not transmitted by the LED to the LASCR, the circuit is open, and no current can get to the light bulb. When sufficient light is blocked from the photoresistor (or darkness increases to the threshold level) its resistance increases. This increases the positive voltage on the base (in effect, the base is pulled up toward +12 V), which turns on the transistor. Current flows through the LED, causing it to emit light, which triggers the LASCR into conduction. The light bulb will now glow, but only at about half brilliance, because the rectifying action of the LASCR lets current flow only during the positive alternation of the cycle.

Resistor R_A (sensitivity adj.) sets the control circuit to turn on at the desired level of darkness.

3-20 TESTING OPTOELECTRONIC DEVICES

Many faulty optoelectronic devices can be detected with an ohmmeter in a similar manner to testing solid-state devices (see Section 2-23).

3-20.1 Testing Photodetectors

To test photodiodes, phototransistors, and photofets, place the positive lead of the ohmmeter to the anode, collector, or drain and the negative lead to the cathode, emitter, or source, respectively. When light strikes the device, the resistance reading should be low, and when light is blocked from the device, the resistance reading should be high. Follow the same procedure for checking photoresistors, except that ohmmeter polarity need not be observed.

Checking LASCRs is a little more difficult. There must be a gate-to-cathode resistor of 100 kΩ to 1 MΩ, depending on the intensity of the light source used. Light must be blocked from the LASCR when connecting the ohmmeter to prevent it from triggering falsely. The resistance reading should be maximum (infinite). When light strikes the LASCR, the resistance will be

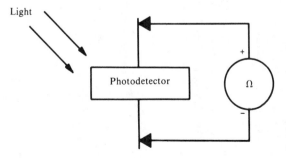

Figure 3-29 Checking photo-detectors with an ohmmeter.

minimum.   The LASCR should latch (remain on even when the light is blocked again), but it may not because the ohmmeter cannot supply enough holding current.

3-20.2 Testing Photocells

Observing polarity, connect a voltmeter (set on the lowest scale) across the photocell.   With no light striking the cell, the meter will read 0 V.   When exposed to light, the meter should read a few tenths of a volt.

3-20.3 Testing Optoisolators

These devices are usually checked in a working circuit, but a simple test circuit could be constructed as shown in Section 3-17.   The +5 V pulse turns on the LED, which turns on the transistor, causing the output voltage to increase.

Figure 3-30 Testing displays: (a) seven-segment common anode LED; (b) seven-segment common cathode LED; (c) 5 × 7 matrix LED; (d) gas-discharge tube.

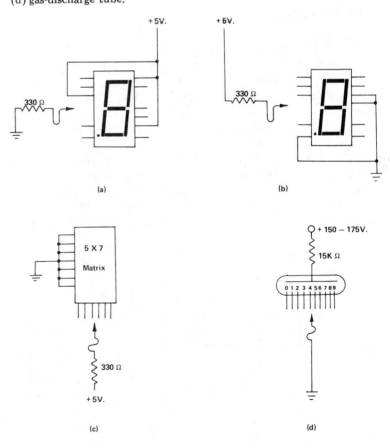

3-20.4 Testing LEDs and IRLEDs

The ohmmeter test for LEDs and IRLEDs is the same as for regular diodes. Place the positive lead of the ohmmeter to the anode and the negative lead to the cathode. The device is now forward biased and should indicate minimum resistance. Switch the leads for reverse bias and the resistance should indicate maximum.

3-20.5 Testing Displays

To test a seven-segment common-anode display, connect all anode pins to +5 V. With a clip lead connected to ground via a 330-Ω resistor, touch all segment pins for a visible indication.

To test a seven-segment common cathode display, connect all cathodes to ground. With a clip lead connected to +5 V via a 330-Ω resistor, touch all segment pins for a visible indication.

A 5 X 7 matrix LED display can be tested by first connecting all cathodes (rows) to ground. Then with a clip lead connected to +5 V via a 330-Ω resistor, touch each anode (column). All seven LEDs should light in each column.

To test a gas-discharge tube display, connect the anode or plate to a +150 to 175 V source via a 15-kΩ resistor. With a clip lead connected to ground, touch each tube pin to light the desired numeral.

Electronic Vacuum Tubes

4-1 INFORMATION ON VACUUM TUBES

Solid-state devices are used in most electrical functions previously performed by vacuum tubes. However, there still exist some vacuum-tube circuits in operation, and some applications (radio-TV transmission and large current control circuits) are better suited for the vacuum tube.

Electron flow is produced in a vacuum tube by heating a filament made of tungsten or thoriated tungsten. The filament may be directly or indirectly heated, where the filament heats a cathode that is electrically isolated from the heater voltage. Electrons boil off the filament or cathode and form a cloud of electrons referred to as the *space charge*. This process is called *thermionic emission*. A conducting metal cylinder, the anode or plate, is placed around the cathode and attracts the electrons when a positive voltage is placed on it. In some vacuum tubes, grids placed between the cathode and anode control the flow of electrons when voltages are impressed upon them. The entire structure is placed in a glass or metal envelope for protection and a vacuum is introduced to permit current flow and proper operation.

Pins protruding from the bottom of the tube connect to the various elements within the envelope. These pins are numbered counterclockwise when viewed from the bottom.

Vacuum tubes require higher voltages and are more fragile than their solid-state counterparts.

Tube parameters help to describe or predict how a vacuum tube performs. The amplification factor (μ) indicates the tube's amplifying capabilities

124

Figure 4-1 Typical vacuum-tube envelopes.

and is expressed as the ratio of a small change in plate voltage (Δe_b) to a small change in grid voltage (Δe_g) while plate current (i_b) remains constant. This is given by the formula

$$\mu = \frac{\Delta e_b}{\Delta e_g} \times i_b \text{ constant}$$

The ac plate resistance r_p shows the actual tube's resistance under operating conditions and is expressed as

$$r_p = \frac{\Delta e_b}{\Delta i_b} \times e_c \text{ constant}$$

The transconductance g_m of the tube is the opposite of ac plate resistance, or the ease with which it conducts, and is expressed as

$$g_m = \frac{\Delta i_b}{\Delta e_c} \times e_b \text{ constant}$$

These three parameters are related as expressed by the formula

$$\mu = g_m r_p$$

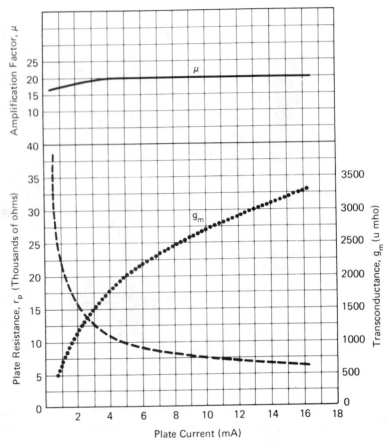

Figure 4-2 Relationships of μ, g_m, and r_p.

4-2 VACUUM-TUBE DIODE

The vacuum-tube diode consists of a directly or indirectly heated cathode, which is connected to the most negative part of the circuit. The heater voltage is labeled supply A (V_A). The anode or plate is connected, usually through a current-limiting resistance (R_L), to a positive voltage, labeled supply B (V_B). Electrons are attracted from the cathode to the plate, and flow through R_L and supply B back to the cathode to complete the circuit. This current flow is termed plate current (I_p or I_B) and is directly proportional to the plate voltage (V_B). If the plate voltage is increased, the plate current will increase. When V_B is increased to a level where no more available electrons will flow from the cathode, the tube is said to be in *saturation*. Vacuum tubes are not normally operated in this saturation region.

Vacuum-tube diodes used in power supplies require heater voltage which is usually supplied from a low-voltage winding on the same transformer

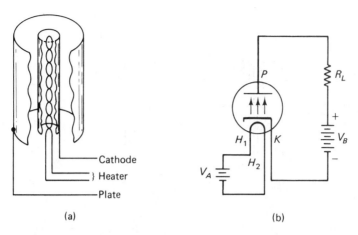

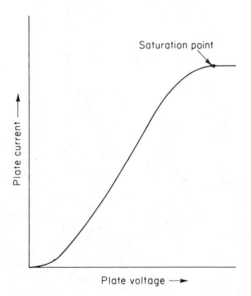

Figure 4-3 Vacuum-tube diode: (a) physical structure; (b) schematic symbol indicating circuit electron flow.

Figure 4-4 Vacuum-tube diode plate characteristic curve.

that supplies the anode voltage. Dc voltage is produced the same way by vacuum-tube diodes as with solid-state diodes and the output voltage is labeled B+. A full-wave power supply may use a duo-diode vacuum tube.

4-3 VACUUM-TUBE TRIODE

The vacuum-tube triode is constructed with a helix type of grid placed between the cathode and anode. This grid, called the *control grid*, limits or controls the amount of plate current through the tube. Normally, the con-

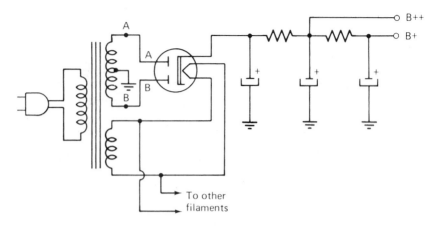

Figure 4-5 Vacuum-tube duo-diode, full-wave power supply.

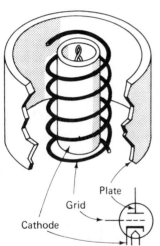

Figure 4-6 Vacuum-tube triode
structure and schematic symbol.

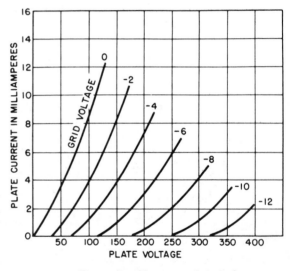

Figure 4-7 Vacuum-tube triode
plate characteristic curves.

trol grid is biased negatively with respect to the cathode. A negative grid
repels electrons coming from the cathode; the more negative it is, the less
plate current flows. When less plate current flows, it appears as though the
plate resistance has increased, which is how the triode amplifies an input
signal.

With a class A triode amplifier, current flows through the tube constantly. The voltage drop V_K across the cathode resistor R_K establishes the
bias for the tube. If V_K is +3 V and the grid is at 0 V, the tube is said to be
biased at -3 V. Capacitor C_3 stabilizes this bias voltage and the tube is in
the quiescent state. Let the plate voltage V_B to ground be about +125 V.

The positive alternation of the input signal makes the bias go more positive (perhaps −2 V), which increases the plate current. Plate resistance decreases and the voltage at the plate also decreases (e.g., to +50 V). On the negative alternation of the input signal, the bias is made more negative (e.g., −4 V), which decreases the plate current. Plate resistance increases and the voltage at the plate increases (e.g., to +200 V). For a 2-V change on the grid, there is a corresponding 150-V change on the plate for a stage gain of 75 as found by the formula

$$A_v = \frac{v_{out}}{v_{in}}$$

Notice that the output signal is 180° out of phase with the input signal.

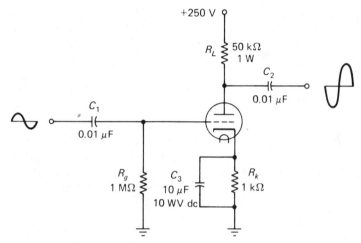

Figure 4-8 Vacuum-tube triode amplifier.

4-4 TETRODE

Vacuum-tube triodes have limited gain at high frequencies because of the interelectrode capacitance between elements, especially the plate and control grid. These electrodes, separated by the vacuum dielectric, form a capacitance, which allows the input signal to "leak" into the output, causing distortion.   The interelectrode capacitance can be minimized by a fourth electrode, the screen grid placed between the control grid and plate, and acts as an electrostatic shield. It is operated more positive than the cathode and less negative than the anode. However, its presence accelerates the electrons so much that when they hit the plate, other electrons are dislodged to form what is called a *secondary emission*. This results in a negative feedback action as shown by the "hump" in the plate characteristic curve. Operating the plate at very high voltages can reduce this effect somewhat. In a vacuum-

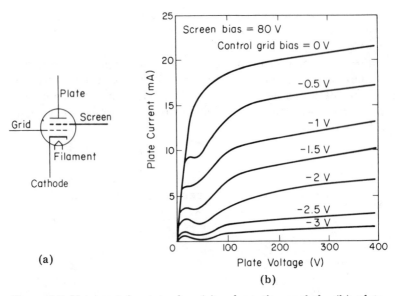

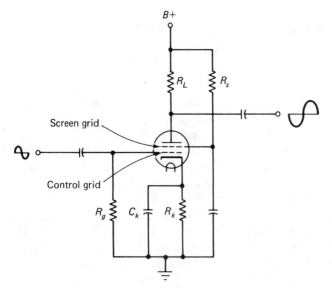

(a)

Figure 4-9 Vacuum-tube tetrode: (a) schematic symbol: (b) plate characteristic curves.

tube *tetrode* amplifier, the screen grid is held a little less positive than the plate by resistor R_s. Any ac variations are bypassed to ground by a capacitor connected from the screen grid to ground. This amplifier is not as practical as other types, but the tetrode does find some application as an oscillator operating within the feedback region.

Figure 4-10 Vacuum-tube tetrode amplifier.

4-5 PENTODE

The problem of secondary emission in the tetrode is reduced or eliminated in the vacuum-tube pentode with a grid called the *suppressor grid*. This grid is placed between the screen grid and plate. The suppressor grid is usually connected to the cathode (very often it is internally connected during manufacturing) and is very negative with respect to the plate. Secondary emission electrons are repelled by the suppressor grid back to the plate, where they contribute to the total output of the tube. Because of this action and the effect of the screen grid acting as a virtual anode, the plate voltage has little effect on plate current. The control grid is then the main control of current through the tube. Generally, *pentodes* can produce more

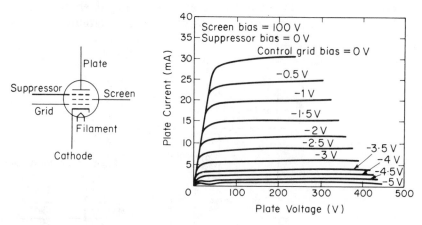

Figure 4-11 Vacuum-tube pentode: (a) schematic symbol; (b) plate characteristic curves.

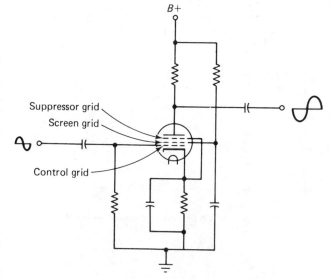

Figure 4-12 Vacuum-tube pentode amplifier.

usable current than triodes and have a high plate resistance. Pentodes are also given one of two classifications, remote cutoff or sharp cutoff. A remote-cutoff pentode requires a large negative grid bias to cut off or stop the current flow of the tube. A sharp-cutoff pentode will reach cutoff at a very small negative grid bias.

4-6 MULTIGRID TUBES

The pentode has the greatest number of grids used in normal amplifying tubes. However, there are other types of tubes having more than three grids. These tubes are used in special applications having functions other than merely amplification. One such tube, the pentagrid converter, has five grids. This tube is used in converting a radio frequency to an intermediate frequency for a radio receiver. In a typical converter circuit, a tunable oscillator

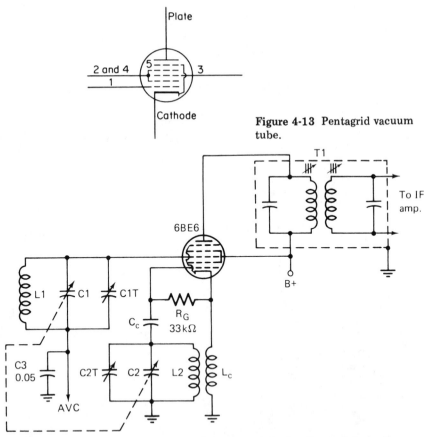

Figure 4-13 Pentagrid vacuum tube.

Figure 4-14 Pentagrid vacuum-tube converter circuit used in radio receivers.

frequency is placed at the cathode and first grid. The incoming radio frequency is placed at the third grid. These frequencies mix within the tube
and the difference or sum frequency is taken from the plate and grids 2
and 4.

There are other multigrid tubes, such as the hexode, with four grids; the
heptode, with five grids; and the octode, with six grids.

4-7 BEAM POWER TUBE

Beam power tubes are specially designed tetrodes and pentodes used to provide large amounts of current. Beam-forming plates are placed partially
around the cathode and grids and between the plate. These plates are connected internally to the cathode. Since the plates are much less negative
with respect to the plate, the electrons are focused somewhat into a beam as
they flow toward the plate. This flow of electrons is so concentrated that
secondary emission is reduced and a large plate current is available for a

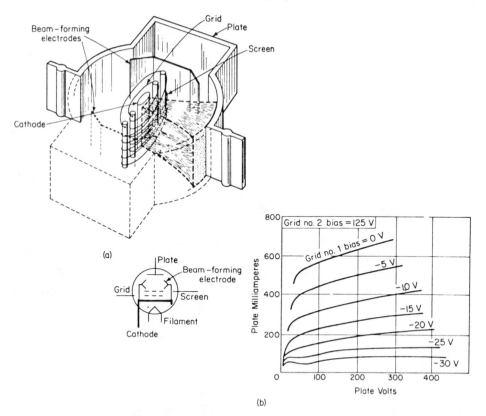

Figure 4-15 Beam power vacuum tube: (a) structure; (b) schematic
symbol and plate characteristic curves.

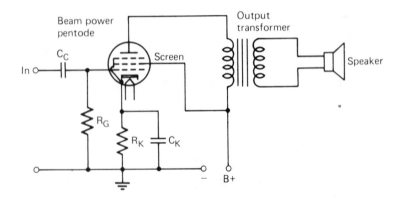

Figure 4-16 Beam power pentode tube used as single-ended output power amplifier.

relatively low value of plate voltage. The cathode and grids are oblong shaped to provide a flat plane perpendicular to the plate, which enhances the beam formation.

Beam power tubes are more efficient than regular tetrode or pentode tubes because they dissipate less heat. Their applications include functions requiring large amounts of current, such as motor control, solenoid control, and particularly, audio output stages. A single-ended output power amplifier stage resembles a normal class A amplifier, except that B+ is applied to the tube through a step-down transformer, which drives a speaker.

4-8 GAS-TYPE VACUUM TUBES

Gas-type vacuum tubes require less vacuum, produce more current for proportionate size, operate cooler, and generally are more efficient than regular vacuum tubes. The gas is usually neon, argon, nitrogen, or mercury vapor. Gas-type tubes are usually designated by a small dot in the schematic drawing tube envelope. Current flow through the tube depends on the ionization of the gas atoms, which produces more electrons (and holes) than the conventional tube with a high vacuum.

Cold-cathode gas-type tubes do not require a heater element, but will conduct when the firing voltage for which they are designed is exceeded. Once in conduction, they maintain the voltage across their terminals. These

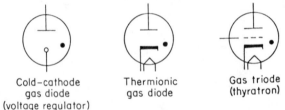

Cold-cathode
gas diode
(voltage regulator)

Thermionic
gas diode

Gas triode
(thyratron)

Figure 4-17 Gas-type vacuum tubes.

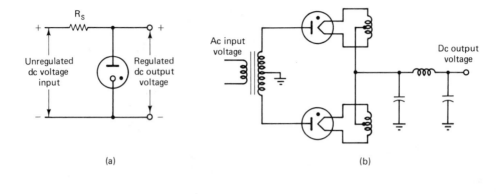

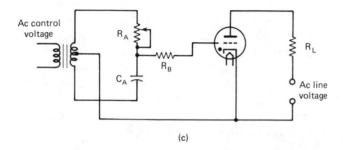

Figure 4-18 Gas-type vacuum-tube applications: (a) cold-cathode voltage regulator; (b) gas-tube full-wave power supply; (c) thyratron control circuit.

gas-type tubes are used as dc voltage regulators and are typically 75, 105, and 150 V. Its solid-state counterpart is the zener diode.

A *thermionic* (requiring a heater) gas diode is the same as a normal vacuum-tube diode, but is capable of handling larger currents and is used for high-power rectifiers. These tubes usually have a 15-V drop across them in the forward-biased condition and are referred to as mercury-vapor rectifiers.

The *thyratron*, or gas triode, operates similar to a solid-state SCR. The control grid voltage determines the ionization level at which the thyratron will fire, but once the tube is conducting, the control grid loses control and cannot cut off the current flow. Grid control can be regained only by stopping the current through the tube and then reapplying a plate voltage lower than the ionization level.

4-9 MULTIPURPOSE TUBES

Multipurpose tubes have two or more separate tubes in a single envelope. Often, these separate tube elements share one common heater and/or cathode. These special tubes were designed and manufactured to save space and cost.

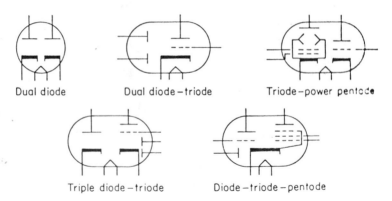

Figure 4-19 Multipurpose vacuum tubes.

It is incorrect to call a *duo-diode* a "tetrode" or a *duo-diode-triode* a "pentode." The separate tubes within a single envelope may be used in the same circuit or totally different circuits. For instance, a duo-diode could be used in a full-wave rectifier, whereas a duo-diode-triode would use the diode section for rectifying and detecting an AM signal, while the triode section would be used as an audio voltage amplifier to drive a power output stage. On schematic diagrams, the separate tubes are designated as V_{1A} for one section, V_{1B} for another section, and so on.

4-10 TESTING VACUUM-TUBE CIRCUITS

The first step in troubleshooting vacuum-tube circuits is to make a visual inspection to determine if all the tubes are lighted. An open filament may cause one tube or several tubes to be out, depending on whether they are wired in series or parallel. Replacing a suspected defective tube is relatively easy, but finding faulty components in the associated circuitry usually involves making dc voltage readings. Using a triode amplifier stage as an example, current flows from ground up through R_K, the vacuum tube, and R_p to the +100-V supply. The normal operating voltage drops are then V_p = +60 V, V_g = 0 V, and V_K = 2 V (all voltage measurements are in reference to ground). The bias on the tube, V_{GK}, is –2 V. The tables show the change in voltages at these points for various defective components that most likely would occur.

If R_K opens, there is no current flow through the tube, no IR drop across R_p, and the plate is at $B+$. The voltage increase at V_K is due to the charge held by C_K. If R_K increases in value, less current flows and there is less IR drop across R_p; therefore, V_p must increase. When R_p opens, no voltage is applied to the plate, no current flows, and there is no IR drop across R_K. If R_p increases in value, it drops more voltage, there is less

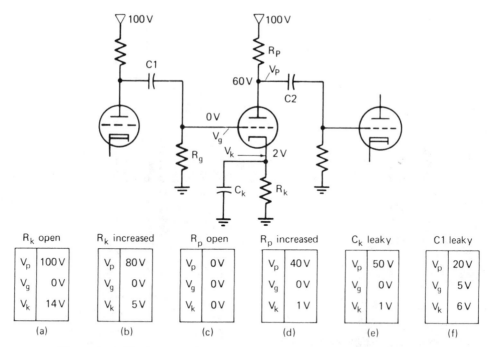

Figure 4-20 Testing a vacuum-tube triode amplifier stage by voltage analysis.

current flow, and V_p and V_K have to decrease. A leaky C_K decreases the effective resistance of R_K and the bias decreases. This allows more current to flow, and the IR drop of R_p increases, causing V_p to decrease. If a coupling capacitor (C_1) becomes leaky, it places a positive dc voltage on the grid of the tube. The tube conducts more, causing V_p to decrease and V_K to increase. An increase in V_K appears to increase the bias, which should limit current through the tube, but this is not the case since the grid is positive.

5

Power Supplies

5-1 INFORMATION ON POWER SUPPLIES

There are many types of electrical and electronic power supplies, providing various ac and dc voltages for equipment operation. Dc voltages are needed for most electronic circuits, and generally an electronic power supply is considered as a device that converts ac into dc.

The ac voltage from the power lines can be rectified directly or passed through an isolation transformer (a turns ratio of 1:1). Depending on the value of dc voltage needed, the transformer may be of either a step-up or step-down type. After the transformer, the ac is rectified into pulsating dc by diodes in the form of a half-wave rectifier, a full-wave rectifier, or a bridge (full-wave) rectifier. The pulsating dc is then filtered or smoothed out by capacitors, inductors, and resistors so as to produce a constant dc output voltage.

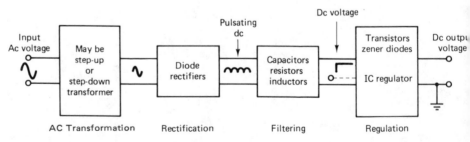

Figure 5-1 Block diagram of basic power supply.

Variations in line voltage and changing loads (circuits that are being operated from the power supply voltage) can cause the output voltage to vary, which introduces electrical problems in equipment. A voltage regulator is usually used as the last stage of a power supply, to regulate the output voltage.   The regulator may consist of zener diodes, transistors, and/or integrated circuits.

5-2 HALF-WAVE RECTIFIER

With *half-wave rectification*, one alternation of the ac input voltage is allowed to pass to the load, while the other alternation is blocked, producing pulsating dc.

The voltage from point A to point B (ground reference) is the ac voltage at the secondary of the transformer. During the positive alternation the diode's anode (point A) is more positive than its cathode (point C). The diode is forward biased and current flows from point B to point A. The voltage drop appearing across the resistor will be the peak value of the positive alternation less the forward voltage drop of the diode (usually about 0.6 V).   When the negative alternation appears, point A is negative with respect to point C and the diode is reverse biased. No current flows through the circuit and the voltage drop across the resistor is zero. The output voltage across the resistor is positive pulsating dc. The peak output voltage is equal to the peak input voltage less the forward voltage drop of the diode V_F (usually considered 0.6 V):

$$V_{\text{out}} = V_{\text{peak}} - V_F$$

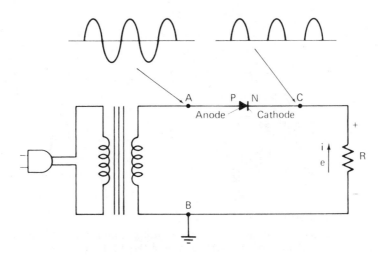

Figure 5-2 Half-wave rectifier.

The average output voltage is one-half the average full-wave value, since the diode conducts only on half-alternations:

$$V_{av} = \frac{0.637}{2} \; V_{out} = 0.3185 \; V_{out}$$

This half-wave pulsating voltage is more difficult to filter to produce pure dc, and the result is usually a large ripple voltage factor.

If the diode is turned around in the circuit, the output voltage will be negative pulsating dc.

5-3 FULL-WAVE RECTIFIER

A basic *full-wave rectifier* uses two diodes and a center-tapped transformer. The positive dc voltage is seen across the load resistor from point C to ground. Ac voltage appears across the secondary winding at points A and B. When point A is positive, point B is negative. The anode of D_1 is more positive than its cathode and it conducts allowing current to flow through the

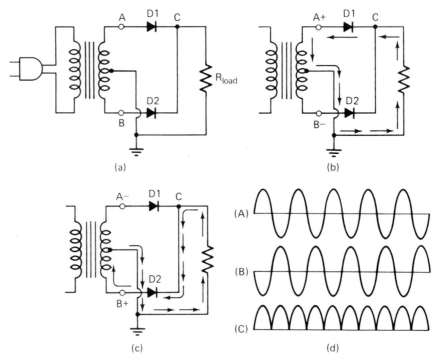

Figure 5-3 Full-wave rectifier using center-tapped transformer: (a) complete circuit; (b) positive alternation; (c) negative alternation; (d) waveforms at points A, B, and C.

center tap to ground, up through the resistor, and back to the diode. At this time the anode of D_2 is negative with respect to its cathode and it does not conduct. On the next half cycle, point A is negative and point B is positive. The anode of D_2 is now more positive than its cathode and it conducts, allowing current to flow through the center tap to ground, up through the resistor, and back to the diode. At this time the anode of D_1 is negative with respect to its cathode and it does not conduct. The output voltage is a continuous series of dc pulses occurring at 120 Hz (twice the input frequency). The peak output voltage is equal to the peak input voltage less the forward voltage drop of the diode V_F (usually considered 0.6 V).

$$V_{out} = V_{peak} - V_F$$

The average output voltage is found by the formula

$$V_{av} = 0.637\ V_{out}$$

The full-wave pulsating voltage is easier to filter to produce pure dc and the ripple voltage factor is less.

If the diodes are reversed in the circuit, the output voltage will be negative at point C with respect to ground.

5-4 BRIDGE (FULL-WAVE) RECTIFIER

The *bridge rectifier* produces a full-wave dc output with the use of four diodes, but does not require a center-tapped transformer. When the positive half-cycle appears, the anode of diode 1 is positive and it conducts, and the cathode of diode 2 is negative and it conducts. Current flows from the ac source through diode 1, through R_L, through diode 2, and back to the source. Diodes 3 and 4 are reversed biased and no current flows through them. They appear as open circuits. On the negative half-cycle the anode of diode 4 is positive and it conducts, and the cathode of diode 3 is negative and it conducts. Current flows from the ac source through diode 4, through R_L, through diode 3, and back to the source. Diodes 1 and 2 are reverse biased at this time and appear as open circuits. The current always flows through R_L in the same direction to produce full-wave rectification. The voltage output V_o is the peak voltage less two times the V_F across the diodes (two diodes are conducting at a time):

$$V_o = V_{peak} - 2V_F$$

The average output voltage is the same as for any full-wave rectifier, and reversing all diodes in the bridge circuit will produce a negative voltage output.

Full-wave bridge rectifier modules containing four diodes in a single unit are commercially available.

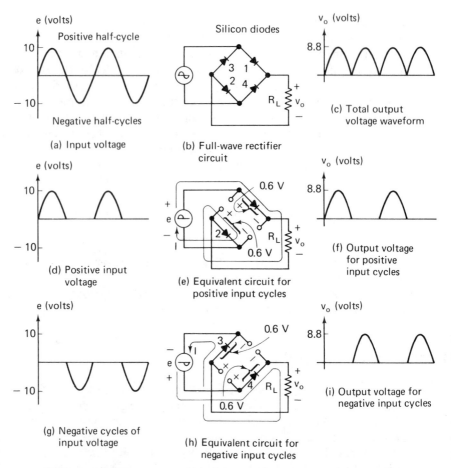

Figure 5-4 Bridge (full-wave) rectifier: (a)–(c) complete rectification; (d)–(f) positive alternation; (g)–(i) negative alternation.

5-5 FILTERING

Capacitors are used in power supplies to smooth out pulsating dc and to produce steady dc voltage. A simple capacitor filter is connected across the output of a rectifier. The capacitor will charge up to the peak voltage of the output. When this peak voltage begins to decrease, the stored electrons in the negative side of the capacitor will discharge and flow through the load in an attempt to keep the voltage constant across the capacitor. Before the capacitor discharges too far, another dc pulse arrives to recharge the capacitor back to the peak voltage. The slight charging and discharging of the capacitor

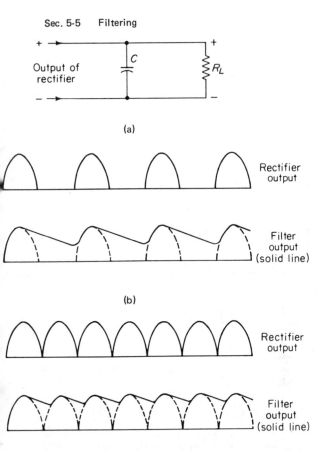

(a)

(b)

(c)

Figure 5-5 Simple capacitor filter: (a) circuit; (b) half-wave filtering; (c) full-wave filtering.

produces a ripple voltage (ac component) that is superimposed on the top of the steady dc. Therefore, full-wave rectification is easier to filter than half-wave rectification. The percentage of ripple determines the quality of filtering: the less the better. The ripple factor can be found thus:

$$\% \text{ of ripple} = \frac{v_{ac(rms)}}{V_{dc(av)}} \times 100$$

where $v_{ac(rms)}$ is the effective voltage of ripple and $V_{dc(av)}$ is the average voltage of the dc output.

The addition of inductors in filters helps to smooth out changes in current and to reduce the ripple factor even more. An LC pi (Π)-type filter could be used where the load draws considerable current. In solid-state circuits where current loads are minimum, an RC pi-type filter works sufficiently and is used in many applications.

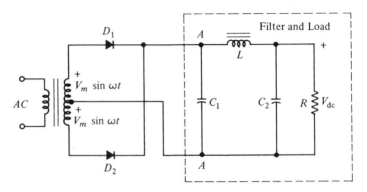

Figure 5-6 *LC* pi-type filter.

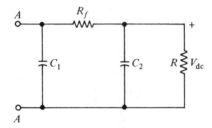

Figure 5-7 *RC* pi-type filter.

5-6 VOLTAGE REGULATOR

Other circuits are added to power supplies to keep their output voltage constant when changes occur in the loads or line voltage. The simplest of these devices is the zener diode (see Section 2-3). However, a circuit also using transistors is more sensitive to voltage changes and is more efficient. A typical *voltage regulator* circuit uses a zener diode as a basic stable reference voltage to bias a voltage-sensing transistor (Q_2). This transistor, in turn, controls the bias on a series-pass transistor (Q_1), which allows more or less current to flow through the load (R_L), thus keeping the output voltage across the load constant. Resistor R_3 forms a voltage divider across the load and can be adjusted to provide the full desired voltage at ideal load conditions.

If the voltage across R_L decreases, the voltage at the base of Q_2 also decreases, which reduces the current through Q_2. The voltage at the collector of Q_2 increases and is applied to the base of Q_1. The current through Q_1 increases, which is the same current through R_L; therefore, the voltage drop across R_L tends to remain constant.

When the voltage across R_L increases, the voltage at the base of Q_2 increases, which increases the current through Q_2. The voltage at the collector of Q_2 decreases and is applied to the base of Q_1. The current through Q_1 decreases, which tends to lower the voltage drop across R_L.

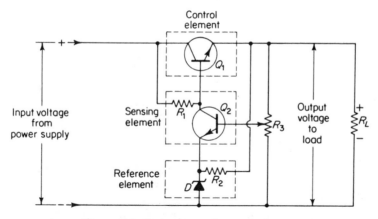

Figure 5-8 Series-pass voltage regulator.

Variations of input voltage from the power supply will affect the circuit in a similar manner. The transistors will conduct more or less depending on what is required to maintain the voltage constant across the load.

5-7 VOLTAGE DOUBLER

Voltage multipliers are inexpensive circuits that produce higher voltage without the use of step-up transformers. However, to be of any quality, they are limited to very light current loads.

In a half-wave *voltage-doubler* circuit, when the negative alternation of the input ac is at C_1, diode D_1 is forward biased and capacitor C_1 is allowed to charge up to peak voltage. When the positive alternation is at C_1, D_1 is reverse biased and looks like an open circuit. Diode D_2 is now forward biased and allows capacitor C_2 to charge up to the peak voltage plus the charge that is stored on C_1. Therefore, the dc output voltage across R_L is nearly twice the peak input voltage. This circuit has a high ripple factor and R_L is constantly discharging C_2.

With a full-wave voltage doubler, the ripple factor is reduced and proves to be more efficient in maintaining a steady dc output voltage. When the positive alternation of the input ac appears at D_1, this diode is forward

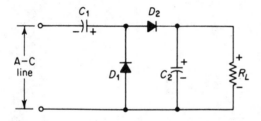

Figure 5-9 Half-wave voltage doubler.

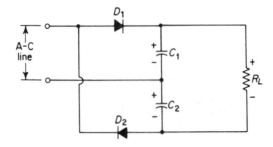

Figure 5-10 Full-wave voltage doubler.

biased and allows C_1 to charge up to peak voltage. Diode D_2 is reverse biased at this time. When the negative alternation is at D_1, it is reverse biased, but D_2 is forward biased, allowing C_2 to charge up to the peak voltage. Since C_1 and C_2 are in series, their stored charges added up to be about twice the input peak voltage, which is seen across R_L.

Remember that R_L is constantly discharging the output voltage, and these circuits are generally used in low-current applications. (In other words, the load resistance should be as high as possible.)

5-8 VOLTAGE TRIPLER AND VOLTAGE QUADRUPLER

A *voltage tripler* consists of a half-wave voltage doubler and an extra half-wave rectifier connected in series. Capacitor C_2 of the voltage doubler charges up to twice the peak input ac voltage, and capacitor C_3 of the half-wave rectifier charges up to the peak ac voltage. These capcitors are in series and their stored charges add up to three times the peak input ac voltage. This tripled dc voltage is applied to R_L.

The *voltage quadrupler* uses two half-wave voltage doublers connected in series. Capacitors C_2 and C_4 are each charged up to twice the peak input ac voltage. Since these capacitors are in series across the load (R_L), their combined dc output voltage is four times the peak input ac voltage.

These circuits will also be efficient if the load current requirements are kept low.

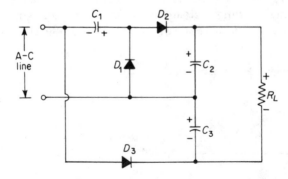

Figure 5-11 Voltage-tripler circuit.

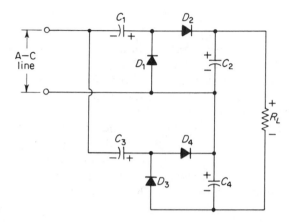

Figure 5-12 Voltage-quadrupler circuit.

5-9 DUAL-VOLTAGE POWER SUPPLY

A *dual-voltage power supply* consists of a positive voltage output and a negative voltage output in relation to ground. The bridge rectifier can be used with a center-tapped transformer and provide a fairly stable dual voltage power supply (minimum interaction between the two supplies). Each power supply requires its own filter. The pi-type filter components typically are 100 Ω for the resistors and 1000 to 4000 μF for the capacitors. With a load of approximately 10 kΩ for each supply, the dc output is about 1.3 times the value of the secondary rms voltage,

$$\text{dc output voltage} = 1.3 \times \text{rms of input voltage}$$

For example, with a secondary voltage of 24 V rms, the dc output voltage would be approximately 31.2 V. This would allow for a +15.6-V supply and a –15.6-V supply.

Dual-voltage power supplies are often used for op-amp circuits and high-power solid-state audio amplifiers that require a balanced power supply.

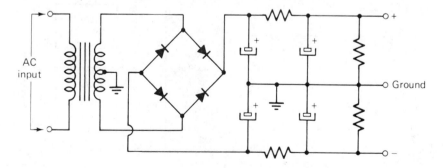

Figure 5-13 Dual-voltage power supply.

5-10 MULTIVOLTAGE POWER SUPPLY

Some circuits and equipment require various dc voltages from a power supply for proper operation. The simplest method is to use a resistive voltage divider, which follows the filter. Current flowing from $-B$ up to $+B_1$ develops voltage gradients along the resistor. The taps on the resistor provide different voltages (single resistors are often used also). The various voltages provided might be: $-B$ = ground, $+B_3$ = +9 V, $+B_2$ = +24 V, and $+B_1$ = +100 V. This type of *multivoltage power supply* is complex to design and may prove unstable, because of interaction between the various circuit loads.

Some electronic circuits require various voltages with a high degree of regulation. Integrated-circuit (IC) regulators have become very popular

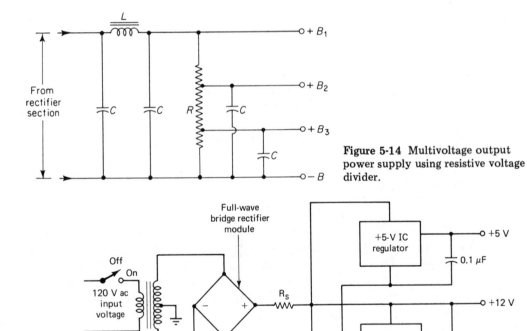

Figure 5-14 Multivoltage output power supply using resistive voltage divider.

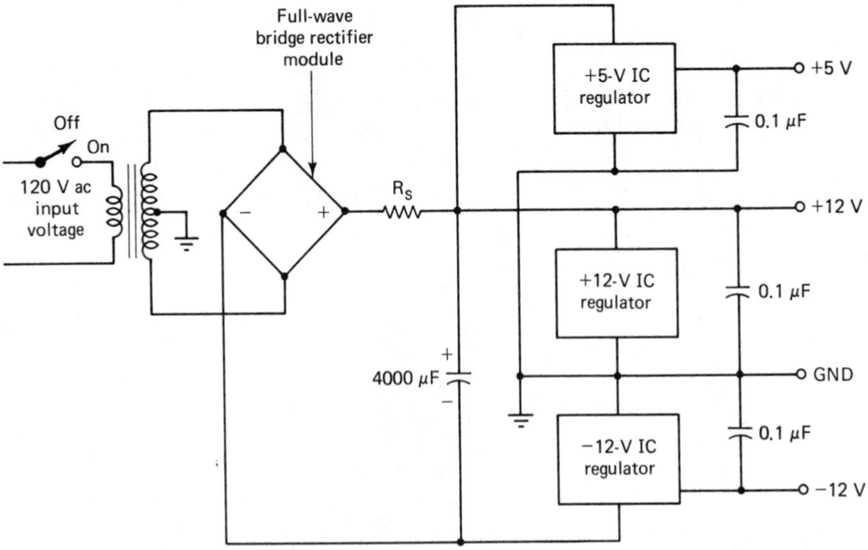

Figure 5-15 Multivoltage power supply using IC regulators.

for such applications, because they are easy to use and are reasonable in cost for the quality they provide. A multivoltage power supply using IC regulators may provide several positive and/or negative voltages, each with a separate regulator. Filtering is usually less critical because the regulators provide such a high degree of regulation. Power transistors and associated circuitry can be added to IC voltage regulators for higher current requirements. Power supplies such as these are used in op-amp, microprocessor, and other electronic circuit applications that require stringent voltage regulation.

5-11 DC-TO-DC CONVERTER

Dc-to-dc converter power supplies are finding numerous applications in solid-state circuitry, particularly for powering op amps and digital-to-analog (D/A) or analog-to-digital (A/D) converters on logic boards. These types of circuits convert a lower dc voltage into a higher dc voltage.

Transistors Q_1 and Q_2, together along with a special pulse transformer, form a square-wave oscillator, which is generally designed to operate at a rate of 8 to 20 kHz. The square wave is stepped up, rectified, and filtered to produce a dc voltage, which is then regulated to the desired output level.

A dual-voltage output can also be obtained by rectifying and filtering the negative level of the secondary square wave.

Manufacturers offer small dc-to-dc converters in a 24-pin DIP rated at 1 W. The input voltages are +5 or +12 V, with single outputs of +5, +12, and +15 V and dual outputs of ±12 and ±15 V.

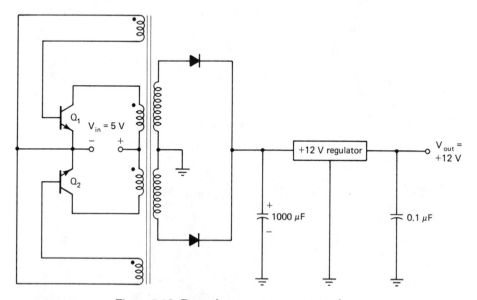

Figure 5-16 Dc-to-dc converter power supply.

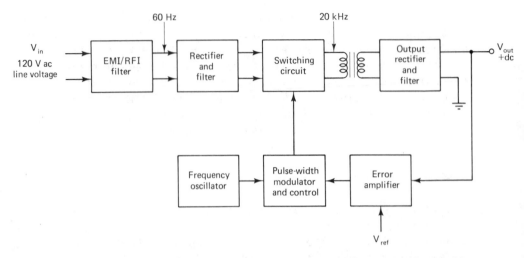

Figure 5-17 Block diagram of a pulse-width-modulated switching regulated power supply.

5-12 SWITCHING POWER SUPPLY

Switching power supplies are more efficient and smaller in size than normal linear power supplies. For these reasons, more of these units are finding applications in solid-state circuits.

Usually, the ac line voltage is filtered from electromagnetic interference (EMI) and radio-frequency interference (RFI), rectified, and filtered to produce steady dc voltage. This steady dc voltage is then chopped by the switching circuit at a fixed frequency, usually 20 kHz, produced by a frequency oscillator. The chopped dc produces a square wave, which is transformed, rectified, and filtered to the desired output voltage level. This voltage is monitored by an error amplifier that compares its value against a stable fixed reference voltage (V_{ref}). Any changes in the output level, due to line or load variations, are amplified by the error amplifier, which controls a pulse-width modulator. The pulse-width modulator varies the conduction period of the switching circuit accordingly, keeping the output voltage constant.

TABLE 5-1 COMPARISON OF LINEAR AND SWITCHING POWER SUPPLIES

Parameter	Linear	Switching
Line and load regulation	Very good, 0.1%	Same
Output noise and ripple	Very good, 3 mV p-p	Fair, 50 mV p-p
Efficiency	Low, 35–45%	High, 70–80%
Size	Unfavorable, 0.5 W/in.^{3}	Favorable, 2.5 W/in.^{3}
Weight	Unfavorable, 10 W/lb	Favorable, 50 W/lb

Since the switching power supply operates in the switching mode, on or off, rather than the linear mode, the absence of voltage and current on alternative half cycles represents no power loss.

Table 5-1 compares a linear power supply to a switching power supply.

5-13 CROWBAR VOLTAGE PROTECTION

Some solid-state circuits cannot withstand an overvoltage.   The *crowbar overvoltage protection* circuit is fast acting to protect the load R_L.   The zener diode sets the inverting input of V_{ref} of +2.5 V.   The noninverting input is adjusted to +2.5 V by the trip adjust pot.   Differential input voltage (V_D) is zero; therefore, the op-amp output is zero and the SCR is open.   If the 5-V power supply increases, the voltage at the noninverting input increases.   The op-amp output increases, causing the SCR to fire.   The SCR appears as a short across the load (crowbar effect) and activates a fuse, circuit breaker, or current limiter.   Once the problem is cleared, the reset switch is used to disable the SCR, allowing full voltage to the load.

In some applications it is desirable to shut down a power supply if the voltage drops below a specific level.   The crowbar under voltage protection circuit uses the noninverting input for V_{ref}.   The trip adjust pot sets the inverting input at +2.5 V.   Differential input voltage is zero, the op-amp output is zero, and the SCR is open.   When the power supply voltage drops below +5 V, the voltage at the inverting input decreases, the op amp output voltage increases, and the SCR fires.   Voltage across the load decreases and an alarm or indicator is activated.   The capacitor across the zener diode temporarily holds the noninverting input at a lower voltage than the inverting input during initial turn-on.   This allows the zener diode time to conduct and prevents false triggering of the SCR.

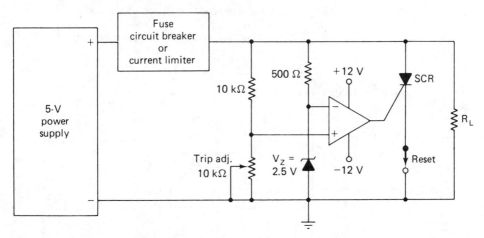

Figure 5-18 Crowbar overvoltage protection.

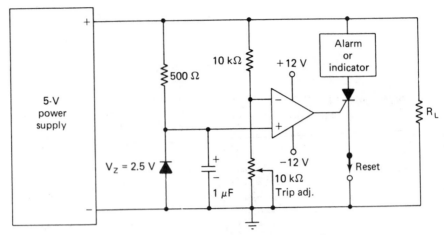

Figure 5-19 Crowbar undervoltage protection.

5-14 TROUBLESHOOTING POWER SUPPLIES

To troubleshoot any electronic circuit, you must know the normal operating voltages of the circuit. In the figure, point F is the desired dc voltage to power other circuits. Point E should be a little more positive because of the added IR drop of R_2. These two points can be measured in reference to ground with a dc voltmeter. Points C and D represent the ac voltage of the transformer secondary. Points A and B is the ac line voltage when S_1 is closed. These points can be measured with an ac (rms) voltmeter or an oscilloscope.

The normal procedure for troubleshooting a power supply is to check the desired dc output voltage and then work back toward the input line voltage until a good reading is obtained. Usually, the faulty component is located between a bad reading and a good reading. Table 5-2 shows the voltages at various points of the power supply and indicates the faulty component that might cause incorrect voltage readings.

Figure 5-20 Power supply troubleshooting.

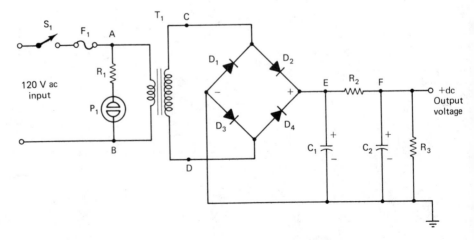

TABLE 5-2 PROBABLE CAUSES OF INCORRECT VOLTAGE READINGS

F	E	C–D	A–B	Ripple	Defective component
+15	+17	12.6 rms	120 rms	Minimum	None (normal operation)
+17	+17	12.6 rms	120 rms	Minimum	R_3 open
0	+17	12.6 rms	120 rms	0	R_2 open
+15	+17	12.6 rms	120 rms	Increase	C_2 open
Less	+17	12.6 rms	120 rms	Increase	C_2 leaky
0	+16	12.6 rms	120 rms	0	C_2 short
+12	+14	12.6 rms	120 rms	Increase	C_1 open
+14.6	+16.6	12.6 rms	120 rms	Slight increase	C_1 leaky
0	0	12.6 rms	120 rms	0	C_1 short
Normal reading or zero		12.6 rms	120 rms	Increase	Diodes open [depends on which one(s)]
+15	+17	12.6 rms	120 rms	Increase	Diodes short
0	0	0	120 rms	0	Transformer secondary open or shorted
0	0	0	0	0	Transformer primary open
0	0	0	0	0	F_1 or S_1 faulty

If the load on the power supply becomes open from the output voltage, point F should increase to the same value as point E, since there is no voltage drop across R_2.

If R_2 opens, there will be no voltage at point F. Capacitor C_2 will discharge through R_3.

If C_2 opens, the dc voltages may remain normal, but the ripple voltage will increase because of less filtering.

A leaky capacitor will lower the voltage it is trying to filter and the ripple voltage will increase.

A shorted capacitor will short out the voltage and little or none will be available for the load.

A single diode being "open" will produce half-wave rectification. All voltages may appear normal, but the ripple voltage will increase. The indications will be the same if two aiding diodes are open. If two opposing diodes "open," there will be no rectification and no dc voltage available.

When a single diode "shorts," the dc voltage may remain normal, but the ripple voltage will increase. There will be an increase in current through the transformer. If two aiding diodes "short," there will be a decrease in the dc output and an increase in ripple. When two opposing diodes "short," all dc voltages will go to zero and there will be an excessive amount of current flow through the transformer.

No voltage at points C and D could mean that the transformer secondary is open or shorted. Also, the transformer primary could be open.

No voltage at points A and B could mean that F_1 or S_1 is faulty.

Amplifiers

6-1 AMPLIFIER CLASSES AND TYPES

An amplifier is usually considered as a device that enlarges or increases a small input voltage or current. This is accomplished by the small input signal (voltage or current) controlling a larger voltage and current to produce an increased output signal.

Amplifiers have many classifications, one of which depends on the amount of time the circuit is conducting. A *class A* amplifier is biased in the linear region and will conduct for 360° of the input cycle. This amplifier has a distortionless output. A *class AB* amplifier is biased near cutoff, so that it conducts less than 360° but more than 180° of the input cycle. A *class B* amplifier is biased at cutoff. It must be turned on by the input signal and will conduct for 180° of the input cycle. A *class C* amplifier is biased beyond cutoff. It must be turned on by the input signal and will conduct for a time much less than 180° (perhaps less than 90°) of the input cycle.

Amplifiers are also classified as to their application, such as:

DC amplifier: directly coupled amplifier, used to amplify dc voltages and currents

Audio amplifier: used to amplify audio frequencies

IF amplifier: used to amplify intermediate frequencies in radio circuits

RF amplifier: used to amplify radio frequencies

Video amplifier: used to amplify picture information in television systems

154

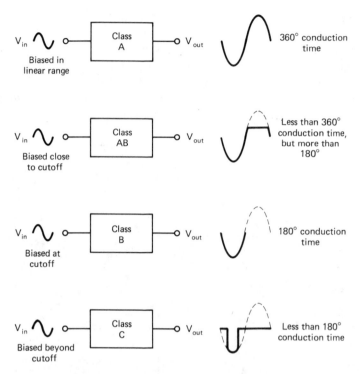

Figure 6-1 General classes of amplification.

Pulse amplifier: specially designed to amplify pulses

Small-signal amplifier: used to increase greatly the amplitude of a small signal (also referred to as a preamplifier)

Driver amplifier: a large-signal voltage or current amplifier used to drive power amplifiers

Power amplifier: a large amplifier capable of producing high currents to operate loads, such as speakers, motors, relays, etc.

Amplifiers are also referred to the configuration of their active-device elements, as shown in Table 6-1.

TABLE 6-1 AMPLIFIERS CLASSIFIED BY ACTIVE DEVICE

Vacuum tubes	Bipolar transistors	FETs
Common cathode	Common emitter	Common source
Common anode	Common collector	Common drain
(cathode follower)	(emitter follower)	(source follower)
Common grid	Common base	Common gate

An *active device*, such as a vacuum tube, bipolar transistor, or FET, when operated at the extremes of cutoff or saturation, can be thought of as a switch in series with a load (R_L). When reverse biased, the device is cutoff, the circuit is open, and V_{out} will equal the supply voltage. When it is forward biased, the device is turned on (saturated) and the circuit is closed with the output nearly at O V.

At other times, when the device is biased to operate within the two extremes, it can be thought of as a variable resistance in a simple equivalent series circuit. The total voltage will divide proportionately across each

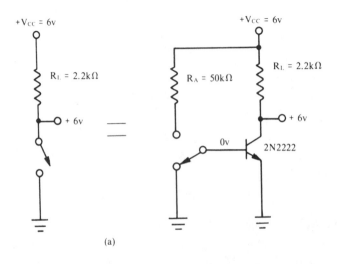

(a)

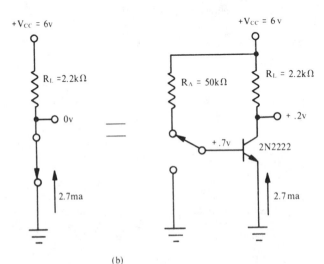

(b)

Figure 6-2 The transistor as a switch: (a) open circuit; (b) closed circuit.

resistance. With V_{out} connected between the resistances, any change in one of the resistances will cause V_{out} to change. The resistance of the active device is controlled by circuit bias (not shown) and input signals. When the resistance of the active device is equal to the load resistor R_L, the output will be one-half of the supply voltage with reference to ground. If the bias is adjusted to allow more current flow through the active device (to turn it on more), its resistance decreases. Therefore, the voltage drop across the active device decreases and V_{out} is less. Remember, Ohm's law for voltage drop across a resistor is $E = IR$. The voltage is directly proportional to current and resistance. If the resistance decreases, so does the voltage drop. Since R_L is fixed, more voltage is dropped across it, while less is dropped across the active device. When the bias is adjusted to restrict current flow through the active device (to cut it down more), its resistance increases. The voltage drop across the active device has now increased and V_{out} is greater.

A small input signal varying the bias on an active device will produce a change in the current flow and in the resistance of the device and produce a varying V_{out} of greater amplitude.

If the active device becomes "open," V_{out} will equal the supply voltage. On the other hand, if the device becomes "shorted," V_{out} will be about zero.

This basic approach can be applied to most circuits using active devices.

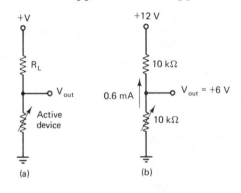

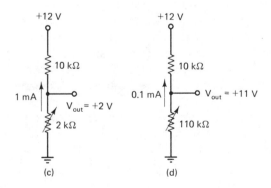

Figure 6-3 Analysis of an active device: (a) equivalent circuit; (b) equal resistance; (c) decreased resistance; (d) increased resistance.

6-3 BIPOLAR NPN TRANSISTOR COMMON-EMITTER CONFIGURATION

The input to a *common-emitter* amplifier is between base and ground (effectively base to emitter) and the output is between collector and ground (effectively collector to emitter). Electron current flows from emitter to collector, with a slight amount leaving the base terminal, $I_C = I_E - I_B$ (refer to Section 2-13 for circuit nomenclature). A voltage divider is formed by R_L, the transistor, and R_E. Resistor R_E stabilizes the transistor due to I_{CO} and temperature effects. Capacitor C_E stabilizes the bias voltage at the quiescent operating (Q) point. Degenerative feedback occurs across R_E without C_E and v_{out} is greatly reduced. Resistors R_A and R_B form a voltage divider to forward bias the transistor. For class A operation V_B will be about 0.6 V more positive than V_E. For example, V_E might be +1.0 V and V_B will be +1.6 V. Voltage at the collector V_c will be about one-half of V_{cc} for maximum undistorted swing of the output signal (v_{out}).

When V_B goes positive, I_B increases, I_C increases, the internal resistance of the transistor decreases, and the voltage at the collector (V_C) decreases. When V_B goes negative, I_B decreases, I_C decreases, the internal resistance of the transistor increases, and V_C increases. Therefore, the output voltage (v_{out}) is $180°$ out of phase with the input voltage (v_{in}).

The amplifier voltage gain is found by dividing v_{out} by v_{in}, $A_v = v_{out}/v_{in}$.

The maximum power dissipation curve of the transistor means that the circuit should never be operated above a given value of $V_c \times I_c$. For example, if the maximum power rating is 200 mW and $V_c = 20$ V, then I_c should not exceed 10 mA. ($P_{max} = V_c \times I_c = 20$ V $\times 10$ mA $= 200$ mW). Similarly, if $V_c = 10$ V, then I_c should not exceed 20 mA.

6-4 BIPOLAR PNP TRANSISTOR COMMON-EMITTER CONFIGURATION

The PNP common-emitter amplifier has the same characteristics as its NPN counterpart and the circuit components perform the same function. The difference between the two is the power supply voltage, and the bias voltage is reversed. Also, the majority current carrier in the PNP transistor is hole current (refer to Section 2-14). When the signal at the base goes positive, I_B decreases, I_C decreases, the resistance of the transistor increases, and V_c increases toward $-V_{cc}$. When the signal at the base goes negative, I_B increases, I_C increases, the resistance of the transistor decreases, and V_C decreases toward ground (O V).

Often, PNP transistors are used with NPN transistors in the same circuitry. Their emitters can be connected toward $+V_{cc}$ and the collectors connected toward ground. In this case, when the signal at the base goes

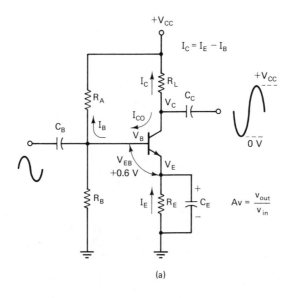

$I_C = I_E - I_B$

$Av = \dfrac{v_{out}}{v_{in}}$

(a)

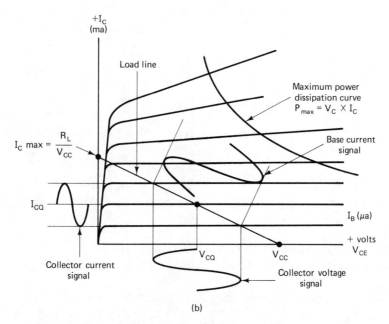

(b)

Figure 6-4 NPN transistor common-emitter amplifier: (a) schematic diagram; (b) load line showing waveforms.

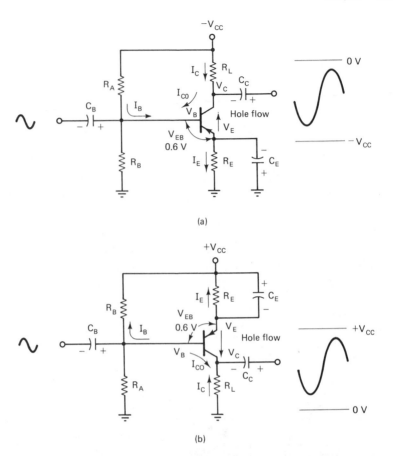

(a)

(b)

Figure 6-5 PNP transistor common-emitter amplifier: (a) using a negative power supply; (b) using a positive power supply.

positive, I_c decreases and V_c decreases toward ground. Then when the signal at the base goes negative, I_c increases and V_c increases toward $+V_{cc}$.

6-5 BIPOLAR NPN TRANSISTOR COMMON-COLLECTOR (EMITTER-FOLLOWER) CONFIGURATION

The *common-collector* circuit is similar to the common-emitter circuit except that there is no resistor in the collector circuit and the output is taken across an unbypassed emitter resistor (R_L). Resistors R_2 and R_1 establish the bias at the base and set the Q point. Capacitor C_1 is a decoupling capacitor to reduce power supply variations and may or may not be needed. Capacitors C and C_c couple the signal from one circuit to another, but block dc from each stage. When the incoming signal at the base goes positive, I_B

increases, I_E increases, and the voltage drop across R_L increases (the output voltage goes positive). The voltage at the emitter has followed the input signal voltage and is the reason why this circuit is referred to as an "emitter follower." When the signal at the base goes negative, I_B decreases, I_E decreases, and the voltage drop across R_L decreases. The overall bias voltage has changed very little, if any, and even though there is a current gain ($A_i = \Delta I_E / \Delta I_B$), the output voltage will always be less than the input voltage. ($A_v = v_{out}/v_{in}$, less than 1.) There is no signal phase inversion. Because the emitter-base junction is not completely forward biased, the input impedance is high for this circuit and the output impedance is essentially the value of R_L (usually low resistance). This circuit then becomes excellent for high-to-low impedance matching between circuits and to serve as a buffer circuit.

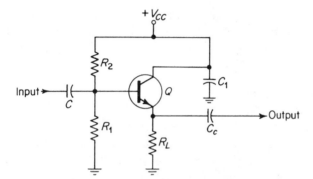

Figure 6-6 NPN transistor common-collector configuration.

6-6 BIPOLAR NPN TRANSISTOR COMMON-BASE CONFIGURATION

The input to a *common-base* circuit is at the emitter (effectively emitter to base) and the output is taken at the collector (effectively collector to base). While voltage gain is accomplished with this circuit, the current gain, called *alpha* (α), is less than 1. The input current flows from the emitter to the collector with a small amount flowing out of the base circuit; therefore, $I_C = I_E - I_B$, and alpha is collector current divided by emitter current, $A_i = I_c/I_E$. The actual circuit for a common-base can be similar to a common-emitter circuit. Resistors R_1 and R_2 forward bias the base of the transistor.

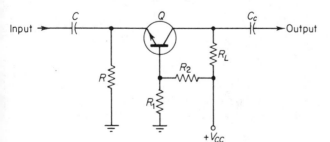

Figure 6-7 NPN transistor common-base configuration.

The collector circuit contains the load resistor and the resistor in the emitter circuit (the input) is not bypassed with a capacitor. Capacitors C and C_c are used for signal coupling.

There is no signal phase inversion with the common base circuit. When the input signal goes positive, I_E increases, causing V_E to increase, which decreases the forward bias on the EB junction and I_C decreases. The resistance of the transistor has increased, which causes V_c, the output voltage, to go positive. When the input signal goes negative, I_E decreases, causing V_E to decrease, which increases the forward bias on the transistor. Collector current I_c increases, but the resistance of the transistor decreases and the output voltage goes negative.

The input impedance is low (effectively the value of the emitter resistor) and the output resistance is high, which makes this circuit applicable for low-to-high impedance matching. There is less capacitive feedback (output to input) for this configuration, which increases its high-frequency response, making it well suited for RF circuits.

6-7 COMPARISON OF BIPOLAR TRANSISTOR AMPLIFIERS

A transistor is able to amplify when a signal is placed between two of its terminals and the output is taken from two of its other terminals. Since there are only three terminals on a transistor, one terminal has to be common to both input and output circuits. Therefore, there are three possible circuit configurations with the transitor [i.e., common emitter, common collector (also called emitter follower), and common base].

The type of configuration can be recognized by identifying where the input and output occur:

> Common emitter
> > Input at base
> > Output at collector
> Common collector
> > Input at base
> > Output at emitter
> Common base
> > Input at emitter
> > Output at collector

These circuit arrangements have special characteristics, which have optimum advantages in specific applications.

The common-emitter circuit is the most used in the electronics industry. The common-collector circuit ranks second in use, and the common-base

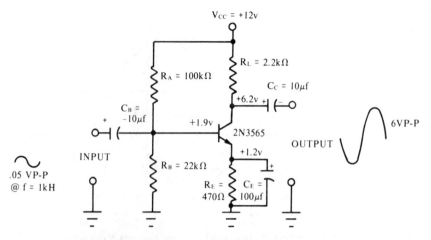

Current Gain: Yes, A_1 (beta) = I_C/I_B
Voltage Gain: Yes, $A_v = \Delta V_c/V_b$ = 6vp-p/.05vp-p = 120
Power Gain: Yes, $A_p = A_1 \times A_v$. Highest compared to other circuits
Input Impedance: Moderate (500 – 1kΩ) compared to other circuits
Output Impedance: Moderate ($\approx$ 50kΩ) compared to other circuits
Signal Phase Inversion: Yes (180°)
Basic Uses: Switch, regulator, amplifier, and oscillator

Figure 6-8 Common-emitter amplifier.

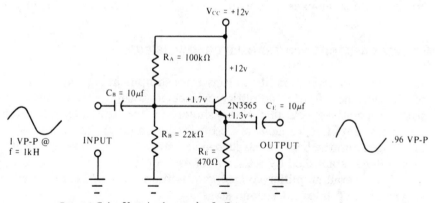

Current Gain: Yes, A_1 (gamma) = I_F/I_B
Voltage Gain: No, $A_v = \Delta V_c/\Delta V_b$ (less than one $\approx$ .90 – .99)
Power Gain: Yes, $A_p = A_1 \times A_v$ lowest compared to other circuits
Input Impedance: Highest (20k – 300kΩ) compared to other circuits
Output Impedance: Low ($\approx$ 300 – 500Ω) compared to other circuits
Phase Inversion: No
Basic Uses: Impedance matching and buffer amplifier

Figure 6-9 Common-collector (emitter-follower) amplifier.

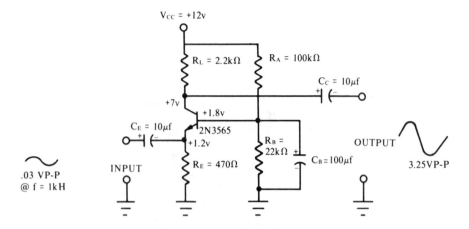

Current Gain: No, A_1 (alpha) = I_C/I_E (less than one $\approx$ .90 – .99)
Voltage Gain: Yes, $A_v = \Delta V_c/\Delta V_e$ = 3.25 vp–p/.03 vp–p = 108
Power Gain: Yes, $A_p = A_1 \times A_v$ Moderate compared to other circuits
Input Impedance: Lowest (50 – 500Ω) compared to other circuits
Output Impedance: Highest ($\approx$ 300k – 1MΩ) compared to other circuits
Phase Inversion: No
Basic Uses: Used least, but serves well for impedance matching and RF amplifiers

Figure 6-10 Common-base amplifier.

circuit is the least used. Understanding the comparisons among the circuit configurations enables a person to better understand their application.

6-8 TWO-STAGE CASCADED NPN TRANSISTOR AMPLIFIER

An often used arrangement in electronics equipment is to have several stages of the common-emitter amplifier connected together by means of RC (re-sistive–capacitive) coupling. As shown in the *two-stage cascaded* NPN transistor amplifier, capacitors C_c block the dc potential of the collector of one stage from the base of the succeeding stage, which prevents upsetting the bias voltage established by resistors R_1 and R_2.

The overall amplifier voltage gain is the product of the gain of the first stage and the gain of the second stage:

$$A_{v_{\text{overall}}} = A_{v_{\text{first stage}}} \times A_{v_{\text{second stage}}}$$

Remember, voltage gain is $A_v = v_{\text{out}}/v_{\text{in}}$.

Signal phase relationships are also important to consider in an amplifier. When the signal at the base of Q_1 is positive, the signal at the collector of Q_1 will be negative. If the effects of C_c are considered negligible, the base of Q_2 will also be negative, while the collector of Q_2 will be positive. Notice

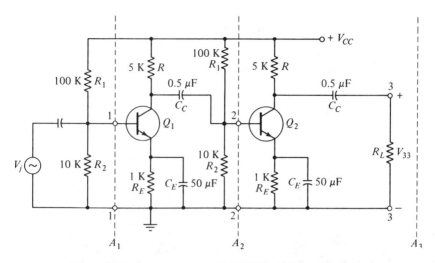

Figure 6-11 Two-stage cascaded NPN transistor amplifier.

that there is a 360° phase relationship (in phase) between the input signal and output signal.

Many amplifiers utilize negative feedback (often called degenerative feedback) to reduce distortion, stabilize gain, increase the bandwidth, and/or adjust the input and output resistances.

In a series voltage feedback arrangement, a portion of the signal at the collector of Q_2 is fed back to the unbypassed emitter resistor of Q_1. The signal is in phase at both of these points, which reduces the gain of Q_1.

With the shunt current feedback method, a portion of the signal at the unbypassed emitter of Q_2 is fed back to the base of Q_1. These signals are about 180° out of phase, thereby reducing the input signal being applied to Q_1.

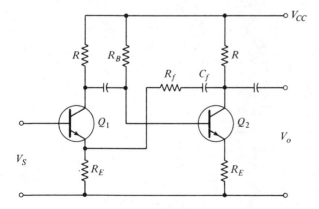

Figure 6-12 Series, voltage feedback.

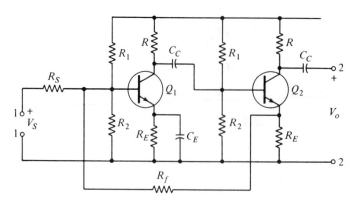

Figure 6-13 Shunt, current feedback.

6-9 JFET AMPLIFIERS

The same circuit configurations and operations for bipolar transistors also apply to JFET amplifiers, with a similar name description [i.e., common source, common drain (source follower), and common gate]. Biasing arrangements can be simpler for a JFET amplifier, since it is a depletion-type device (see Section 2-17).

In a common-source amplifier, self-bias is developed across the source resistor R_S. As electrons flow through the circuit, the top of R_S (at the source) becomes positive, say +0.5 V. The gate is at 0 V by way of R_G connected to ground. The gate is then more negative than the source and V_{GS} is 0.5 V. An input signal has the same effect as with the bipolar transistor.

Figure 6-14 JFET common-source amplifier.

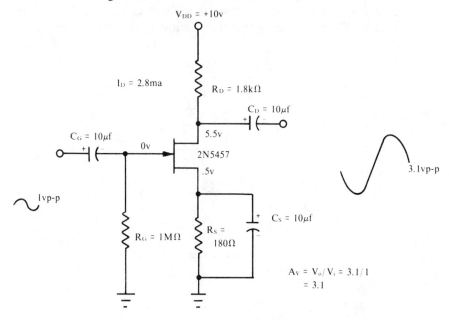

The main advantages of this circuit over bipolar types is a higher input impedance, and it can be used with higher operating voltages for its comparable size.

The common-drain amplifier has high input impedance and low output impedance, making it desirable for impedance matching, especially in the use of test equipment.

A common-gate amplifier has low input impedance and high output impedance. Like its counterpart, the common-base amplifier, its uses are usually limited.

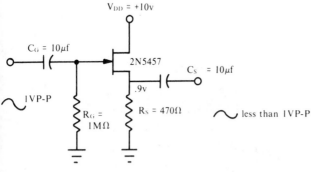

Figure 6-15 JFET common-drain (source-follower) amplifier.

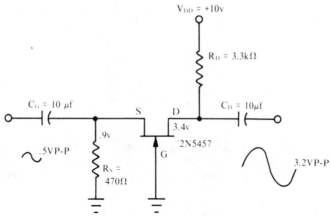

Figure 6-16 JFET common-gate amplifier.

6-10 DEPLETION-TYPE MOSFET AMPLIFIERS

Depletion-type MOSFET amplifiers function and operate the same as bipolar transistor and JFET amplifiers. The three types of configurations produce the same results, and all have the advantage of high gate resistance compared to other types of devices. The depletion-type MOSFET has the additional advantage of being operated with positive, negative, or zero bias voltage (see Section 2-18).

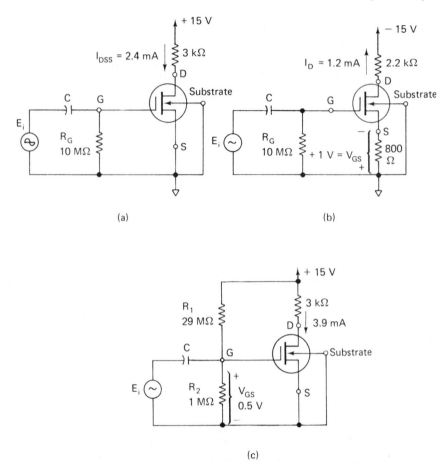

Figure 6-17 Depletion-type *n*-channel MOSFETs showing biasing arrangements: (a) *n*-channel MOSFET zero gate bias; (b) *n*-channel MOSFET depletion-mode bias; (c) *n*-channel enhancement-mode bias.

With zero bias, the source is usually connected directly to ground. The gate is connected to ground via R_G and V_{GS} = O V. When negative bias is used, the gate can be connected directly to a negative voltage, or a source resistor can be used to develop self-bias. The source is then more positive than the gate and in effect V_{GS} is negative. The circuit is then said to be operating with depletion-mode bias.

A positive bias arrangement can be achieved by returning the gate directly to a positive voltage or using a voltage-divider resistive network. The resistors are selected so that V_G is more positive than V_S. The circuit is now said to be operating with enhancement-mode bias.

6-11 ENHANCEMENT-TYPE MOSFET AMPLIFIERS

The *enhancement-type* MOSFET amplifier functions the same as the bipolar, JFET, and depletion-type MOSFET transistor amplifiers. The primary consideration for this amplifier is establishing forward bias, since it is normally off and must be turned on for operation (see Section 2-19).

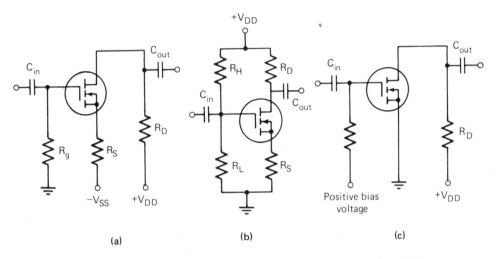

Figure 6-18 Enhancement-type *n*-channel MOSFETs showing biasing arrangements: (a) source at fixed negative voltage; (b) resistor voltage divider; (c) gate at fixed positive voltage.

For an *n*-channel enhancement-type MOSFET, the gate must be more positive than the source for conduction.

One method of accomplishing this is to connect the gate to ground via R_g and connect the source to a fixed negative voltage.

A resistor voltage divider may be used where the resistors are selected, so that V_G is more positive than V_S.

When the source is connected to ground, forward bias can be achieved by connecting the gate to a fixed positive voltage.

With *p*-channel enhancement-type MOSFET amplifiers, the voltage supplies will usually be reversed or the biasing arrangement will be designed to have V_G more negative than V_S.

6-12 DUAL-GATE MOSFET AMPLIFIER

Dual-gate MOSFETs are primarily used for frequency converters or mixers in radio circuits. Applying two different frequencies to gate G_1 and gate G_2 will produce a sum and difference frequency at the drain. However, biasing

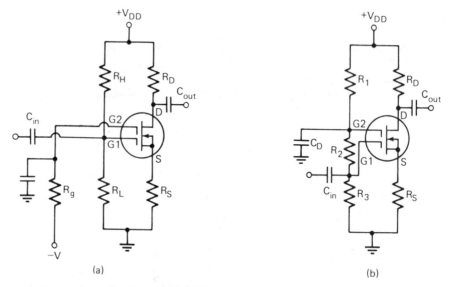

Figure 6-19 Dual-gate MOSFET (depletion-type n-channel) amplifiers showing biasing arrangements: (a) fixed voltage bias; (b) resistor voltage-divider bias.

arrangements are still needed for proper circuit operation. Similar biasing techniques used for the depletion-type MOSFET are also applicable for the dual-gate MOSFET. For an n-channel device, the gate must be biased negative with respect to the source. Nearly always, the input signal is applied to gate G_1, while gate G_2 has a dc gain control voltage or is connected to a fixed voltage. A series resistive voltage divider can be used to bias the dual-gate MOSFET.

In RF applications, where only gate G_1 is used, gate G_2 is held at RF signal ground by a capacitor connected from gate G_2 to ground.

With some applications where only one gate is needed, the two gates can be connected together and the dual-gate MOSFET responds as a normal single gate MOSFET.

6-13 DIFFERENTIAL AMPLIFIER

The *differential* amplifier is one of the most practical and versatile circuits used in electronics. It forms the basis of integrated circuits used in audio, video, IF, tuned, and operational amplifiers.

A basic circuit consists of two common-emitter transistor circuits in parallel (Q_1 and Q_2), supplied by a constant-current source (Q_3) in the emitter circuit. Resistors R_{C_1} and R_{C_2} should be equal and transistors Q_1 and Q_2 must be matched. The circuit has two inputs and two outputs. Voltage at input 1 (V_{i_1}) is applied to Q_1 and voltage at input 2 (V_{i_2}) is

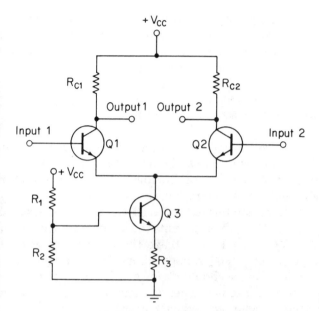

Figure 6-20 Differential amplifier.

applied to Q_2. Output voltage can be taken from output 1 to ground (V_{o_1}), output 2 to ground (V_{o_2}), or between the two outputs ($V_{o_1} - V_{o_2}$), which is the difference between V_{o_1} and V_{o_2}, hence the name "differential amplifer."

When both inputs are at ground, both Q_1 and Q_2 are cut off and each output is at $+V_{cc}$. The difference between them is zero. If one of the transistors is turned on by a positive voltage to its input, the corresponding output will decrease (go low). This results in a voltage difference between the two outputs. If both inputs are turned on the same amount (input voltage the same), the outputs will decrease the same amount and there will be no difference between them. If the transistors are not turned on the same amount, the outputs will be different and a voltage difference will exist between them.

Table 6-2 helps to show this relationship.

TABLE 6-2 INPUT–OUTPUT RELATIONSHIPS IN DIFFERENTIAL AMPLIFIER

Input 1 (V_{i_1})	Input 2 (V_{i_2})	Output 1 (V_{o_1})	Output 2 (V_{o_2})	Difference between outputs
0	0	V_{cc}	V_{cc}	0
Positive	0	Low	V_{cc}	$V_{o_2} - V_{o_1}$
0	Positive	V_{cc}	Low	$V_{o_1} - V_{o_2}$
Positive (same)	Positive	Low	Low	0
Most positive	Positive	Lowest	Low	$V_{o_2} - V_{o_1}$
Positive	Most positive	Low	Lowest	$V_{o_1} - V_{o_2}$

If V_{o_2} is set as a reference point, then when V_{i_1} is more positive than V_{i_2}, the output at V_{o_1} will go negative. Input 1 then behaves like an inverting input. If V_{i_2} becomes more positive than V_{i_1}, the output at V_{o_1} will go positive. Input 2 then behaves like a noninverting input.

6-14 BASIC DESIGN FOR CONSTRUCTING AN NPN CE AMPLIFIER

There are many factors to consider when designing an efficient class A amplifier: the impedance of the devices before and after the amplifier, input and output impedance, current requirements of the load, ambient temperature and changes, desired gain, and frequency response. However, a dc operating point (quiescent point) must be established for each circuit. It is desirable to have V_c at half the value of V_{cc}, which permits the greatest ± swing of the output voltage. Resistor R_E is used to stabilize the circuit, but also decreases the available output swing. The voltage across R_E can range from 8 to 20% of V_{cc}. The remaining voltage should divide evenly across the transistor and R_L. Highest stability is attained when R_A and R_B are low in value but they are usually selected higher to reduce current drain on the power supply.

The beta of a specific transistor is usually variable, depending on the values of I_B and I_C. However, it is very important to find the amount of I_E

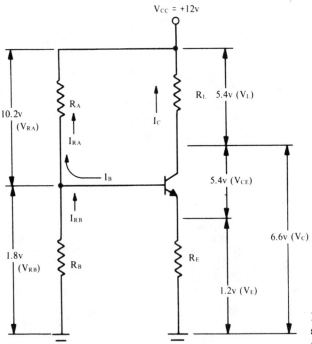

Figure 6-21 Circuit for constructing NPN common-emitter amplifier.

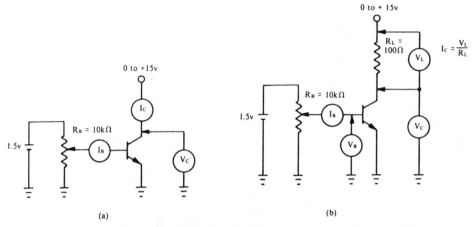

Figure 6-22 Circuits for measuring I_C and I_B: (a) using two ammeters; (b) using one ammeter and calculating I_C.

that produces a given amount of I_C. A simple, practical approach is to measure these currents by setting V_{CE} for one-half of V_{CC} and adjust I_B with a 10K to 100K potentiometer for the desired I_C. Once the value of I_B is found to produce a given I_C, the following procedure for determining the dc load and biasing resistors may be used.

6-14.1 Procedure for Determining Load and Biasing Resistors

1. Select or determine: V_{CC}
2. Find: $V_E = 0.10V_{CC}$
3. Find: $V_L = \dfrac{V_{CC} - V_E}{2}$ (note: $V_L = V_{CE}$)
4. Find: $V_C = V_L + V_E$
5. Select: desired I_C
6. Find: $R_L = V_L/I_C$ (note: $I_C \approx I_E$)
7. Find: $R_E = V_E/I_C$
8. Let: $V_{EB} = 0.6$ V for silicon transistors and 0.1 V for germanium transistors
9. Find: $V_{RB} = V_{EB} + V_E$
10. Let: $I_{RB} = 0.1$ mA (selected low to limit current drain on power supply)
11. Find: $R_B = V_{RB}/I_{RB}$
12. Find: I_B that produces I_C from previous test
13. Find: $I_{RA} = I_{RB} + I_B$
14. Find: $V_{RA} = V_{CC} - V_{RB}$
15. Find: $R_A = V_{RA}/I_{RA}$

Use commercially available resistors that are close to the calculations
The voltage measurements for the circuit, V_C, V_E, and V_B should be within
a few tenths of a volt of the calculated values.

The dc gain of the circuit can be approximated by dividing R_L by R_E
($A_{V_{dc}} = R_L/R_E$). If a bypass capacitor (C_E) is added across R_E, the ac gain
will be much larger but will also be dependent on the coupling capacitor
and the load to which the amplifier is connected.

6-15 TROUBLESHOOTING A TRANSISTOR AMPLIFIER

The most common approach to troubleshooting a transistor amplifier is to
check the dc voltages at the collector, base, and emitter (V_C, V_B, and V_E
respectively). Two important points generally should be remembered: i
$V_C = V_{CC}$, the transistor is not conducting or if V_C is lower than the norma
indication, the transistor is conducting more heavily. The transistor itsel
may have developed a problem; an internal short, leakage between elements
or an open. Another cause may be a defective load resistor or biasing
resistor. Usually, resistors increase in value or become open.

Any one of several defects can stop conduction through the transistor
causing $V_c = V_{CC}$. If R_E opens, V_C will rise to V_{CC}, and V_B and V_E wil
rise slightly, depending on the voltage-divider biasing of R_1 and R_2. If R,

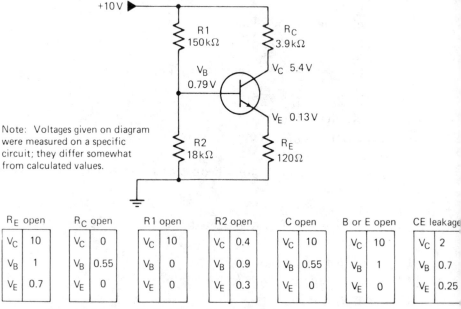

Note: Voltages given on diagram
were measured on a specific
circuit; they differ somewhat
from calculated values.

R_E open		R_C open		R1 open		R2 open		C open		B or E open		CE leakage	
V_C	10	V_C	0	V_C	10	V_C	0.4	V_C	10	V_C	10	V_C	2
V_B	1	V_B	0.55	V_B	0	V_B	0.9	V_B	0.55	V_B	1	V_B	0.7
V_E	0.7	V_E	0	V_E	0	V_E	0.3	V_E	0	V_E	0	V_E	0.25

Figure 6-23 Troubleshooting a bipolar transistor amplifier using voltage
analysis.

becomes open, R_2 electrically pulls the base of the transistor to ground, V_B = O V, and there is no IR drop across R_E. Therefore, V_E = O V and V_C rises to V_{CC}. If the collector opens internally, V_C will rise to V_{CC} and V_B and V_E will decrease. If the base or emitter opens internally, V_C rises to V_{CC}, V_E decreases, and V_B will increase to the voltage-divider reference point.

One condition may indicate that the transistor is saturated, V_C = O V, when in fact it is not conducting. When R_C becomes open, no voltage is applied to the collector, and V_B and V_E will also decrease.

Two defects can cause an increase in IC or even saturation of the transistor. A collector-to-emitter leak may develop which decreases the resistance of the transistor. The base voltage remains about the same, but V_C decreases while V_E increases slightly. If R_2 becomes open, the base of the transistor is pulled electrically toward V_{CC} via R_1. The forward bias increases, V_B increases, and V_E increases while V_C decreases.

7

Oscillators and Multivibrators

7-1 INFORMATION ON OSCILLATORS AND MULTIVIBRATORS

Oscillators convert dc to ac or other forms of pulsating dc. The waveforms can be sine wave, square wave, triangle wave, sawtooth wave, or other combinations of these basic types. Multivibrators are a group of oscillators that produce square-edge waveforms, delay and reshape pulses, and temporarily store electrical information.

A universal amplifier has a 180° phase shift from input to output. If the output is shifted another 180° and fed back to the input, the amplifier will oscillate providing that the amplitude of the feedback is sufficient. The signal is in phase and is referred to as *positive* or *regenerative feedback*. Negative feedback decreases gain and tends to stabilize an amplifier. Positive feedback increases gain, causes instability, and the amplifier tends to oscillate. It can be said that an oscillator is an amplifier that provides its own input signal.

Most sine-wave oscillators utilize an LC tank circuit. An LC tank circuit can oscillate when connected to a battery and switch. When the switch is closed, C charges to the battery potential. When the switch is opened, C discharges through L, which creates an electromagnetic field around L. When C is discharged, the field collapses, which continues to supply current to charge C in the opposite direction. Now, C must discharge in the reverse direction and the tank circuit continues this action in the form of oscillations. The oscillations are damped and eventually die out, or stop, because

176

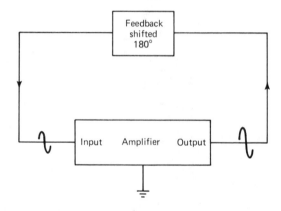

Figure 7-1 Positive feedback for oscillation.

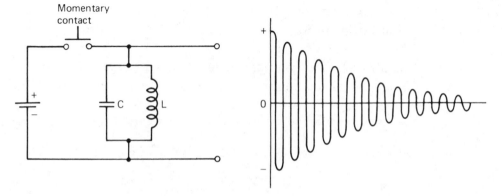

Figure 7-2 Tank circuit action.

of the resistance in the circuit. However, if after one cycle, the switch could be momentarily closed to add more energy to the circuit, the oscillations would continue at a fixed amplitude.

In an oscillator, the switch is an active device, such as a vacuum tube, transistor, or other solid-state device.

The resonant frequency f_r of the tank circuit can be found by the formula

$$f_r = \frac{1}{2\pi \sqrt{LC}}$$

Oscillators and multivibrators are found in radio and television circuits, magnetic recorders (tape, disk, etc.), computers and microprocessors, electronic entertainment devices, control circuits, test equipment, and other industrial electronic applications.

General oscillator requirements.

1. All types require a dc power supply.
2. Sine-wave types require positive feedback.
3. Sine-wave types require RC or LC components.
4. Relaxation types (sawtooth wave) require a capacitor for charging and a discharge path.
5. Multivibrators usually have two complementary outputs. When one output is high, the other output is low.

Oscillator optimum features.

1. A stable output frequency.
2. A constant-output amplitude.
3. Low distortion of waveform.
4. A fixed-frequency output.
5. A variable-frequency output (usually limited to a span of a decade).
6. Various frequency ranges by switching components to extend the variable-frequency output.
7. Perhaps, some method of external triggering for synchronizing purposes.

Also refer to Sections 2-4, 2-10, 2-12, 4-6, and 13-10 through 13-17.

7-2 *RC* PHASE-SHIFT OSCILLATOR

An *RC phase-shift oscillator* is suited for sine-wave frequencies below 10 kHz. The basic circuit is a universal bias common-emitter amplifier, including components Q_1, R_C, R_E, C_E, R_A, and R_B. The circuit is forward biased by R_A and R_B. The three-section RC network of C_1–R_1, C_2–R_2, C_3–R_3, and R_F provide the positive feedback path from the output (collector) to the input (base). Each RC section shifts the feedback voltage about 60°; hence three sections will produce a phase shift of 180°. The values of R and C for each section are usually the same. This circuit is used for a single fixed frequency, since changing R and C would have to be done for all three sections simultaneously to achieve the 180° phase shift. The output is seen across R_L. Resistor R_F is made variable to be able to adjust sufficient feedback voltage to cause oscillation.

When power is applied to the circuit, current flow through the transistor branch causes a voltage change at the collector. This voltage change is applied to the base via the 180° phase-shift network, where it provides the positive feedback to sustain oscillations. The approximate frequency can be determined by the formula

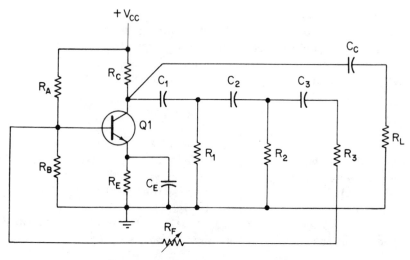

Figure 7-3 RC phase-shift oscillator.

$$f \approx \frac{1}{18RC}$$

For a frequency of about 1 kHz, the following component values can be used:

R_A = 100 kΩ C_E = 500 μF

R_B = 22 kΩ C_C = 0.1 μF

R_C = 1 kΩ $C_1 = C_2 = C_3$ = 0.01 μF

R_E = 470 Ω Q_1 = 2N3565

R_L = 100 kΩ Let $+V_{CC}$ = 9 to 12 V

$R_1 = R_2 = R_3$ = 4.7 kΩ

R_F = 1 kΩ

7-3 WIEN-BRIDGE OSCILLATOR

A *Wien-bridge oscillator* uses RC components to generate a low-distortion sine wave with a typical frequency range of 5 Hz to 500 kHz. A block diagram is useful in understanding its operation. The output signal is already positive feedback (the amplifier may consist of two stages). Components R_1-C_1 and R_2-C_2 form a frequency-selective network that will provide the necessary amount of feedback to sustain oscillations at a single frequency, designated as f_0. Frequencies below f_0 will in effect be blocked by C_1, by not having

enough gain to cause oscillation. Frequencies above f_0 will in effect be
shorted to ground by C_2. The two capacitors are ganged together to provide
a fine frequency adjustment. Other capacitors may be switched into the
circuit to extend the frequency range of the oscillator. When $R_1 = R_2$ and
$C_1 = C_2$, the approximate frequency can be calculated by the formula

$$f_0 \approx \frac{1}{2\pi RC}$$

In an actual circuit some negative feedback is used to stabilize the
oscillator. Thermistor R_3 and fixed resistor R_4 form a voltage divider which
is connected to the emitter of Q_1. The positive feedback applied to the base
of Q_1 via C_C must be greater than the negative feedback applied to the
emitter in order for the circuit to oscillate.

The Wien-bridge oscillator is often found in laboratory sine-wave
generators, because of its low distortion.

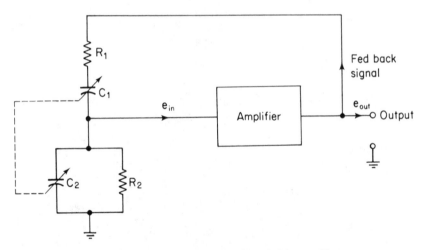

Figure 7-4 Block diagram of Wien-bridge oscillator.

Figure 7-5 Schematic diagram of Wien-bridge oscillator.

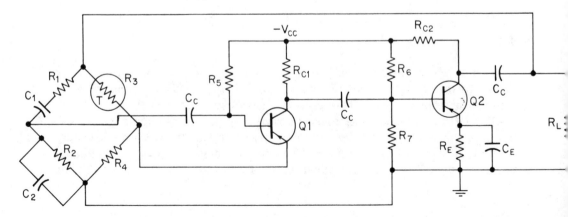

7-4 ARMSTRONG OSCILLATOR

An *Armstrong oscillator* uses an LC tank circuit to create its sine-wave oscillations. Forward bias is established on the transistor by R_1 and R_2. Capacitor C_2 bypasses ac around R_1. Components R_E and C_E stabilize the transistor and prevent thermal runaway. The tank circuit consists of L_1 and C_1. A "tickler coil" L_2 is inductively coupled (aiding) to L_1 and provides positive feedback voltage to the transistor. The output is seen across L_3, which is also inductively coupled to L_1. The transistor acts as a switch which supplies energy to the LC tank circuit to keep it oscillating at a constant amplitude.

When power is applied to the circuit, the initial rise of current provides a sufficient change of energy from L_1 to L_2. This energy is applied to the base of the transistor, where it is amplified and causes a higher amplitude in the tank circuit. The tank circuit now begins a "flywheel" effect to produce oscillations. These oscillations are coupled back to the base of the transistor as feedback and alternately drive it into saturation and cutoff. The transistor, acting as a controlled switch by the feedback, applies energy to the tank circuit to maintain its oscillations at a constant amplitude.

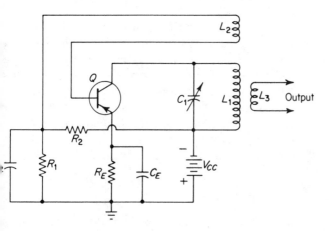

Figure 7-6 Armstrong (tickler coil) oscillator.

7-5 HARTLEY OSCILLATOR

The *Hartley oscillator* produces a sine-wave output with the use of an LC tank circuit. It is easily recognizable, because the inductor of the tank circuit is tapped to provide the required positive feedback. The lower section of L_1 is the feedback coil. The tap may be adjustable to provide the necessary feedback voltage for oscillation. As shown, resistors R_A and R_B develop a slight forward bias on Q_1. The radio-frequency choke (RFC) inductor passes the dc collector current, but due to its inductive reactance, it limits oscillator frequency current that reaches the power supply. Positive feedback is passed by C_2, while C_3 passes the energy from Q_1 to the tank

181

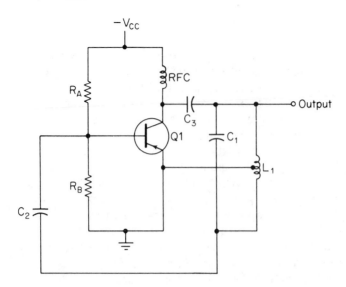

Figure 7-7 Hartley oscillator.

circuit, L_1-C_1. Since no dc current flows in the tank circuit, the oscillator is said to be shunt fed.

When power is applied to the circuit, the collector goes slightly positive (toward ground). There is a $180°$ phase difference across L_1; therefore, a negative voltage (in phase) is applied to the base of Q_1, which drives the transistor harder. When saturation is reached, C_1 stops charging and begins to discharge through L_1, which immediately reverses the polarity of L_1 and cuts off the transistor. The "flywheel" effect of the tank circuit alternately turns on and off Q_1, which supplies the needed energy to the tank circuit to sustain oscillations at a constant amplitude.

The approximate frequency can be found by the formula

$$f_r \approx \frac{1}{2\pi\sqrt{L_1 C_1}}$$

Either C or L of the tank circuit can be variable for fine frequency adjustments.

7-6 COLPITTS OSCILLATOR

The *Colpitts oscillator* uses an LC tank circuit to produce a sine-wave output. It is easily recognizable, because of its split capacitors in the tank circuit. As shown, C_1, C_2, and L_1 comprise the tank circuit, where C_2 provides the positive feedback for oscillation. Components R_A and R_B provide forward bias of Q_1. The radio-frequency choke (RFC) inductor passes the dc collector current, but due to its inductive reactance, it limits the oscillator frequency current that reaches the power supply. Positive feedback is passed

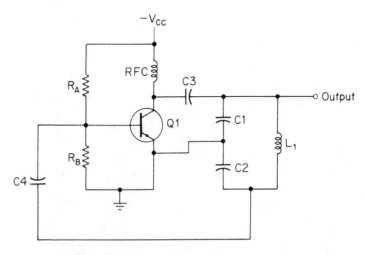

Figure 7-8 Colpitts oscillator.

by C_4 to the base of Q_1, while C_3 passes the energy from Q_1 to the tank circuit. Since no dc current flows in the tank circuit, the oscillator is said to be shunt fed.

When power is applied to the circuit, the collector goes slightly positive (toward ground). There is a 180° phase difference across L_1; therefore, a negative voltage (in phase) is applied to the base of Q_1, which drives the transistor harder. When saturation is reached, C_1 and C_2 stop charging and begin to discharge through L_1, which immediately reverses the polarity of L_1 and cuts off the transistor. The "flywheel" effect of the tank circuit alternately turns on and off Q_1, which supplies the needed energy of the tank circuit to sustain oscillations at a constant amplitude.

The approximate frequency can be found from the formula

$$f_r \approx \frac{1}{2\pi \sqrt{L_1 \, C_T}}$$

where $C_T = C_1 C_2 / C_1 + C_2$.

Both C_1 and C_2 or L_1 can be variable for fine frequency adjustments.

7-7 CLAPP OSCILLATOR

The *Clapp oscillator* is a variation of the Colpitts oscillator (see Section 7-6), but with improved frequency stability. Components C_1 and C_2 are made large enough to "swamp" out or cancel the transistor's inherent junction capacitances. This effect negates junction capacitance changes due to transistor variations with temperature. A capacitor (C_3) placed in series with

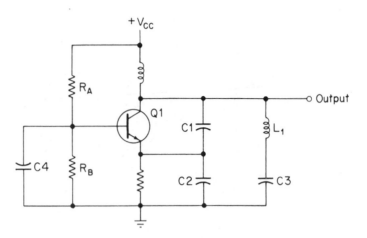

Figure 7-9 Clapp oscillator.

L_1 produces a series resonant frequency. However, C_1 and C_2 must be much greater in value than C_3.

The approximate frequency can be found by the formula

$$fr \approx \frac{1}{2\pi \sqrt{L_1 C_3}}$$

Components L_1 or C_3 can be variable for fine frequency adjustment; however, this oscillator does not have as large a range of frequency adjustment as the Hartley and Colpitts oscillators.

7-8 CRYSTAL OSCILLATOR

Crystal oscillators are used in circuits that require a high degree of frequency stability. The crystal is usually cut from quartz, where the type of cut and thickness determines its operating frequency. The natural frequency of a crystal can range from 10 kHz up to 10 MHz. A crystal operates on the principle of the piezoelectric effect. When the crystal is put under physical stress, it will generate a slight voltage difference. Similarly, if a voltage is placed on it, a physical stress results. When the voltage is momentarily applied, the crystal will oscillate at its natural resonant frequency. Crystals have a high Q (in the order of 20,000); however, they are susceptible to frequency changes with temperature changes. For this reason, they may be mounted in a temperature-controlled oven to maintain an oscillator's accurate operating frequency. Crystals may be used to replace the inductor in LC oscillators, although at higher frequencies, the case in which it is mounted provides the capacitance for a complete tuned tank circuit.

Components C and L of the tank circuit, shown in the figure, serve as output waveform filters, while the exact oscillator frequency is set by the

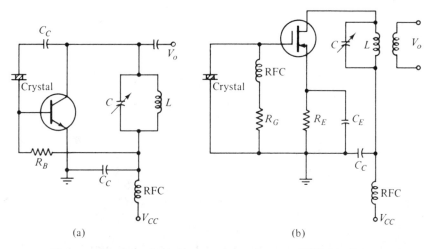

Figure 7-10 Crystal oscillator: (a) using an NPN bipolar transistor; (b) using an *n*-channel depletion-type MOSFET.

crystal. In the NPN bipolar transistor circuit (a), the crystal provides the positive feedback path to turn the transistor off and on. The MOSFET circuit (b) uses the crystal as a tank circuit at the gate. The oscillations are amplified by the MOSFET and passed on to the *LC* filter.

Tunable oscillators using crystals are usually varied by plugging in crystals with different natural resonant frequencies.

7-9 BLOCKING OSCILLATOR

A *blocking oscillator* does not produce a complete cycle oscillation, but rather a pulse, to be used for special applications. The pulse is considerably less than 360°. A basic circuit resembles a regenerative "tickler" coil oscillator (see Section 7-4).

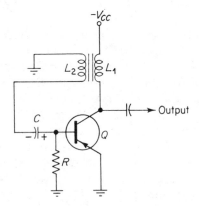

Figure 7-11 Blocking oscillator.

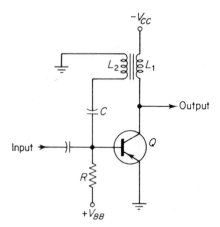

Figure 7-12 Triggered blocking oscillator.

Initially, when power is applied, Q conducts and the collector current begins to rise. This rising current through L_1 induces a voltage in L_2. The coupling between L_1 and L_2 is arranged so that the bottom of L_2 becomes negative and the top (grounded end) becomes positive. This positive feed back voltage fed to the base of Q via C drives the transistor into saturation. When there is no change in collector current, there is no induced voltage to apply to the base of Q. However, C has built up a positive charge, which drives Q into cutoff. The transistor will remain cut off and the circuit will be in a state of relaxation until the charge on C discharges through R to the point where forward bias is resumed. Then the entire cycle is repeated.

With a triggered blocking oscillator, R is not returned to ground, but connected to $+V_{BB}$, which keeps Q normally cutoff. A negative pulse of sufficient amplitude to overcome the reverse bias on the base will cause Q to go through a conductive cycle and then return to cutoff.

Blocking oscillators are commonly found in television circuits.

7-10 RELAXATION (SAWTOOTH) OSCILLATOR

Relaxation oscillators produce non-sine-wave outputs, such as sawtooth rectangular, and square-edge pulse. Their operation depends on the charge and discharge of the capacitor. The rate or frequency at which they operate depends basically on the RC time constant ($t = RC$). For a sawtooth wave the charging time constant must be larger than the discharge time constant.

A gas-discharge tube can be used as a switch to provide a long charge time constant and a short discharge time constant. When power is first applied to the circuit, the gas in the tube is un-ionized and appears as an open circuit. Capacitor C begins to charge up exponentially through R toward the power source. At a certain point, the gas tube "fires" and the gas ionizes providing a low resistance path for C to discharge. The voltage across

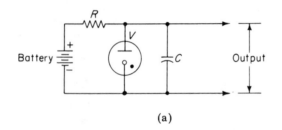

(a)

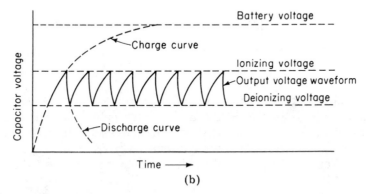

(b)

Figure 7-13 Gas-discharge-tube relaxation oscillator: (a) schematic diagram; (b) output voltage waveform.

the capacitor decreases to the point where the gas deionizes and the tube again becomes open. The capacitor then begins charging up to complete another cycle.

A Shockley diode can be used as the switch in a relaxation oscillator. The voltage across the capacitor rises exponentially until the voltage break-over point of the diode occurs. The capacitor discharges through the diode until the current falls below the minimum holding current of the diode. The diode then opens (stops conducting), and the process starts over.

The SCR can be used to produce a sawtooth waveform. Resistor R_1 must be large enough to keep the current through the SCR just below the minimum holding point. This will allow the SCR to open after the capacitor has discharged. Resistor R_2 should be in the megohm range and is used to set the circuit into oscillations. It can also vary the frequency within a limited range. The capacitor can be any value over a wide range, depending on the frequency desired. When power is first applied to the circuit, the

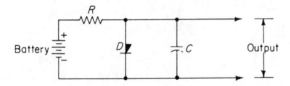

Figure 7-14 Relaxation oscillator using a Shockley diode.

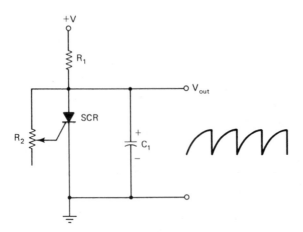

Figure 7-15 Relaxation oscillator using an SCR.

capacitor begins to charge up expotentially. The voltage will continue to rise at the anode of the SCR, until enough gate current is developed to "fire" the SCR. The capacitor will now discharge back through the SCR. The SCR then "opens" and the process begins again.

For other relaxation oscillators, refer to Sections 2-11 and 2-12.

7-11 ASTABLE MULTIVIBRATOR

An *astable* (also referred to as free-running) *multivibrator* can produce sawtooth and square-edge waveforms. In digital circuits it may be referred to as a clock. The waveforms should be symmetrical if corresponding components are equal (i.e., $R_{C1} = R_{C2}$, $C_1 = C_2$, $Q_1 = Q_2$, $R_{E1} = R_{E2}$, $C_{E1} = C_{E2}$, and $R_{B1} = R_{B2}$).

When power is applied to the circuit, both transistors begin conducting, but one will conduct more than the other. Assume that Q_2 conducts more and its collector starts going negative. This negative-going voltage is trans-

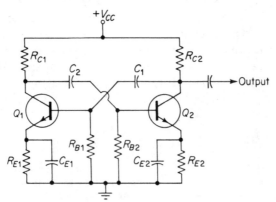

Figure 7-16 Astable multivibrator.

ferred to the base of Q_1 via C_1. Transistor Q_1 becomes cut off and its collector goes to the $+V_{CC}$ level and this positive-going voltage is applied to the base of Q_2, turning it on harder. Capacitor C_2 is charged up to voltage at the collector of Q_1 and will keep the circuit in this condition until it discharges through R_{B2} and begins to decrease the forward bias on Q_2. As the forward bias decreases, so does the current through Q_2, and its collector voltage begins to be positive. This positive voltage is applied to the base of Q_1 via C_1. Transistor Q_1 is turned on and its collector voltage begins to go negative. This negative voltage is applied to the base of Q_2 via C_2, making it go into cutoff. The circuit will remain in this condition until C_1 discharges to a point to decrease the current through Q_1. The cycle then repeats and a constantly changing output is available. The major RC timing components, which determine the frequency of the multivibrator, are C_1, C_2, R_{B1} and R_{B2}.

7-12 MONOSTABLE (ONE-SHOT) MULTIVIBRATOR

A *monostable multivibrator* must be triggered by an input pulse to produce a single pulse (or *one-shot* as it is often called) output. The values of the circuit components determine the length of time the circuit is in operation. Other trigger pulses occurring during the time the multivibrator is in an operation cycle will not affect the circuit. The multivibrator can be used as a time delay circuit and to reshape square-edge pulses.

When power is applied to the circuit Q_2 begins to conduct and its collector goes to nearly 0 V. The fixed voltage supply $+V_{BB}$ keeps Q_1 cut off. The collector of Q_1 is at $-V_{CC}$ and C charges up to this value. The circuit is at rest. When a negative trigger pulse of sufficient amplitude to overcome the fixed reverse bias of $+V_{BB}$ is present, Q_1 is forced into conduction. The collector voltage of Q_1 goes toward 0 V. Capacitor C attempting to discharge

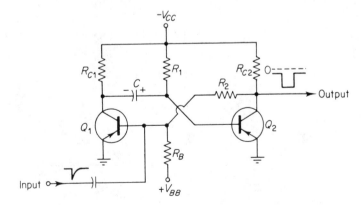

Figure 7-17 Monostable (one-shot) multivibrator.

through R_1 places a positive voltage on the base of Q_2. Transistor Q_2 is cut off and its collector goes to $-V_{CC}$. The $-V_{CC}$ voltage at the collector of Q_2 is applied to the base of Q_1, which keeps it in saturation. The circuit will remain in this condition until C has discharged to the point where Q_2 begins to turn on. At this time the collector of Q_2 goes positive, which decreases the current through Q_1. The action is cumulative and Q_1 becomes cut off, while Q_2 goes into saturation and the circuit is again in the initial condition, at rest. The RC time constant of C and R_1 is the main factor in determining the pulse width of the output. The use of PNP transistors produces a negative-going pulse. If NPN transistors are used and the power supplies are reversed, a positive-going pulse would be available.

7-13 BISTABLE (SET-RESET FLIP-FLOP) MULTIVIBRATOR

The *bistable (set-reset flip-flop) multivibrator* will remain in one of two conditions or states: Q_1 on and Q_2 off or Q_1 off and Q_2 on. It can be forced into either state by inputs A or B. This circuit can be used as a basic memory device, since it will remain (or remember) the state in which it was placed.

Assume that when power is applied, both transistors begin to conduct. However, Q_2 conducts more and its collector voltage begins to go positive. This voltage is applied to the base of Q_1 through R_2. In a cumulative action, Q_1 is cut off, and its collector voltage goes to $-V_{CC}$, which is applied back to the base of Q_2, causing it to go into saturation. Output 1 is now at $-V_{CC}$ and output 2 is at about ground potential. Power supply $+V_{BB}$ is of sufficient voltage to keep Q_1 cut off and the circuit will remain in this state.

A positive pulse to input B will cause the current through Q_2 to decrease and its collector voltage goes toward $-V_{CC}$. This negative-going voltage is applied to the base of Q_1, which starts it conducting. The positive-going

Figure 7-18 Bistable (set-reset flip-flop) multivibrator.

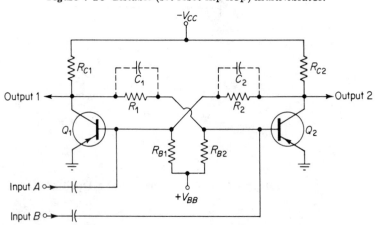

voltage (toward ground) of collector Q_1 is applied back to the base of Q_2, causing it to conduct less. This cumulative action continues until Q_2 is cut off and Q_1 is in saturation. Output 1 will now be at about ground potential and output 2 will be at $-V_{CC}$. The circuit will remain in this state due to $+V_{BB}$. A positive pulse to input A will cause the circuit to go back to the original state.

Negative input pulses can also be used to change the state of the flip-flop. A negative pulse will cause a cutoff transistor to begin conducting.

7-14 BISTABLE (TOGGLE FLIP-FLOP) MULTIVIBRATOR

The *bistable (toggle flip-flop) multivibrator* is similar to the set-reset flip-flop described in Section 7-13, except that steering diodes enable the flip-flop to change states by a single input. This action resembles that of a toggle switch.

Assume that Q_2 is conducting and Q_1 is cut off. Output 2 will be at ground potential, while output 1 will be at $-V_{CC}$. The base of Q_2 is negative ($\approx -V_{CC}$) and diode D_2 is reverse biased. The base of Q_1 is positive (≈ 0 V) and diode D_1 is slightly forward biased, due to the resistive voltage divider of R_1 and R_2. When a negative-going pulse appears at the input, it will pass through conducting diode D_1 to the base of Q_1. Transistor Q_1 starts conducting and the current through Q_2 starts to decrease. The cumulative action of the cross-coupling of the two transistors causes Q_1 to go into

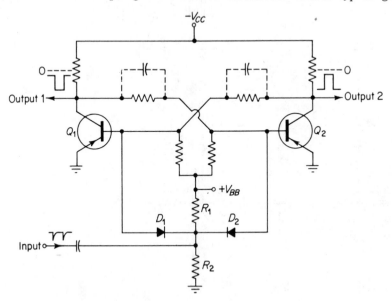

Figure 7-19 Bistable (toggle flip-flop) multivibrator.

saturation and Q_2 to become cut off. Output 1 is now at ground potential, while output 2 is at $-V_{CC}$. At this time diode D_2 is slightly forward biased and diode D_1 is reverse biased. The next negative trigger pulse at the input will cause the circuit to go back to the other state through similar action, where output 1 is at $-V_{CC}$ and output 2 is at ground potential.

7-15 BISTABLE (STEERED SET-RESET FLIP-FLOP) MULTIVIBRATOR

The *bistable (steered set-reset flip-flop) multivibrator* is a combination of the set-reset flip-flop (Section 7-13) and the toggle flip-flop (Section 7-14). It uses three inputs to change its state; however, only two are used at a time.

The output at the collector of Q_2 is labeled Q and the output at the collector of Q_1 is labeled $\overline{Q}$ (the complement of Q). One input is labeled S (set), another R (reset), and the last input T (trigger). The T input is used every time the flip-flop changes states. A positive voltage at input S and a negative-going trigger pulse at input T will turn on Q_1 and turn off Q_2. Similarly, a positive voltage at input R and a negative-going trigger pulse at input T will turn on Q_2 and turn off Q_1.

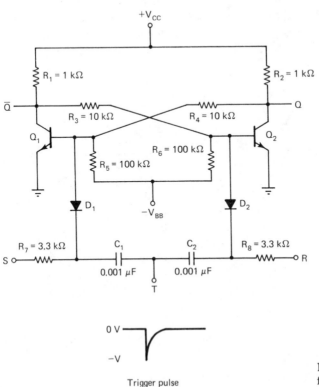

Figure 7-20 Steered set-reset flip-flop.

Assume that Q_1 is cut off and Q_2 is conducting. The positive voltage on the collector of Q_1 is applied to the base of Q_2, which slightly forward biases D_2. The low voltage (≈ 0 V) at the collector of Q_2 is applied to the base of Q_1; however, supply voltage $-V_{BB}$ reverse biases D_1. Placing a positive voltage at input S reverse biases D_1 even more. Input R is at ground potential at this time. When a negative-going trigger pulse is applied to input T, D_2 conducts more, and Q_2 is cut off. The positive-going voltage at the collector of Q_2 causes Q_1 to start conducting. The flip-flop is now in the on (or set) state. To switch the flip-flop back to the original state, input S must be 0 V, input R at a positive voltage and a negative-going pulse at input T. This time D_2 is reverse biased, and D_1 is forward biased, conducts, and turns off Q_1. Transistor Q_2 is then forced on and the flip-flop is back to the original state.

7-16 TESTING OSCILLATOR AND MULTIVIBRATOR CIRCUITS

There exists specialized equipment for testing oscillators that is developed by manufacturers of industrial electronic systems and transmitting and receiving systems. However, three basic types of test equipment will be able to service most oscillator problems: the voltmeter, a wideband oscilloscope, and the digital frequency counter. These test instruments should have high input impedance to minimize any loading effects.

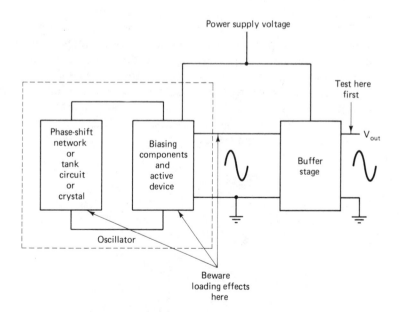

Figure 7-21 Testing oscillator circuits.

Special plastic alignment tools should be used when adjusting *LC*-type oscillators, since metal screwdrivers have inductive properties and can cause a frequency shift.

Care should be taken when testing an oscillator because improper connection of test instruments can cause a shift in frequency, reduce amplitube, or stop oscillations altogether. If a buffer stage follows the oscillator, the test should be made at this point first. Any test procedures given with a particular oscillator should be followed.

Table 7-1 lists four general problems associated with oscillators and possible causes. Since, power supply voltage affects all these problems, the first check should be made for correct voltage.

TABLE 7-1 OSCILLATOR PROBLEMS AND POSSIBLE CAUSES

Problem	Possible cause
No output	Defective transistor, crystal, or connection. Open or shorted component.
Reduced amplitude	Defective transistor. Defective bias resistors. Loading by another circuit.
Frequency error	Loading by another circuit. Defective tank circuit, *RC* network, or crystal. Defective bias resistors.
Frequency unstable	Loading by another circuit. Defective connection. Temperature-sensitive component. Defective crystal or transistor.

Tapping components with an insulated tool can help locate defective connections. Temperature-sensitive components may be located by using a desoldering pencil, to apply heat to a certain area, or commercially available chemical "cool sprays," to reduce the temperature of a selected component.

Multivibrators are tested similar to a standard amplifier, except that the outputs (usually at the collectors) will be complementary (i.e., one high, and one low). If the circuit is monostable or bistable, apply a trigger pulse to determine if it will change states.

Audio Circuits

8-1 INFORMATION ON AUDIO CIRCUITS

An audio system detects a small audio signal from an input transducer such as a microphone, phonograph cartridge, magnetic recording head, or an electronic circuit in a radio or television set. This small audio signal is voltage amplified by several stages, which might include tone control circuits and a volume control. The amplified audio signal is then converted by a power amplifier to a large current signal which is applied to an output transducer in the form of a loudspeaker. In turn, the loudspeaker varies the air on the speaker cone to produce sound waves. The human ear is capable of hearing sounds from about 20 to 20 kHz; therefore, an audio amplifier or system is usually designed to amplify this range of frequencies equally and is referred to as a *flat frequency response*.

However, audio circuits must be able to process complex ac signals characterized by steep complex wavefronts of a transient nature separated by incalculable periods of absolute silence. The tone from a single musical instrument is complex in itself. The physical shape of the instrument forms resonant cavities, which produces harmonic frequencies. These harmonic frequencies are higher multiple frequencies of the fundamental frequency,

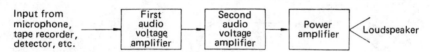

Figure 8-1 Block diagram of an audio system.

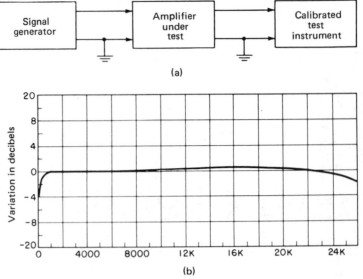

(a)

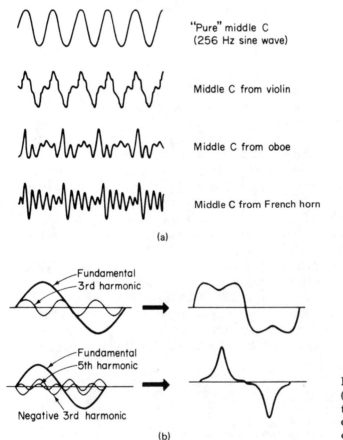

(b)

Figure 8-2 Frequency response of an audio amplifier: (a) block diagram of method of testing amplifier; (b) frequency-response curve.

"Pure" middle C
(256 Hz sine wave)

Middle C from violin

Middle C from oboe

Middle C from French horn

(a)

Fundamental
3rd harmonic

Fundamental
5th harmonic

Negative 3rd harmonic

(b)

Figure 8-3 Complex waveforms: (a) complex harmonic waveforms that determines musical instruments' characteristic sounds; (b) analysis of odd-harmonic waveforms.

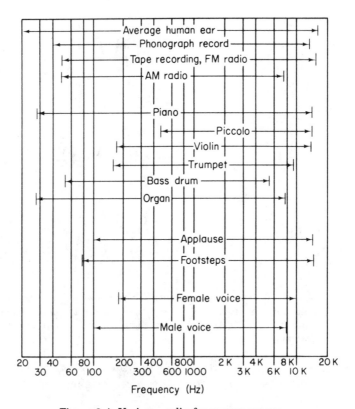

Figure 8-4 Various audio-frequency ranges.

but less in amplitude, which combines to produce the individual characteristic sound of the instrument. A single-note sound from an instrument may be complex in structure, but will have the same reoccurring pattern. Some musical instruments are capable of producing nearly the entire audio range of frequencies, whereas others produce primarily lows or highs.

Figure 8-5 shows a fundamental frequency (A), its in-phase third harmonic of lesser amplitude (B), and the resultant waveform (C) of the combining (adding and subtracting) voltages. A single-frequency sine wave (D) can be used in testing audio circuits for harmonic distortion. Typical complex audio signals are shown in (E) and (F).

8-1.1 The Decibel

The human ear does not perceive sound linearly, but logarithmically. This means that doubling the power of a sound level does not result in the ear feeling the sensation as twice as intense, but in reality much less. The unit of measurement to express this phenomenon is called the *decibel* (1/10 of a

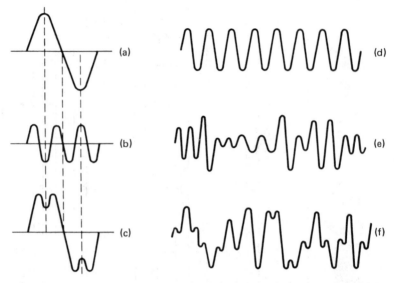

Figure 8-5 Waveform analysis: (a) fundamental frequency; (b) third-harmonic frequency; (c) resultant waveform of fundamental and third harmonic; (d) single frequency; (e) and (f) typical complex audio frequencies.

bel) and is abbreviated dB. A decibel then shows the relationship or ratio of two powers, voltages or currents, as given by the formulas

$$dB = 10 \log_{10} \frac{P_{out}}{P_{in}}, \quad dB = 20 \log_{10} \frac{V_{out}}{V_{in}}, \quad dB = 20 \log_{10} \frac{I_{out}}{I_{in}}$$

Since voltage and current are proportional (increasing one increases the other), the factor 20 is used in the voltage and current formulas for decibels.

Doubling the power of an amplifier means a 3-dB gain, whereas a power loss of one-half means a 3-dB loss. Doubling the voltage or current of an amplifier means a 6-dB gain and halfing the voltage or current results in a 6-dB loss.

Reference levels are established in audio work. A 1-mW reference level is referred to as zero decibels and is indicated as 0 dBm.

8-2 TYPES OF INTERSTAGE COUPLING

There are four standard methods used to transfer a signal voltage from one circuit to another. *Direct coupling* transfers the signal from the output of one stage directly to the input of a following stage. There is little or no problem with frequency reactive components and very little signal loss at various frequencies, but the type of biasing method needed may involve complex resistive circuits.

Type of Sound	Relative Intensity (in dB)
Reference level	0
Threshold of average hearing	10
Soft whisper; faint rustle of leaves	20
Normal whisper; average sound in home	30
Faint speech; softly playing radio	40
Muted string instrument; softly spoken words (at a distance of 3 ft)	50
Normal conversation level; radio at average loudness	60
Group conversation; orchestra slightly below average volume	70
Average orchestral voume; very loud radio	80
Loud orchestral volume; brass band	90
Noise of low-flying airplane; noisy machine shop	100
Roar of overhead jet-propelled plane; loud brass band close by	110
Nearby airplane roar; beginning of hearing discomfort	120
Threshold of pain from abnormally loud sounds	130

Figure 8-6 Decibel comparisons of some common sounds.

Resistive–capacitive (RC) coupling, also called *resistive coupling*, is the most used, least expensive, and is lightweight. A capacitor transfers the signal from the output of one stage to the input of the next stage. This capacitor blocks the higher dc voltage of the first stage from upsetting the bias voltage at the input of the following stage. However, the increased capacitive reactance at low frequencies reduces the signal amplitude applied to the next stage and low frequencies are not amplified as much.

Impedance coupling, used very little in audio circuits, uses an inductor as the load in the first stage. A capacitor is then used to block dc, while passing the ac signal to the next stage. The increased inductive reactance at high frequencies causes the signal amplitude to increase at high frequencies and results in a "peaking" effect. This type of coupling is heavier and more costly.

Transformer coupling may be used in circuits for impedance matching (power amplifier to speaker), phase splitting for push-pull operation, and to increase the interstage gain of an amplifier. A transformer provides dc voltage isolation from one circuit to another and the turns ratio can accomplish some signal gain. A major disadvantage of transformer coupling is the increased weight and cost.

8-3 RIAA EQUALIZATION AMPLIFIER

Basically, a phonograph record is produced by a lateral (side to side) cutting action of the cutting-stylus mechanism within a groove. Low frequencies originating from musical instruments have large amplitudes which would drive the cutting mechanism into adjacent grooves. Higher frequencies from

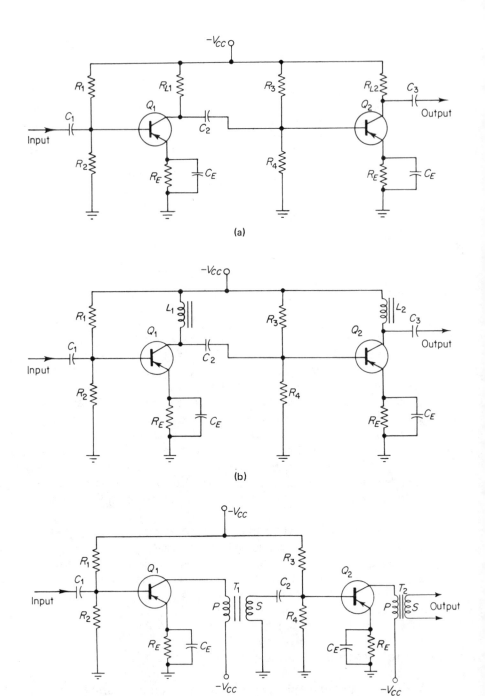

Figure 8-7 Types of interstage coupling: (a) resistive–capacitive (*RC*); (b) impedance; (c) transformer.

the instruments have smaller amplitudes, which will not drive the cutting mechanism sufficiently and result in a poor signal-to-noise ratio when the record is played back. Therefore, during the recording process an electronic procedure known as *equalization* attenuates the amplitude of the lower frequencies and amplifies the amplitude of the higher frequencies.

When a record is played back, the preamplifier of an audio system must reverse the equalization process, which is illustrated by the Record Industry Association of America (RIAA) curve. The preamplifier must have a higher gain for the frequencies from 50 to 500 Hz and have less gain for the frequencies from 2 to 20 kHz.

A preamplifier may consist of discrete transistors or use an integrated circuit; however, the external resistors and capacitors will form the essential feedback characteristics for proper equalization. With the RIAA equaliza-

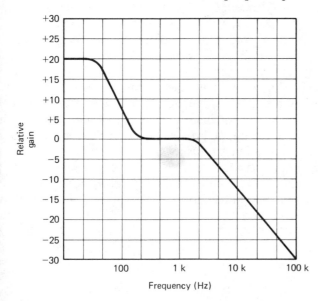

Figure 8-8 RIAA equalization curve.

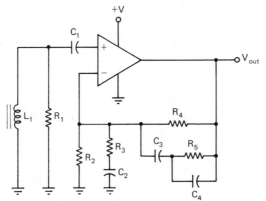

Figure 8-9 RIAA equalization amplifier.

tion amplifier shown, R_1 matches the input impedance to the magnetic cartridge L_1. Resistors R_2 and R_4 establish the dc bias for the circuit. Resistors R_3 and R_5 set the reference gain for frequencies of 500 Hz to 2 kHz. Capacitors C_3 and C_4 determine the amount of gain for various frequencies. At frequencies below 500 Hz, their X_c is large, causing less feedback signal, which results in higher gain for these frequencies. For frequencies above 2 kHz, their X_c is low, allowing more feedback signal, which reduces the gain for these frequencies.

8-4 NAB EQUALIZATION AMPLIFIER

Magnetic tape recording uses a tape head which is an inductive device whose impedance varies directly with frequency. When frequency increases, X_L increases. Therefore, at higher frequencies, X_L is greater and a larger amplitude is produced. Information recorded on magnetic tape then has low amplitudes for low frequencies and high amplitudes for high frequencies.

When a magnetic tape is played back, the tape preamplifier of an audio system must compensate for the level differences during recording by amplifying the low frequencies and attenuating the high frequencies. The National Association of Broadcasters (NAB) equalization curve is a standard which indicates the speed of the tape during playback. One speed is for 7.5 inches per second and the other for 3.75 inches per second.

A tape preamplifier may consist of discrete transistors or an integrated circuit; however, the external resistors and capacitors will form the essential feedback characteristics for proper equalization. With the NAB equalization

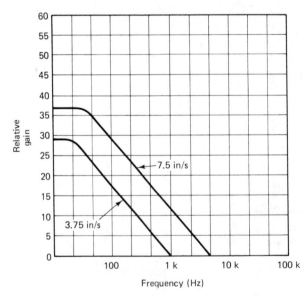

Figure 8-10 NAB equalization curve.

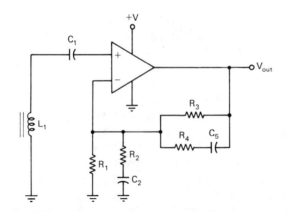

Figure 8-11 NAB equalization amplifier.

amplifier shown, C_1 couples the audio signal from the magnetic playback head L_1 into the amplifier. Resistors R_1 and R_3 establish the dc operating bias for the circuit. The reference gain of the circuit is set by R_2 and R_4. Capacitor C_5 determines the amount of gain for various frequencies. At lower frequencies, the X_C is large, causing less feedback signal, which results in higher gain. At higher frequencies, the X_C decreases, causing more feedback signal to reduce the gain of the amplifier.

8-5 SINGLE-ENDED POWER AMPLIFIER

Single-ended power amplifiers use transformer coupling to match the high impedance of the collector circuit of a transistor to the low impedance of a loudspeaker. The primary winding acts as the collector load, which develops the amplifier output voltage. The voltage induced in the secondary winding

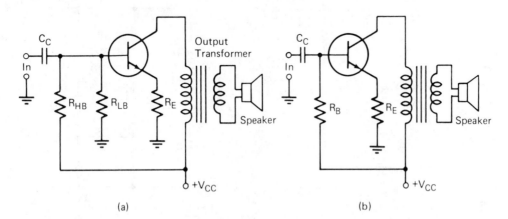

(a)   (b)

Figure 8-12 Single-ended power amplifiers: (a) universal bias arrangement; (b) base bias with emitter-feedback arrangement.

is less because of the step-down ratio to match the impedance of the speaker. Since the power in the primary must equal the power in the secondary ($P_p = P_s$; therefore, $E_p I_p = E_s I_s$), the current in the secondary is increased to drive the speaker. Neither the primary winding nor the secondary winding of the transformer is the transistor load. The actual load is the speaker voice coil. A voice coil may have an ohmic value of 4 Ω, 8 Ω, 16 Ω, or up to 50 Ω. The signal power applied to the voice coil sets up varying magnetic fields, which aid or oppose the fields of the speaker magnet, causing the diaphragm cone to vibrate, which produces audible sounds. Therefore, electrical energy of the power amplifier is converted into acoustical energy.

Because the inductive action of the transformer is based on current, very often the output transistor is of the power type. It may also be connected to a heat sink or metal chassis to dissipate heat. A single-ended audio power amplifier must be operated class A to produce as little distortion as possible.

8-6 PHASE INVERTERS

In a push-pull amplifier the input signal must be split 180° out of phase and applied to each section of the output stage. A center-tapped transformer can be used for this purpose (see Section 8-7), or solid-state phase inverters may be employed.

The simplest *phase inverter* is a single transistor amplifier with an unbypassed (no capacitor) emitter resistor. Since the current flowing in the emitter circuit is essentially the same as the current in the collector circuit resistors R and R_E are equal and produce the same amount of voltage

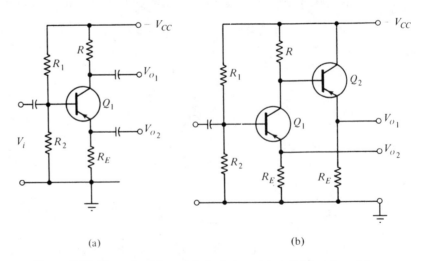

(a)   (b)

Figure 8-13 Phase inverters: (a) single transistor; (b) two transistors.

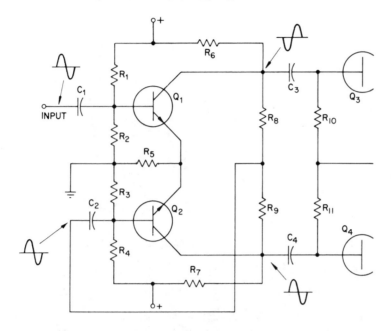

Figure 8-14 Paraphase amplifier.

drop. The signal at V_{O_2} is in phase with V_i, while the signal at V_{O_1} is 180°
out of phase with V_i.

Two transistors can be used to provide emitter-follower circuits for
more balanced impedance matching to the push-pull output stage. The
output of Q_1 is fed to the input of Q_2 and the outputs to the push-pull
circuit are taken from the emitter of each transistor. Output at V_{O_1} will be
180° out of phase with the input signal, while the output V_{O_2} will be in
phase.

Another phase inverter, also called a *paraphase amplifier*, uses two transistors where a portion of the output of Q_1 is fed back to the input of Q_2
via resistor voltage dividers R_8 and R_9. When Q_1 is conducting more, Q_2 is
conducting less, and vice versa. The circuit is self-adjusting because of the
arrangement of R_8 and R_9, and the signal that appears at the input of Q_2
will always be the same amplitude as the input signal applied to Q_1. The
out-of-phase signals are taken from the collectors and applied to the inputs
of a push-pull amplifier.

8-7 TRANSFORMER OUTPUT PUSH-PULL POWER AMPLIFIER

A *push-pull* power amplifier can have nearly double the output power of a
single-ended output amplifier. Theoretically, the output transistors are
biased class B, where only one transistor is conducting at a time. The input

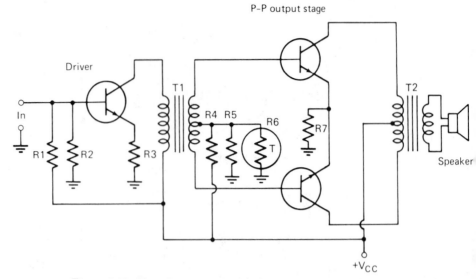

Figure 8-15 Transformer output push-pull power amplifier.

signal is inverted by the center-tapped transformer T_1, which alternately turns each transistor on and off. However, the slight interval of time that both transistors are off during the switching transition causes what is called crossover distortion (see Section 8-13). For this reason the output transistors are biased class AB (slightly on). Resistors R_4 and R_5 provide this slight forward bias. Resistor R_6 is a thermistor, usually mounted close to the output transistors, which helps to compensate for changes in bias due to increases in temperature. If too much forward bias is applied, the circuit will saturate with the input signal, and if there is not sufficient bias, the transistors will cut off. Resistor R_7 provides some negative feedback to reduce noise distortion and help stabilize the circuit. Transformer T_2 is center tapped to supply each output transistor with supply voltage. This step-down transformer converts the output voltage in the primary to a high current in the secondary, which drives the speaker.

The driver stage uses a universal bias arrangement and a transformer as its load also. Resistor R_3 provides negative feedback to decrease noise distortion and help stabilize the circuit. This amplifier must be operated class A.

Phase-inverter circuits, shown in Section 8-6, can be used to drive the push-pull amplifier. In many instances they prove to be better because of less weight and lower cost than the interstage transformer.

8-8 COMPLEMENTARY OUTPUT POWER AMPLIFIER

The *complementary output* power amplifier does not require two input signals 180° out of phase or an output transformer to couple the signal to a speaker. Since no output transformer is used, such circuits are also called *output transformer-less* (OTL) amplifiers. The output transistors (an NPN, Q_2, and a PNP, Q_3) must have characteristics that are matched fairly closely and are referred to as complementary transistors. These transistors are operated class AB push-pull. Resistors R_{E_2} and R_{E_3} are usually less than 5 Ω and provide some stabilization to the output transistors. The output impedance between the two resistors is low, which makes capacitive coupling to the speaker possible. Capacitor C_2 is usually 2000 μF or larger in value to provide the best low-frequency response and at the same time prevents the voice coil of the speaker from dc saturation. In the quiesient (Q) state (no input signal), the output point between R_{E_2} and R_{E_3} should be one-half of $+V_{cc}$. During operation, the transistors alternately conduct to allow the ac signal current to flow through C_2 and the speaker.

Transistor Q_1 is the driver for the complementary transistors and uses a universal-bias arrangement. However, the bias resistors voltage source comes from the output point, which provides feedback to stabilize the entire circuit and keep it at the designed operating Q point. Resistor R_L is the load for Q_1 and R_{CD} is the low-value resistor which develops the voltage to drive Q_2 and Q_3. The circuit is designed so that the collector current of Q_1 is such that the voltage at the midpoint of R_{CD} is nearly equal to one-half of $+V_{cc}$, or neutral to the output point. Therefore, the base of Q_2 is slightly positive to the neutral point, while the base of Q_3 is slightly negative to the neutral point. This keeps the complementary transistors turned on during the quiescent condition and reduces cross-over distortion during operation.

A positive-going input signal will tend to turn on Q_1, which turns off

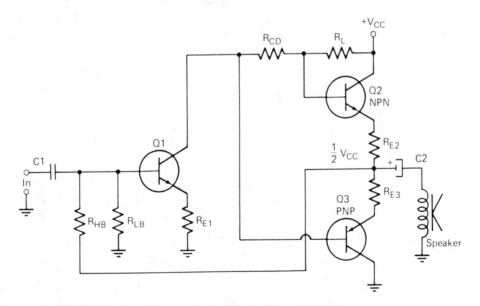

Figure 8-16 Complementary output power amplifier.

Q_2 and turns on Q_3. A negative-going input signal will have the opposite effect, where Q_1 turns off, Q_2 turns on, and Q_3 turns off.

Circuits like this may use a dual $\pm$ voltage supply, which means that the ground connections of R_{L_B}, R_{E_1}, and the collector of Q_3 would be connected to the negative voltage supply. The speaker may be connected to the negative supply or to ground. In this case the output point will be about 0 V with reference to ground during the quiescent state.

8-9 QUASI-COMPLEMENTARY OUTPUT POWER AMPLIFIER

A true complementary output consists of an NPN and PNP transistor in series (see Section 8-8). The characteristics of these transistors should be matched; however, silicon PNP power transistors are expensive and cannot be matched perfectly. In a *quasi-complementary output* stage, Q_3 and Q_5 simulate the PNP counterpart of a true complementary output stage and the effect is the same.

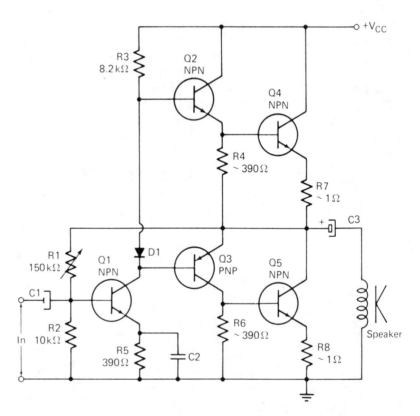

Figure 8-17 Quasi-complementary output power amplifier.

Transistor Q_1 is a common-emitter amplifier using universal bias supplied by feedback voltage from the output point. Resistor R_1 is variable so as to adjust the proper operating point on Q_1. The voltage at the output point should be one-half $+V_{cc}$. Transistor Q_2 is an emitter-follower circuit which drives Q_4. Transistor Q_3 is a common-emitter circuit which drives Q_5. Diode D_1 provides the necessary voltage to keep Q_2 and Q_3 turned on during the quiescent state. Transistors Q_2 and Q_3 operate in push-pull, which causes Q_4 and Q_5 to operate in a similar push-pull fashion. A positive-going input signal will tend to turn on Q_1. This tends to turn on Q_3, which causes Q_5 to turn on while Q_2 turns off and result in Q_4 turning off. A negative-going input signal has the opposite effect, where Q_1, Q_3, and Q_5 turn off, while Q_2 and Q_4 turn on.

A circuit like this may use a dual $\pm$ voltage supply, which means that the ground connection would be connected to the negative voltage supply. The speaker may also be connected here or to ground. The quiescent voltage at the output point would be nearly 0 V.

The quasi-complementary circuit can usually produce more output power than the standard complementary output amplifier.

8-10   TONE CONTROLS

Tone controls are used in audio amplifiers to overcome the unwanted effects of the original sound caused by the system, speaker response, room acoustics, and other factors. But most important is the listener's personal taste, since some persons prefer "bassy" music whereas others prefer it "trebly."

The term *bass* refers to low frequencies or low audible sounds; therefore, a bass tone control circuit controls low-frequency amplification or attenuation. Passive tone controls require "audio taper" (logarithmic) potentiometer, since our hearing ability is also logarithmic. When the wiper is set at the halfway point of rotation, the total resistive element is split into

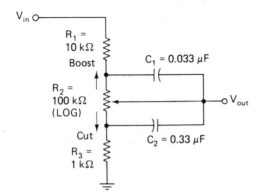

Figure 8-18 Typical passive bass tone control.

two portions, with 90% above the wiper and 10% below the wiper. Basically
when the wiper is placed toward R_1 (bass boost), capacitor C_1 is shorted and
there is more resistance from V_{out} to ground, which produces a larger amplitude. When the wiper is moved toward R_3 (bass cut), there is less resistance
between V_{out} and ground and the lower frequencies are allowed to pass
through C_1 and C_2 to R_3. Therefore, the amplitude is less.

The term *treble* refers to high frequencies or high audible sounds;
therefore, this circuit controls high-frequency amplification or attenuation.
Essentially, the components of the bass circuit have been rearranged for the
treble circuit except that the values of the capacitors are changed, since this
circuit deals with high frequency. When the wiper of R_2 is placed toward
C_1 (boost), there is a larger impedance to ground and more V_{out} is produced.
When the wiper is moved toward C_2 (cut), there is less resistance and the
signal is shunted to ground.

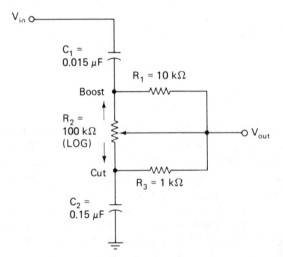

Figure 8-19 Typical passive treble
tone control.

Figure 8-20 Complete passive bass and treble tone control.

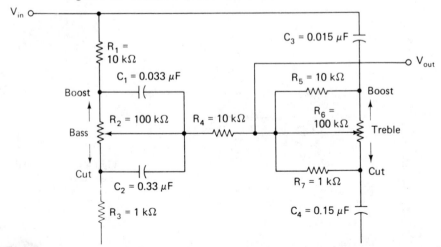

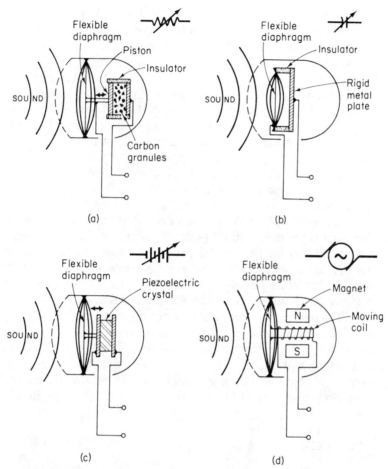

Figure 8-21 Microphones: (a) carbon button; (b) capacitor; (c) crystal; (d) dynamic.

A base and treble tone control can be combined into a complete circuit. Resistor R_4 helps to isolate the two controls and minimizes interaction. Passive tone controls such as these consume power and must be followed by a voltage amplifier. Active controls, on the other hand, are designed to use the feedback characteristics of transistor circuits and will maintain or even amplify the processed tone signals.

8-11 MICROPHONES

A microphone is an input transducer to an audio system that converts acoustical energy into electrical energy. Sound waves strike a diaphragm which is connected to a device that either changes resistance or produces a small voltage output.

In a *carbon button* microphone the diaphragm is connected to a piston which compresses carbon granules. The greater the compression, the lower the output resistance. The microphone is placed in a circuit where the changing resistance of the carbon button generates a small voltage which is applied to an amplifier. This microphone has a poor frequency response, can produce objectionable noise, and is used primarily for voice communication.

The diaphragm serves as one plate in a *capacitance* microphone; the other plate is fixed. A voltage is applied between the two plates. As the diaphragm vibrates to the incoming sound waves, it alternately widens and narrows the air gap, thus varying the dielectric thickness. The changing capacitance produces a changing current which is fed to an amplifier. The capacitor microphone has fairly good frequency response, but a low output amplitude.

A *crystal* microphone uses a crystal of barium titanate mounted between the diaphragm and a fixed plate. It works on the piezoelectric effect, where a stress placed on the crystal produces a slight voltage. Sound waves hitting the diaphram alternately "squeeze" the crystal and produce a varying output voltage, which is applied to an amplifier. The crystal microphone has reasonable frequency response and produces a large output voltage.

The *dynamic* microphone works on the principle of magnetic induction. The diaphragm is connected to a movable coil that is mounted between the poles of a small permanent magnet. When sound waves are present, the diaphragm and coil vibrate and a small current is induced in the coil. The varying current is fed to an amplifier. The dynamic microphone has good frequency response and is probably the most used type in audio systems.

8-12 LOUDSPEAKERS

A loudspeaker, often referred to as just a speaker, is an output transducer from an audio system that converts electrical energy into acoustical energy. The basic construction of the speaker consists of a paper or fabric cone connected to a circular frame by way of a flexible suspension. The inner or small end of the cone is attached to a movable device. With inexpensive speakers, the device is a permanent magnet armature place in front of a fixed voice coil. Current flowing in the voice coil sets up an electromagnetic field that attracts and repels the permanent magnet. This, in turn, vibrates the speaker cone and produces sound. Another type, most often found in high-quality speakers, is the dynamic type, wherein the voice coil moves within a permanent magnet structure. The voice coil is wound on a form of nonmagnetic material. When current flows through the voice coil, the electromagnetic field moves the coil in and out of the permanent magnet which causes the cone to vibrate and produce sound.

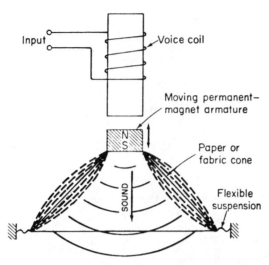

Figure 8-22 Fixed-coil loudspeaker.

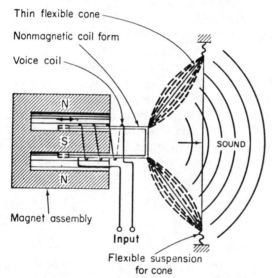

Figure 8-23 Moving-coil loudspeaker.

Some speakers have an extra small cone mounted in the center of the larger cone, which helps to produce high frequencies better. High frequencies are best produced by small speakers, referred to as *tweeters*, whereas low frequencies are best produced by large speakers, referred to as *woofers*. In a coaxial speaker system a tweeter is mounted in front of the woofer and the two input wires power both speakers. A third, medium-size speaker to handle intermediate frequencies may be added to the system; this entire unit is called a triaxial speaker.

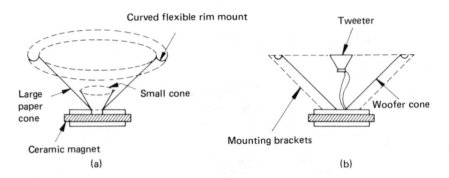

Figure 8-24 Dual-purpose speakers: (a) small cone to enhance high frequencies; (b) coaxial.

8-13 TESTING AUDIO CIRCUITS

Special audio test instruments are available to test complex audio systems; however, basic test instruments can usually be used to locate and correct a malfunction. These instruments consist of an audio signal generator or signal injector, a voltmeter, and an oscilloscope.

Testing audio circuits usually involves localizing the problem in an entire audio system. The most obvious problem is no sound at all. An initial check, as with any electronic device, is to make sure the power supply voltages are normal. Once the power supply is cleared, the next step is to apply a signal to the system and take measurements through the various stages with an oscilloscope. After connecting the signal source to the input of the system, an observation at point A will determine if the input signal is

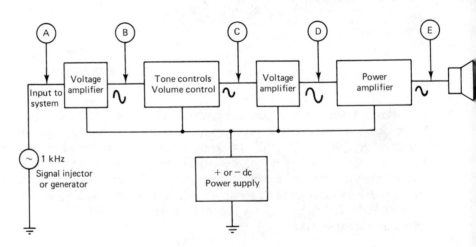

Figure 8-25 Troubleshooting an audio system.

distortion free and of sufficient amplitude (this signal should be very small in amplitude). At point *B*, after the preamplifier or first voltage amplifier, the amplitude of the signal should be greatly increased. The signal may be reduced at point *C* if it has passed through passive tone control circuits. If this point is after the volume control, the signal may be varied (amplitude-wise) by the control. The signal should again be larger after the next voltage amplifier stage at point *D*. At point *E* the normal signal should be small because of the loading effect of the low-impedance speaker. Any stage that does not pass the signal satisfactorily is then inspected and tested more carefully to find the malfunction (see Section 6-15).

Another, faster approach to testing an audio system would be to start at the output and work back through the stages to the input. The test signal would first be placed at point *E*, then *D*, and so on. The sound at the speaker should become louder as the test signal is moved closer to the input. Again, the defective stage could be found between the points, where the signal is correct and where it is distorted or nonexistent.

8-13.1 Waveform Distortion

Waveform distortion can cause objectionable sounds to the listener. These problems are usually more difficult to find and correct. A few of these problems are listed below.

Nonlinear output distortion. Nonlinear distortion, also called *amplitude distortion*, occurs when a portion of the amplitude of the signal is amplified more than the entire signal. A pure sine wave should be symmetrical (the positive alternation, the same amplitude as the negative alternation). A shift of the operating point can cause one alternation to be amplified more than the other. This problem is usually caused by biasing components changing values. Resistors generally increase in value, whereas, capacitors

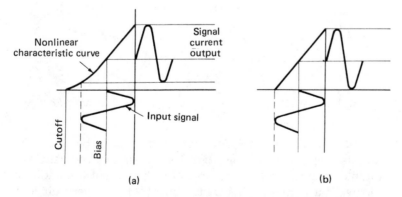

Figure 8-26   Output distortion: (a) nonlinear; (b) ideal circuit.

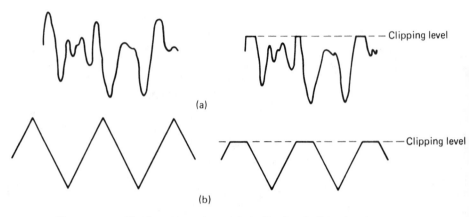

(a)

(b)

Figure 8-27 Clipping distortion: (a) audio signal; (b) triangle-wave test signal.

become leaky. The transistor itself may be defective. All components should be checked when this problem is encountered.

Clipping distortion.   Clipping distortion is a severe case of shifting of the operating point. The circuit may be reaching saturation or cutoff or a combination of both. A triangle-wave test signal found in most function generators is excellent for detecting this type of problem. Here again, all component values should be checked, as well as the beta of the transistor.

Crossover distortion.   Crossover distortion occurs with class AB push-pull amplifiers. Basically, one transistor is not turning on sufficiently before the other transistor is turned off. Often, the matched characteristics of the transistors are no longer symmetrical and the beta of each transistor should be checked. Biasing components, including temperature-compensating devices, should also be checked. Sometimes a biasing adjustment control can easily correct this problem.

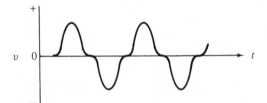

Figure 8-28 Crossover distortion.

Harmonic distortion.   An audio system must be able to correctly amplify complex signals involving countless harmonic frequencies. However, the audible output can become objectionable, when the audio system generates its own harmonics. This particular problem is very difficult to determine

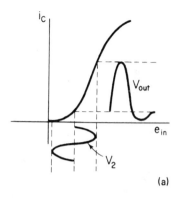

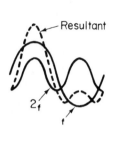

(a)

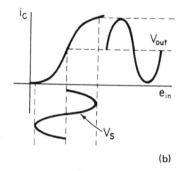

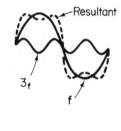

(b)

Figure 8-29 Harmonic distortion:
(a) even-harmonic distortion—
second harmonic in phase with
fundamental; (b) odd-harmonic
distortion—third harmonic in phase
with fundamental.

but feedback networks should be checked, since their function is to reduce
harmonic distortion.

Frequency distortion.   Frequency distortion occurs in an audio system
when certain frequencies are amplified more than others. This is caused by
the capacitive and inductive effects in a circuit, since the impedance of these
reactances varies with frequency. A frequency-response test is usually the
only way to find this problem (see Fig. 8-2).

Phase distortion.   Phase distortion occurs when a frequency component of
a complex input signal is delayed more than the others as it passes through
an amplifier. There may be equal amplification, but the delay may cause
considerable distortion at the output. This problem is exceptionally difficult
to determine and special testing procedures are usually required.

Noise distortion.   Noise distortion is detected at the speaker in the form of
ticks, pops, hisses, and/or scratchy sounds. Noise is extraneous signals of
various amplitude and frequencies, which attach themselves to the primary

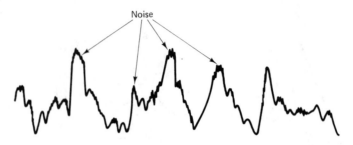

Figure 8-30 Noise distortion on an audio signal.

audio signal. Noise may be caused from loose connecting cables, a defective input transducer, a defective speaker, loose soldering joints, or by the electronic components in a system. Detecting noise is usually done with an oscilloscope stage by stage, while tapping the circuits lightly with an insulated tool.

Basic Radio Systems and Circuits

9-1 INFORMATION ON AMPLITUDE MODULATION (AM)

Less power is required to transmit high frequencies than low frequencies; therefore, it is more efficient to transmit high frequencies containing information over great distances. In *amplitude modulation* (AM) a constant-amplitude RF carrier wave of, say, 800 kHz is produced. The information to be transmitted, perhaps an audio frequency (AF) of 500 Hz, is sent to a modulating circuit, where it varies the amplitude of the RF carrier wave. An AM signal as seen on an oscilloscope will show modulation peaks and modulation valleys. The amount of RF carrier wave variation depends on the amplitude of the audio signal and is called *percentage of modulation*. The loudness of the information at the receiver is a result of the percentage of modulation. If the RF carrier wave is modulated more than 100% (overmodulation), the amplitude of the AM waveform will more than double for one alternation of the AF signal and cut off the RF carrier wave during the second alternation. This produces unwanted distortion and a "garbling effect" of the information at the receiver; therefore, AM transmission is kept below the 100% level of modulation.

During the modulation process other frequencies, called sidetones or sidebands, are produced. Upper-side-tone frequencies are above the RF carrier wave, and lower sidetone frequencies are below the RF carrier wave. If a 800 kHz RF carrier wave is modulated by a 500-Hz AF signal, the upper sidetone $(f_c + f_a)$ would be 800.5 kHz and the lower sidetone $(f_c - f_a)$ would be 799.5 kHz.

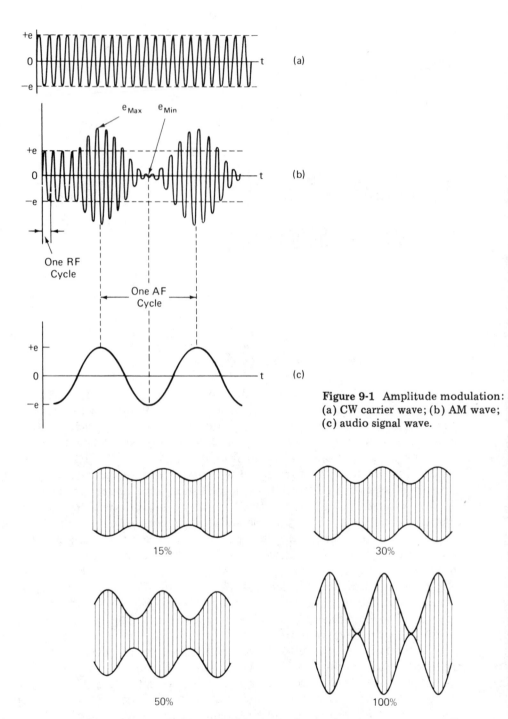

Figure 9-1 Amplitude modulation: (a) CW carrier wave; (b) AM wave; (c) audio signal wave.

15%

30%

50%

100%

Figure 9-2 Amplitude-modulation waveforms showing percentages of modulation.

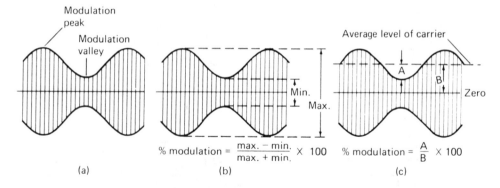

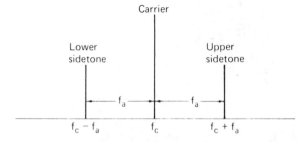

Figure 9-3 Amplitude-modulation definitions: (a) peak and valley; (b) percent modulation; (c) estimating percent modulation.

Figure 9-4 AM sidetones.

Overmodulation produces spurious sidetones that results in distortion to the information contained in the AM signal and may cause interference to other radio stations operating at different RF carrier-wave frequencies.

9-2 THE AM TRANSMITTER

Basically, the *AM transmitter* consists of two sections, one to produce and amplify the RF carrier wave and the other to amplify the AF signal and produce modulation.

The heart of the transmitter is the oscillator (see Chapter 7), which initiates the RF carrier wave. A buffer amplifier follows the oscillator, which isolates it from other circuits that might cause loading and upset the operating frequency. Other stages are then usually included in a practical transmitter such as RF amplifiers that increase the amplitude of the RF carrier wave, and frequency multiplier circuits that increase the original oscillator frequency. The last stage of the RF section is a power amplifier that supplies the AM signal to the antenna, which in turn produces electromagnetic radiation into the atmosphere.

The audio section of the transmitter uses speech or AF voltage amplifiers (see Chapter 8) that drive the modulator circuit. Often, the modulator

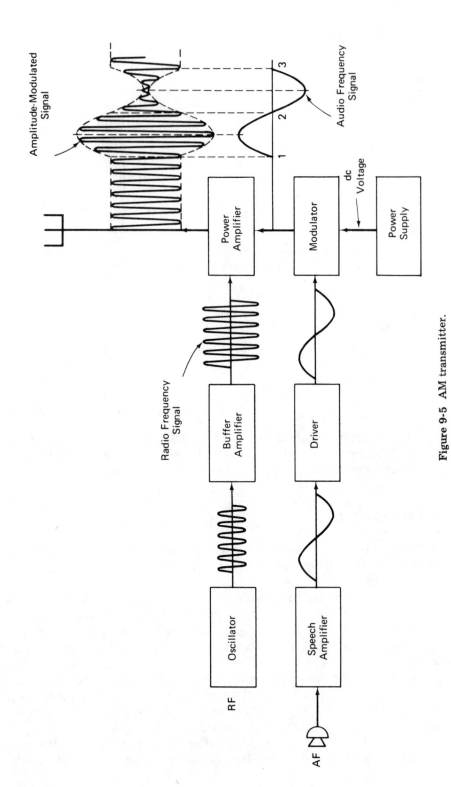

Figure 9-5 AM transmitter.

circuit is in series with the RF power amplifier and a separate power supply is used for this purpose. The supply voltage to the RF power amplifier is varied at an AF rate, which produces the AM signal.

Vacuum tubes are still frequently employed in RF power amplifier and modulating stages with large transmitters, because they are capable of handling large currents and high temperatures. Advances in solid-state power transmitters will someday replace the older vacuum-tube stages.

9-3 VOLTAGE AMPLIFIER

RF voltage amplifiers are used to increase the amplitude of the RF oscillator output and/or the AM wave in a transmitter. If the RF signal contains no modulation, the amplifier can be operated class A, B, or C, whereas an AM signal must use class A or B amplification. An RF voltage amplifier must be able to pass the RF carrier frequency as well as the sideband frequencies. With standard AM broadcasting this bandpass is from 10 to 15 kHz.

Transformer coupling is used in RF amplifiers, where they are arranged as tunable tank circuits. Each tank circuit is adjusted (C_1, C_2, and C_5) for maximum selectivity (the desired frequency) and sensitivity (largest amplitude).

With the common-emitter RF amplifier shown, C_3 places the bottom of the resonant circuit at signal ground. Capacitor C_4 places the emitter at signal ground. Bias voltage is applied through R_1 and L_2 to the base. Resistor R_2 stabilizes the transistor. This type of amplifier tends to self-oscillate, which interferes with the RF signal; therefore, an adjustable neutralizing

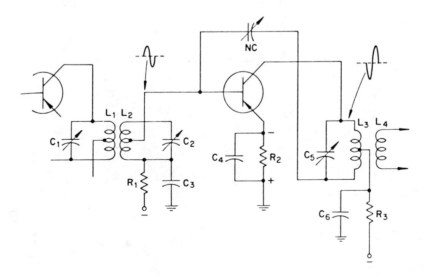

Figure 9-6 Common-emitter RF amplifier.

capacitor (NC) feeds back some of the output signal to cancel this effect. In solid-state circuitry this is called *unilateralization*, because both resistive and capacitive internal coupling between transistor elements is canceled. Collector supply voltage is supplied via L_3 and R_3. Capacitor C_6 is used for power supply decoupling.

In the common-base RF amplifier shown, C_2 and C_3 places the bottom of the resonant circuits at signal ground. The tank circuits use adjustable-core transformers for tuning. A radio-frequency choke (RFC) helps prevent RF from affecting the power supply voltage. This type of circuit usually does not require unilateralization.

The JFET RF amplifier is similar to the common-emitter RF amplifier. The supply voltages are fed to the gate and drain via RFC (L_4) and RFC (L_5), respectively. Capacitors C_2 and C_4 prevent the low resistance of the inductors in the tank circuits from shorting the supply voltages to ground. Components L_3 and C_3 provide unilateralization for the circuit.

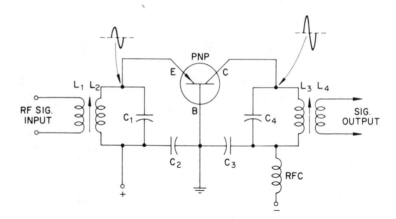

Figure 9-7 Common-base RF amplifier.

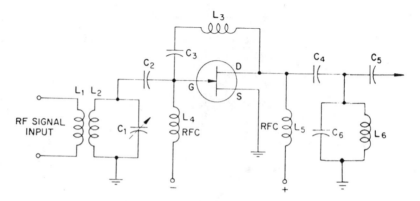

Figure 9-8 JFET RF amplifier.

Electronic communication transmission usually requires high RF operating frequencies, but RF oscillators are usually more stable at lower operating frequencies. Therefore, in many transmitters, frequency multipliers are used to increase the lower oscillator frequency to the required higher RF carrier wave.

A frequency multiplier amplifier is usually operated class C, because of the high order of harmonic generated by operating on the nonlinear portion of the transistor. In a frequency-doubler (X2) amplifier, the fundamental frequency is applied to the input. The output of the circuit consists of a tank circuit which is tuned to twice the fundamental frequency. Since the amplitude of the harmonics decrease as the frequency increases, usually only frequency doublers and triplers are used to obtain the maximum efficiency from the circuits. A frequency-tripler amplifier would have its output tank circuit tuned to the third (X3) harmonic of the input frequency.

To obtain the required high-RF carrier wave, doublers and triplers are cascaded (one after another) until the desired frequency is reached and is then followed by the RF power amplifier. As an example, a transmitter system requiring 6 MHz, but operating with a master oscillator at 1 MHz, could use a doubler, raising the frequency to 2 MHz, followed by a tripler, raising the frequency to 6 MHz.

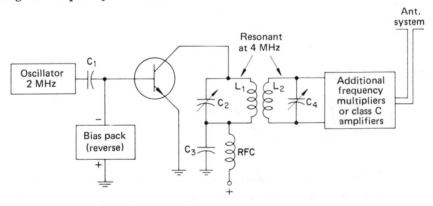

Figure 9-9 Frequency multiplier.

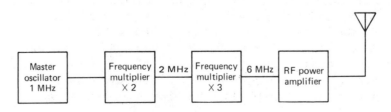

Figure 9-10 Frequency multipliers in a transmitter system.

The *RF power amplifier* may be considered the workhorse of the transmitter, since this is the final stage of the system before the antenna. Maximum
power supply voltages and currents are applied to this stage to enable the
antenna to radiate electromagnetic waves efficiently. Because of the large
currents and heat generated in this stage, the components are usually much
larger in size. Transistors and vacuum tubes require heat sinks (see Section
2-16) and may even require forced-air and water cooling systems. In many
transmitters, a separate power supply is used for the final RF power amplifier. Unilateralization (neutralization) is usually required in this final stage.

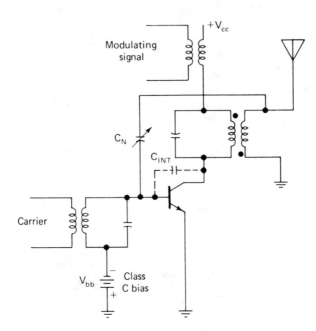

Figure 9-11 RF power amplifier.

For high-power applications, such as standard radio broadcasting, the
highest possible efficiency is obtained by operating the RF power amplifier
class C. However, high-level modulation must be used where the collector
(or plate) circuit is the point which the modulation signal is injected. A
larger modulation power is also required. For example, if 500 W is supplied
to the final power amplifier (FPA) that is 80% efficient, the carrier output
would be 400 W. For 100% modulation, 250 W of modulating power must
be supplied to the FPA since 80% of 250 W = 200 W, the output power
required in the sidebands. If the modulator is 40% efficient, then 625 W of
power must be supplied by the modulator power supply.

High-level modulation is accomplished in the collector (or plate) circuit of the final RF power amplifier. The unmodulated RF carrier wave is applied to the base of Q via T_1. An AF modulating signal is applied to the V_{cc} supply line to the collector of Q through T_3. When the AF modulating signal voltage aids the collector voltage supplied by V_{cc}, the amplitude of the RF carrier increases. When the AF modulating signal voltage opposes the collector voltage, the amplitude of the RF carrier decreases. The AM signal of the collector tank circuit is then coupled to the output (antenna) by T_2. High-level modulation requires large amounts of power (see Section 9-7).

Low-level modulation usually refers to modulating the RF power amplifier at the base or emitter or even a preceding stage. However, with this type of modulation the final RF power amplifier must be operated either class A or B to prevent signal distortion.

In *base modulation*, the RF carrier is applied to the base of Q via T_1. The AF modulating signal is also applied to the base through C. The AF

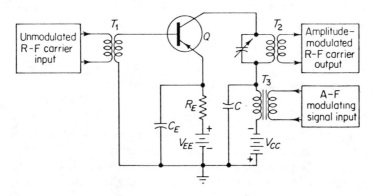

Figure 9-12 Collector modulation.

Figure 9-13 Base modulation.

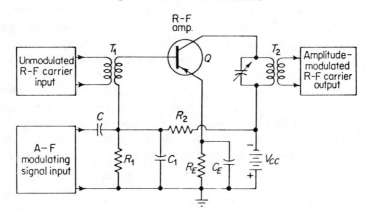

227

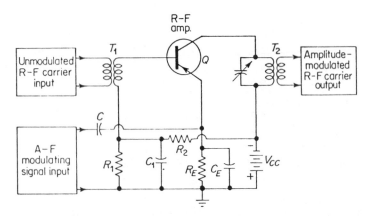

Figure 9-14 Emitter modulation.

modulating signal varies the gain of Q resulting in the AM signal of the collector tank circuit. This AM signal is fed to the output by T_2.

In *emitter modulation*, the RF carrier is fed to the base via T_1. The AF modulating signal is coupled through C to the emitter and varies the gain of the transistor. The AM signal of the collector tank circuit is fed to the output via T_2.

9-7 INFORMATION ON AM RADIO RECEIVER CIRCUITS

There are various types of AM radio receivers, but probably the most popular is the *superheterodyne*. The most common superheterodyne receiver is the standard AM commercial broadcasting type that operates within the band of frequencies from 540 to 1600 kHz. Electromagnetic radio waves cut across the antenna and induce a slight voltage which is fed to the RF amplifier. The RF amplifier has a tank circuit which is tuned to the desired station. At the same time a local oscillator is tuned to 455 kHz above the desired station frequency. The signal from the RF amplifier and the signal from the local oscillator are fed to the mixer stage, where mixing or heterodyning occurs to produce four output frequencies: the station signal, the oscillator signal, and their summation and difference. Only the difference signal of 455 Hz is selected by a tank circuit and is fed to the intermediate frequency (IF) amplifier section. The IF amplifier provides increased selectivity (the ability to select one station from another without interference) for the receiver. The IF signal is fed to the detector stage, where the RF signal is rectified and through an RC time-constant circuit, the audio signal is retrieved. Also, in this stage is an automatic volume control (AVC) circuit that keeps the volume relatively constant regardless of amplitude variations

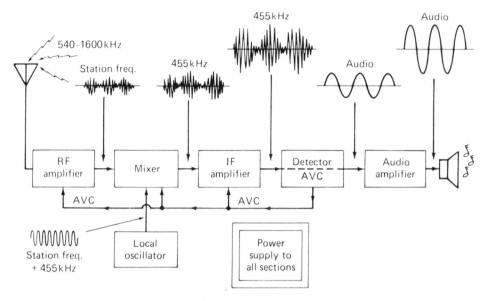

Figure 9-15 AM superheterodyne radio receiver.

in the transmitted radio wave. To accomplish this, the AVC voltage is fed back to the IF amplifier, mixer, and RF amplifier to control their gain. The AF signal from the detector is applied to the audio amplifier section, where it produces sound at the speaker.

9-8 RADIO-FREQUENCY AMPLIFIER

A *radio-frequency* (RF) *amplifier* is an amplifier that amplifies all frequencies other than those within the audio range. In better superheterodyne receivers, the RF amplifier is referred to as the first or preselector stage, even though IF amplifiers are also RF amplifiers. A receiver using an RF amplifier has the advantages of increased signal amplification, increased selectivity (due to an additional tuned circuit), isolation of the local oscillator from the antenna (reducing the possibility that a part of the local oscillator frequency may be radiated into space, causing interference in nearby electronic equipment), and improved image rejection (the ability to select the true station frequency and not some multiple-image frequency).

Permeability tuning is used in the RF amplifier shown, where T_1 and T_2 are ganged together as indicated by the dashed lines, so they will track (tune to the same frequency) as the tuning dial is rotated. Trimmer capacitors C_1 and C_5 are used for more precise alignment. Resistors R_1 and R_2 forward bias the transistor, R_3 and C_3 provide circuit stability, and R_4 is the

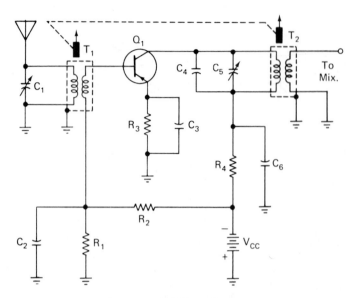

Figure 9-16 RF amplifier.

collector load. Capacitors C_2 and C_6 bypass RF to ground to keep it from affecting the power supply. Capacitor C_4 is part of the collector tank circuit. Class A is used for the RF amplifier to minimize any distortion of the AM signal. The incoming RF signal from the antenna is amplified and coupled to the mixer stage.

9-9 MIXER (CONVERTER) CIRCUIT

The *mixer* or *converter circuit* mixes or heterodynes the incoming RF signal and the local oscillator signal to produce a summation signal and difference signal. Therefore, the output of the mixer contains the selected radio station RF signal, the local oscillator signal, the summation frequency, the difference frequency, and the sideband frequencies.

A single transistor is often used as the RF amplifier, local oscillator, and mixer. The desired radio station is "tuned in" by C_1 and L_1, which is applied to the base of the transistor. Components C_2 and L_2 comprise the local oscillator in the emitter–collector circuit. Capacitor C_2 is ganged to C_1, so the oscillator tracks 455 kHz above the selected station RF carrier wave. Trimmer capacitors (C_{1T} and C_{2T} are used for precise alignment). Resistors R_{HB} and R_{LB} bias the transistor, while R_S and R_E provide circuit stability. Capacitor C_c provides feedback for oscillation. The tank circuit using L_3 is designed to resonate at 455 kHz. Since the output of the mixer contains many frequencies, including the sidebands, this final tank circuit acts as a filter to pass the 455 kHz and its related sidebands. The mixer is

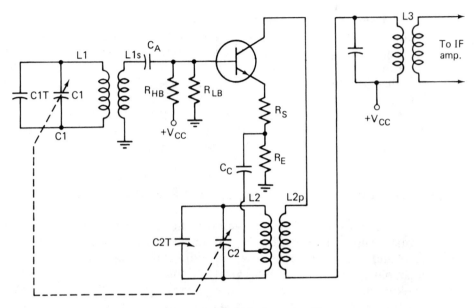

Figure 9-17 AM mixer circuit.

operated class A so as not to distort the AM signal, and its output is fed to the IF amplifier.

9-10 INTERMEDIATE-FREQUENCY AMPLIFIER

The *intermediate-frequency* (IF) *amplifier* improves selectivity in the super-heterodyne receiver. Since it operates at a fixed frequency, it can be de-signed to provide optimum gain and bandwidth. The operating IF for a standard AM broadcast receiver is 455 kHz, with a bandwidth of 10 kHz. A superheterodyne receiver may have one or more IF stages. The more IF stages used (up to a limit, of course), the more selectivity the receiver has.

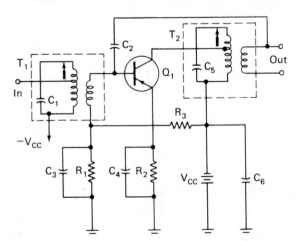

Figure 9-18 IF amplifier.

Tuning is usually done by adjustment of a powered iron-core transformer, as shown. The dashed lines around the tank circuits indicate that they are enclosed in aluminum cans to prevent radiation and interference to or from other circuits. Components T_1 and C_1 are actually the output tank circuit of the mixer stage. Bias on Q_1 is set by R_1 and R_3. Stability of the circuit is improved by R_2 and C_4. Capacitors C_3 and C_6 place the bottom of the tank circuits at signal ground. Capacitor C_2 is used for neutralization to prevent self-oscillations. The IF amplifier is operated class A and feeds the 455 kHz signal to the audio detector stage.

9-11 AUDIO DETECTOR AND AUTOMATIC VOLUME CONTROL (AVC)

The *audio detector circuit* removes the audio component from the AM signal. Audio information is contained in both sidebands, but only one sideband is needed to retrieve the audio and convert it to sound. Therefore, a simple diode (D_1) is used to rectify the AM signal. Transistor Q_1 and the tank circuits are the last IF amplifier stage that feeds the AM signal to D_1. Each negative RF pulse causes D_1 to conduct. Capacitor C_1 charges up to each peak value, but it is allowed to discharge slightly between RF pulses so it can follow the audio rate of the pulses. The slight filtering action of C_1 produces a relatively clear copy of the original audio information, which is coupled by C_c to the 20-kΩ potentiometer (volume control), which in turn is fed to the audio section of the receiver.

Automatic volume control (AVC) is used to amplify weak signals more, and strong signals less, so that the overall output level set by the volume con-

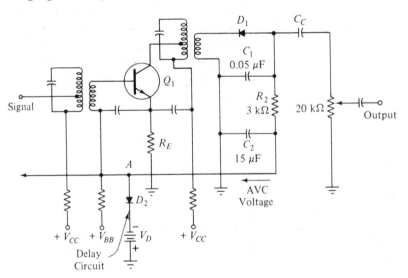

Figure 9-19 AM audio detector and AVC.

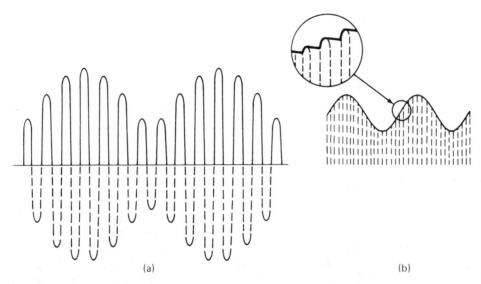

Figure 9-20 Audio-signal detection: (a) rectified AM wave; (b) resultant audio wave.

trol remains fairly constant, even though signals fade and weaken in transmission. Delayed AVC responds slower to amplitude changes and results in a more stable output at the speaker. Diode D_2 and voltage $-V_D$ comprise the delay circuit that negatively biases the preceding stages: IF amplifier, mixer, and RF amplifier. Components R_2 and C_2 have a larger time constant that produces a negative dc voltage relative to the amplitude of the incoming signal. Until the AVC line reaches $-V_D$, D_2 conducts and places $-V_D$ on the AVC line, which is the desired bias level for all amplifiers. When the AVC voltage from D_1 becomes more negative than $-V_D$, D_2 opens and the varying AVC bias controls the gain of the amplifiers. This results in a high gain for weak signals and an optimum AVC characteristic for large signals.

9-12 INFORMATION ON FREQUENCY MODULATION (FM)

Frequency modulation (FM) is the process of transmitting information over an RF carrier wave, where the amplitude is constant, but the modulation signal applied to the carrier changes the frequency of the transmitted signal. The frequency of the modulating signal determines the rate the RF carrier will shift above and below the resting or center frequency, whereas the amplitude of the modulating frequency determines the amount it is shifted on each side of the center carrier frequency. For example, a 100-MHz RF carrier wave is modulated by an AF 1-kHz tone of low amplitude that may cause a ±25-kHz carrier shift. During the positive alternation of the AF signal the RF carrier would shift to 100.025 MHz, while during the negative

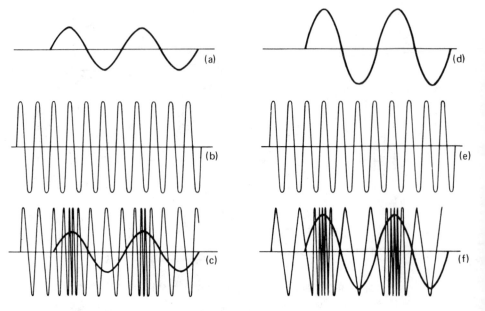

Figure 9-21 Frequency modulation: (a) audio signal; (b) RF carrier wave; (c) FM wave; (d) audio signal with greater amplitude; (e) RF carrier wave; (f) differently modulated FM wave.

alternation, the RF carrier would shift to 99.975 MHz. The rate at which the RF carrier would swing between these two extremes would be 1 kHz. If the amplitude of the AF signal was increased to cause a ±75-kHz shift (defined as 100% modulation), the upper deviation would be 100.075 MHz, while the lower deviation would be 99.925 MHz.

The relationship between the RF carrier deviation and the modulating AF signal is expressed as

$$\text{modulation index} = \frac{\text{RF frequency deviation}}{-\text{AF modulating frequency}} \; ; \quad \text{example:} \; \frac{25 \text{ kHz}}{1 \text{ kHz}} = 25$$

The *deviation ratio* is the ratio of the maximum allowable deviation (75 kHz) to the highest AF modulating frequency (15 kHz); therefore,

$$\text{deviation ratio} = \frac{75 \text{ kHz}}{15 \text{ kHz}} = 5$$

The standard FM radio broadcast band is from 88 to 108 MHz, accounting for 100 channels. A radio station is assigned a channel of 200 kHz, including a maximum frequency deviation of ±75 kHz or 150 kHz, plus two guard bands of 25 kHz each to help prevent interference from any adjacent channels.

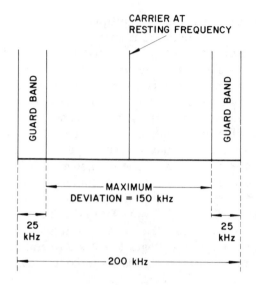

Figure 9-22 FM channel allocation.

9-13 THE FM TRANSMITTER

In many typical *FM transmitters* the direct system modulation occurs in stages of low power levels. An audio amplifier (often called a speech amplifier) drives a reactance modulator circuit, which controls the deviation of a

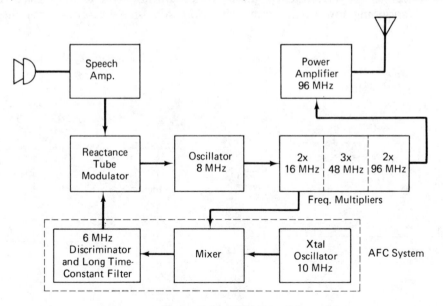

Figure 9-23 FM transmitter.

master oscillator (8 MHz).   Frequency multipliers then increase the FM signal to the station's allocated operating frequency (96 MHz). At the same time, the original oscillator deviation is increased to ±75 kHz for the required 100% modulation.   The final power amplifier, consisting of one or more stages, increases the output power to meet the licensed requirements for transmission.   This final stage is coupled to the antenna to produce electromagnetic radiation into the atmosphere.   The antenna is usually mounted high atop a building, mountain, or radio transmitting tower.

Even with no modulation signal present, the reactance modulator circuit can cause the center frequency of the oscillator to shift. To keep the oscillator close to its center frequency, an automatic frequency control (AFC) circuit is used.   A crystal-controlled oscillator (10 MHz) is fed to a mixer with a portion of the RF carrier wave from the first frequency multiplier.   The difference frequency (6 MHz) is then filtered by a long time-constant filter to produce quasi-dc voltages that tend to stabilize the master oscillator.   An AFC circuit such as this is required to maintain a center frequency stability of ±2 kHz for broadcast stations.

9-14 PREEMPHASIS AND DEEMPHASIS

Noise at high frequencies is most objectionable in radio transmission systems.   High-frequency audio tones are also less in amplitude than corresponding low audio tones; therefore, the signal-to-noise ratio is greater at

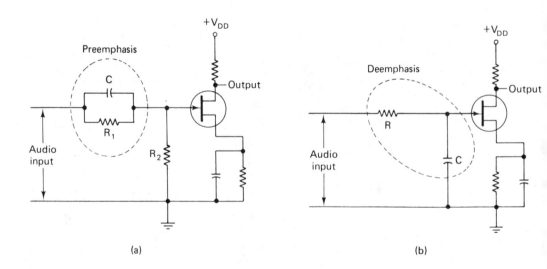

Figure 9-24 Emphasis circuits; (a) preemphasis; (b) deemphasis.

high frequencies.   An FM system is basically used to reduce noise inter-ference; however, a low-amplitude modulating signal means a small deviation and results in a lower capability of the system to reduce noise. A process called *preemphasis* boosts the high frequencies before modulation takes place. A typical preemphasis circuit uses a parallel RC circuit (R_1 and C) in series with an input resistor R_2 to an audio voltage amplifier. As the input frequency increases, X_c decreases, which produces a larger voltage drop across R_2, resulting in more amplification at high frequencies.

In the transmitted FM signal, there exists an unnatural balance between high- and low-frequency tones; therefore, a *deemphasis* circuit is used in the FM receiver to correct this defect. A typical deemphasis circuit would use a resistor in series with a capacitor to the input of an audio voltage amplifier. The higher frequencies would decrease X_c, causing them to be shorted to ground, resulting in less gain. For the original signal to be reproduced, the time constants of the preemphasis circuit and the deemphasis circuit should be equal to each other.

9-15 REACTANCE MODULATORS

A *reactance modulator* is usually an active-device circuit used to vary the reactance of the master oscillator tank circuit (L_1 and C_1). In a transistor reactance modulator (power supply voltage not shown), an input modula-ting signal varies the collector voltage. When the voltage increases, the

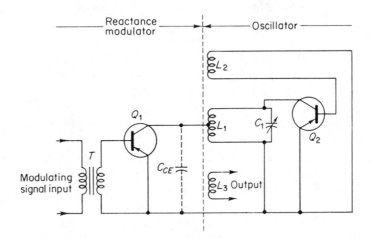

Figure 9-25 Transistor reactance modulator.

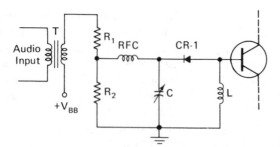

Figure 9-26 FM using a varicap.

collector–emitter capacitance (C_{CE}) decreases, and when the voltage decreases, the capacitance C_{CE} increases. Therefore, the value of C_{CE} varies with the modulating frequency. This changing capacitance is applied to the tank circuit of the oscillator, which in turn causes a frequency deviation. The amplitude of the AF modulating signal determines the amount of frequency deviation.

A VVC diode (see Section 2-5) can be used as a reactance modulator. In a typical circuit the VVC diode (CR_1) is in series with the tuning capacitor (C). Resistors R_1 and R_2 form a voltage divider to reverse bias CR_1, so that its capacity will be in the proper range with respect to the capacitor. The AF signal superimposed by transformer (T) on this reverse bias voltage increases and decreases the capacity of the VVC diode. The total capacity of the tank circuit is affected, which varies the oscillator above and below its center frequency. The amount of frequency deviation is directly proportional to the amplitude of the AF modulating signal. High-frequency energy is blocked from the reverse-biased network by RFC.

9-16 INFORMATION ON FM RADIO RECEIVER CIRCUITS

The basic FM receiver operates on the superheteodyne principle and is similar to the AM receiver. The RF amplifier, mixer, and IF amplifiers have a wider bandpass (200 kHz) and of course operate at higher frequencies (88 to 108 MHz). The local oscillator tracks with the tuning and operates 10.7 MHz (the IF frequency) below the incoming FM signal. The amplified FM signal from the RF amplifier is fed to the mixer along with the local oscillator frequency. Several frequencies are produced by the mixer stage: the FM signal, the oscillator frequency, the summation frequency, the difference frequency, and the sideband frequencies. Only the difference frequency of 10.7 MHz is filtered out and sent to the IF amplifier. From the IF amplifier, the amplified FM signal is sent to a limiter circuit, which is used to clip any AM modulation from the FM signal, which reduces noise and other transmission interference attached to the signal. The FM signal is then processed by the discriminator, which is an FM detector circuit that derives the AF amplitude variations from the frequency variations of the

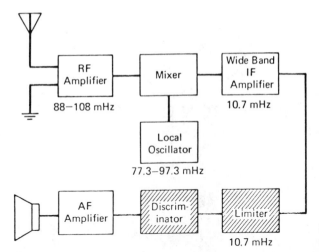

Figure 9-27 FM receiver block diagram.

signal. The AF signal is then applied to the audio section, which contains a deemphasis network required to restore the high-frequency information back to the proper amplitude relationship with the lower frequencies. The AF signal is then applied to a speaker, where it produces sound.

9-17 LIMITERS

Frequency-modulated signals suffer some amplitude modulation during transmission through the atmosphere. Natural and human-made static combine to produce this AM effect of the FM waveform, which is amplified as the signal passes through the receiver. The purpose of the *limiter* is to remove the AM component while maintaining the frequency variations of the original signal.

A bipolar transistor stage can produce limiting action by operating the stage as an overdriven amplifier. With a low collector voltage, the stage can be overdriven by a reasonably small amplitude signal. During the positive

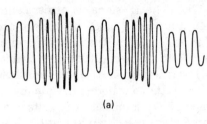

(a)

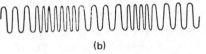

(b)

Figure 9-28 Limiting action on FM waveform: (a) without limiting; (b) with limiting.

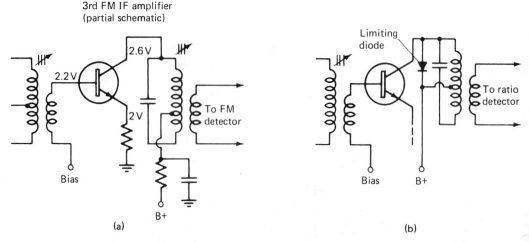

Figure 9-29 Limiters: (a) limiting action via biasing; (b) limiting diode.

input peaks, the transistor saturates, and when the negative peaks occur, the transistor is completely cut off. The tank circuit in the collector circuit restores the true sine wave to the signal.

Another method of limiting is to connect a diode across a portion of the tank circuit. The diode will conduct at high signal levels and the junction capacitance will change, which lowers the Q of the circuit and some limiting action occurs.

9-18 DISCRIMINATORS

The function of the *discriminator* is to convert the frequency deviations of the FM signal into intelligible audio frequencies. The output of the limiter is coupled by L_1 and L_2-L_3 to the discrimination circuit that consists of matched components: diodes D_1 and D_2, capacitors C_4 and C_5, and resistors R_2 and R_3. The center tap of L_2-L_3 is connected to the point between C_4-C_5 and R_2-R_3 via L_4. Each tank circuit is tuned to the IF center frequency (10.7 MHz). When the IF signal is on its center frequency, both diodes conduct alternately but evenly, and the voltage drops across R_2 and R_3 oppose each other (as indicated by the arrows) and the output is zero. Detection occurs because of the phase differences that appear in the resonant circuits when the IF carrier deviates above and below the center frequency. When the carrier shifts to a higher frequency, the circuit appears more inductive, the voltage of D_1 increases, the voltage of D_2 decreases and similarly, the IR drop of R_2 increases, while the IR drop of R_3 decreases, which produces an AF positive alternation at the output. When the carrier shifts to a lower frequency, the process is reversed, the circuit appears more capacitive, the voltage of D_1 decreases, the voltage of D_2 increases, the IR

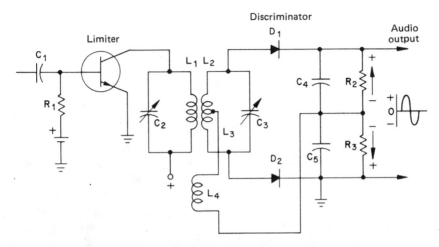

Figure 9-30 FM discriminator.

drop of R_2 decreases while the IR drop of R_3 increases and an AF negative alternation is produced at the output. In this manner the original modulating signal is detected from the RF carrier.

9-19 RATIO DETECTOR

The *ratio detector* is another type of FM demodulator and is similar to the discriminator (see Section 9-18), except that one of the diodes (in this case, D_1) is turned around. Therefore, electron flow through the output resistors, R_3 and R_4, is in series and the sum of the two voltages appears across the

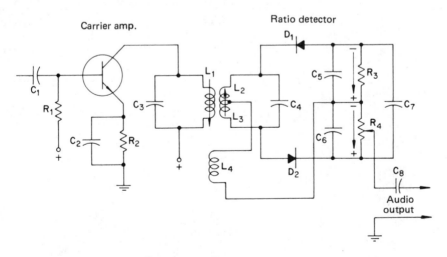

Figure 9-31 FM ratio detector.

output even when the IF carrier is at its center frequency. This dc voltage is blocked from the following stage by C_8. When the carrier deviates, one diode will conduct more than the other, and the IR drops across the output resistors will change accordingly. For instance, when the IF carrier is at center frequency, each resistor may have a 2-V drop across it. As the carrier frequency deviates and D_1 conducts more than D_2, the IR drop of R_3 may rise to 3 V and the IR drop of R_4 would go to 1 V. Similarly, if D_2 conducts more than D_1, the IR drop of R_4 may go to 3 V and the IR drop of R_3 to 1 V. Although the voltage across both resistors remains essentially the same, the voltage across each resistor varies at an audio rate. Therefore the output is tapped off R_4 and fed to the audio section.

Capacitor C_7 is a large value (usually several microfarads) that tends to maintain the dc voltage across R_3 and R_4, and minimizes any sudden changes in the total voltage of the carrier caused by static or AM during transmission. For this reason, a limiter is seldom used with a ratio detector.

9-20 FM STEREO MULTIPLEX TRANSMITTER

Stereophonic, often referred to simply as *stereo*, is the process of multichannel sound that gives a feeling of depth or directional characteristics to the various sounds that are broadcast or played back from recorded material. The most common source of stereo utilizes two channels.

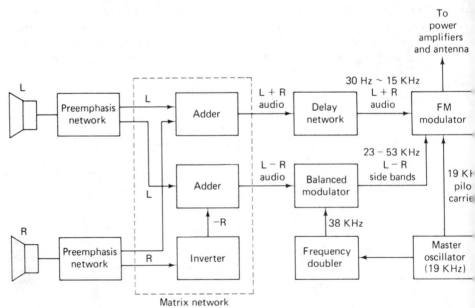

Figure 9-32 FM stereo transmitter.

In a basic *FM stereo multiplex transmitter*, two microphones and audio channels are used: the left channel (L) and the right channel (R). Both channels are fed through a separate preemphasis network. The two channels are added together to produce the L + R signal. The R channel is inverted and added to the L channel to produce the L – R signal. A 19-kHz master oscillator is fed to the FM modulator and also a frequency doubler. In turn, the frequency doubler (38 kHz) is applied together with the L-R signal to produce an AM signal whose sidebands are from 23 to 53 kHz. With this process the 38-kHz frequency is supressed (not sent to the modulator for transmission). Since this initial modulation on the L – R signal takes time, the L + R signal is applied to a delay network so that both signals reach the FM modulator at the same time. The three signals, L + R (30 to 15 kHz), L – R (23 to 53 kHz), and the 19-kHz pilot carrier, are then processed by the FM modulator and sent on to the power amplifiers and antenna for transmission into the atmosphere. All the frequencies of the composite modulating signal lie well within the 200 kHz allocated for an FM radio channel. The 19-kHz pilot subcarrier is transmitted to provide a synchronizing signal at the radio receiver in conjunction with the transmitter. The term "multiplex" refers to the process of the two audio signals sharing the same FM carrier frequency.

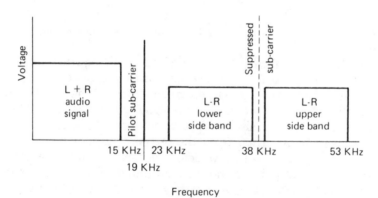

Figure 9-33 Composite modulating signals for FM.

-21 FM STEREO MULTIPLEX RECEIVER

The basic *FM stereo multiplex receiver* is similar to the FM monophonic (mono) receiver, in that the tuner, IF amplifiers, and FM detector are the same. After the composite modulating signal containing the L + R signal (30 to 15 kHz) the 19-kHz pilot subcarrier and the L – R signal (23 to 53 kHz) has been detected by the FM detector it is applied to a 19-kHz amplifier and a 19-kHz bandstop filter. The 19-kHz amplifier passes only the

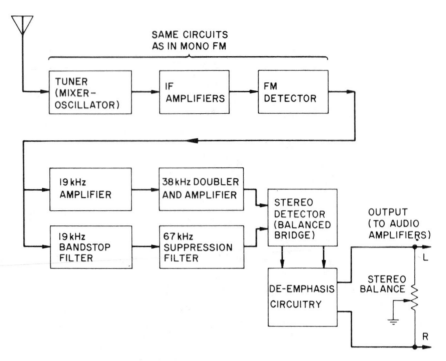

Figure 9-34 FM stereo multiplex receiver.

19-kHz pilot subcarrier, which is fed to a frequency-doubler stage, produ
ing 38 kHz. This frequency is applied to a stereo detector to demodulate th
L – R signal. The 19-kHz bandstop filter rejects the 19-kHz subcarrier bt
passes the rest of the composite signal on to the stereo detector. A 67-kH
supression filter is also used to reject the SCA (Subsidiary Communicatior
Authorizations) signal from the receiver output. This SCA signal is used b
some radio stations to broadcast background music or other private servic
for paying subscribers and may be found on the composite modulating si
nal, usually with a narrow bandwidth of 59.5 to 74.5 kHz.

In the stereo detector the L and R channels are separated into the
original signals, deemphasized, and send on to their respective audio outp
channels.

9-22 BALANCED-BRIDGE DETECTOR

One type of stereo detector to demodulate the modulating composite sign
is the *balanced-bridge detector*. The amplified 38-kHz sub-carrier is fed
the detector via the tank circuits L_1–C_1 and L_2–C_2. This process is som
times called carrier reinsertion, since the carrier was suppressed during trar
mission of the FM signal. The composite modulating signal, less the 19-kH

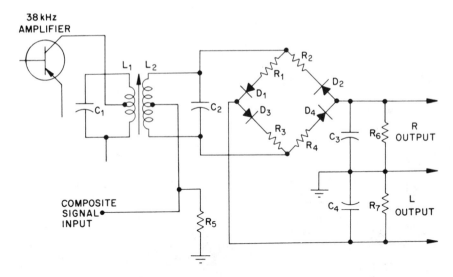

Figure 9-35 Balanced-bridge detector.

subcarrier and any SCA signal, is placed at the center tap of L_2. The four diodes comprise a bridge rectifier circuit that processes the 38-kHz subcarrier in relation to the composite signal. Since the composite signal is placed at the center tap of L_2, the top and bottom of the bridge rectifier will be in phase at any instant of time. However, the 38-kHz subcarrier is out of phase with these same points and the phase relationships occurring between the signals allow the diodes to conduct at the proper times so the currents for the R channel and L channel are applied to the respective output. Capacitors C_3 and C_4 smooth out the rectified pulses to the audio-modulating changes in frequency and amplitude contained in the original right and left channels.

9-23 TYPICAL AM, FM, FM-STEREO RECEIVER

With a typical *AM, FM, FM-stereo receiver*, the selector switch for these different modes has various contacts that select the proper circuits for receiving and reproducing the desired signal. In the FM stereo mode as shown, the power supply activates the FM tuner circuits at point A. The AM tuner does not have any power supply voltage at this time and is inoperative. The 19 kHz is allowed to amplify the 19-kHz subcarrier and each audio channel (L and R) is connected to the modulating composite signal decoder. A selected radio station signal is processed by the FM RF amplifier, FM mixer, and FM oscillator, resulting in the 10.7-MHz FM IF signal. This IF signal is amplified by the IF stages and the FM detector demodulates the AM composite signal. The composite signal is then sent through the SCA trap, the

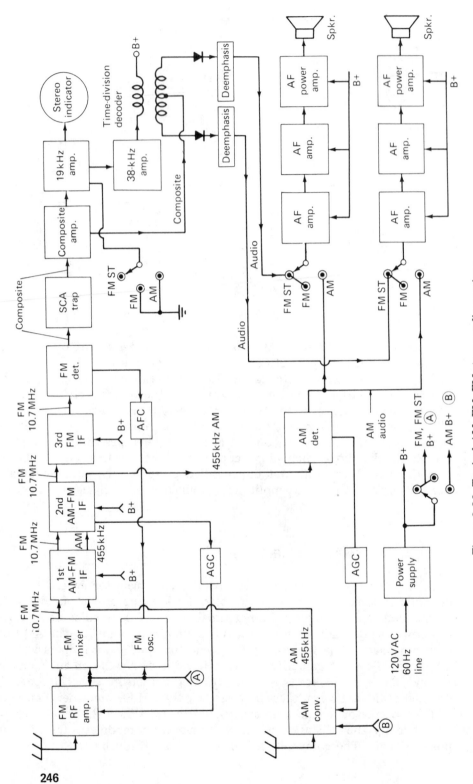

Figure 9-36 Typical AM, FM, FM-stereo radio receiver.

composite amplifier, and the decoder, where each demodulated L and R signal is sent to its respective audio channel.

When the selector switch is set to mono FM, the signals travel the same path, except that the 19-kHz amplifier is grounded and does not affect the detected FM signal as it passes through the decoder to both of the audio channels. (Note: The same signal appears at each channel.)

Placing the selector switch in the AM position removes the power supply from the FM tuner circuits and applies it to the AM converter at point B. Also, the 19-kHz amplifier remains grounded and the two audio channels are connected together to the AM detector stage. The signal from a selected AM radio station enters the AM converter and results in the 455-kHz IF signal, which is sent to a couple of IF amplifier stages. This IF signal is then fed to the AM detector and the demodulated signal is applied to both audio channels.

Notice that each section (AM and FM) has its own AGC (automatic gain control) circuit and the FM shows an AFC (automatic frequency control) circuit, which may or may not be used in modern receivers.

9-24 OVERVIEW OF TROUBLESHOOTING RADIO CIRCUITS

Radio circuits are combined into electronic systems to provide a functional use. In order to find and correct a malfunction, the "systems approach" to troubleshooting is generally used. A malfunction is most often detected at the output of a system. From this point, troubleshooting techniques of observation with test equipment generally proceed back into the system until a stage or circuit with no trouble is detected. Usually, the last stage or circuit where the trouble appears is at fault; however, since many circuits are interconnected by common power supply lines, feedback networks, and control circuits, other circuits seemingly unrelated to the problem should not be overlooked.

The following list is a general systems approach to troubleshooting.

1. Have proper equipment schematic diagram and/or service manuals, if possible.
2. Become familiar with the system's theory of operation to be able to recognize the difference between a malfunction and normal operation.
3. Adjust variable controls or manufacturer's suggestions to correct malfunction, where applicable.
4. Observe the unit under test with the physical senses: hearing, sight, smell, and touch. Become familiar with the sounds of common malfunctions. Look for broken wires and circuit connections or burnt components. Often, the smell of something burning can detect a faulty component. Overheating of components can be detected by touching,

but be careful not to get burned. An insulated tool can be used to push on components and printed circuit boards to detect loose connections and/or noise.

5. Check for proper power supply (dc) operating voltages.

6. Usually, but not always, start at the output of the system and work back to initial stages.

7. Compare voltage and waveform readings with those shown on schematic diagrams or equipment manuals to determine faulty circuits and components.

8. Locate and isolate defective stage or circuits.

9. Locate and test defective components.

10. Clean up residue from soldering and make an overall performance check on system.

9-24.1 Transmitter General Troubleshooting Tips

Many transmitters have built in meters and test equipment; however, common basic test equipment is still a must for accurate troubleshooting procedures. This includes volt-ohm-current meter (the digital type is fine), wideband oscilloscope, audio generator, RF generator, and RF frequency counter. Other specialized test equipment is also used in transmitter testing: dummy load (to replace antenna while testing), wattmeter, dip meter, field strength meter, FM deviation meter, and various analyzers for more specific areas.

AM transmitter. With an AM transmitter (see Section 9-2) the adjustable controls would be set in an attempt to produce a distortionless AM signal output. The output would be checked for the correct power transmission and the signal checked for overmodulation or undermodulation. The RF carrier wave should be the proper frequency and amplitude before entering the modulation process. The AF modulating signal should be distortionless (no noise, no phase distortion, etc.) and of the proper amplitude before entering the modulating process.

FM transmitter. A similar approach to testing an AM transmitter would be used with an FM transmitter (see Section 9-13). In this case different specialized equipment might be used since the type of modulation is a different process.

9-24.2 Radio Receiver Troubleshooting

Troubleshooting a radio receiver would follow the same general procedures as with any system (see Section 8-13) except that special test instruments are available to test the RF sections. These instruments include an RF detector

Symptom	Probable cause
1. Dead receiver.	Speaker circuit; power supply; audio-output stage; earphone jack.
2. Turn-on plop; very little static.	AF amplifiers; IF amplifiers; detector.
3. Turn-on plop; lots of static.	Front end; IF amplifiers.
4. 60-Hz hum with volume at minimum setting.	Power supply filter capacitors; audio output stage.
5. Distorted audio.	Audio section; power supply.
6. Loss of sensitivity.	Front end; IF amplifiers; power supply.
7. Insufficient volume; sensitivity normal.	Audio section; power supply.
8. Oscillations.	Decoupling or neutralizing capacitors; front end; IF amplifier; alignment.
9. Noisy reception.	Defective transistor; poor circuit connections on printed circuit board.
10. Intermittent.	Depends upon symptom when malfunction occurs; usually problem is a transistor or poor printed circuit connection.

Figure 9-37 Typical solid-state receiver.

Symptom	Probable cause or defective section
1. Dead receiver.	Power supply, audio output stage, speaker.
2. FM normal; no AM.	AM converter; AM detector; power supply.
3. Distorted audio.	Audio amplifiers; power supply.
4. 60-Hz hum.	Power supply filters; leakage on B+ line.
5. FM normal, insufficient AM sensitivity.	AM converter; AM IF alignment.
6. FM normal, oscillations in AM mode.	AM converter, AM IF section misaligned.
7. Motorboating.	Open decoupling capacitor (possibly filter capacitor in power supply).
8. FM normal, AM intermittent.	Mechanical or thermal intermittent in the AM sections.

Figure 9-38 AM, audio and power supply sections of an AM-FM receiver.

Symptom	Probable cause or defective section
1. AM normal, no FM.	FM front end; FM IF stage; FM detector; power supply for FM sections.
2. No AM or FM, pop or hum from speaker.	Common IF stages; power supply; audio amplifiers.
3. AM normal, insufficient FM sensitivity.	FM front end; FM IF section; misalignment.
4. Insufficient AM and FM sensitivity.	Common IF amplifier.
5. AM normal, oscillation on FM.	FM front end; FM IF misalignment; neutralizing capacitors.
6. AM normal, FM distorted.	Misalignment, FM detector.
7. AM normal, noisy FM.	FM front end; FM IF stage.

Figure 9-39 FM circuits and common stages of an AM-FM receiver.

probe to be used with a meter or oscilloscope. A signal tracer instrument has an audio detector and amplifier for testing an incoming signal through the various circuits. A signal injector may also be part of the tracer instrument or a separate device. Normally, a receiver may not need to have the master oscillator and IF stages aligned; however, RF generators or a sweep/marker generator should be used for this process.

The following tables list some common trouble symptoms and probable causes for radio receivers. Refer also to Section 9-7, 9-16, 9-21, and 9-23.

Symptom	Most likely cause
1. One channel inoperative on all functions.	Audio amplifier section of the defective channel; foreign object in headphone jack.
2. AM normal; FM and FM stereo inoperative.	Any FM-only stages: composite amplifier; FM muting; FM detector; FM front end.
3. One channel inoperative in FM-stereo mode.	Stereo-decoder circuitry: matrixing resistors; multiplex detector; PLL chip.
4. Stereo separation completely lacking.	Stereo-decoder circuit: 19 kHz or 38 kHz amplifier; bandpass filter; VCO frequency control voltage.
5. Nonfunctioning stereo indicator.	19 kHz source; stereo indicator driver stage; stereo indicator output of IC decoder.

Figure 9-40 Stereo decoder section of an FM-stereo receiver.

10

Basic Television Systems and Circuits

10-1 INFORMATION ON TELEVISION TRANSMISSION

A *television* (TV) system consists of two parts: the *picture* or video signal and the *sound* or audio signal. The TV transmitter actually consists of two separate transmitters. The video signal is produced by the TV camera and together with synchronizing signals is amplitude modulated onto an RF carrier, while the audio signal produced by the microphone is frequency modulated onto another RF carrier. Both modulated signals are combined by a diplexer, which prevents interaction of the transmitters and is sent to the antenna to be radiated into the atmosphere. The TV receiver detects the combined signals and separates them into the AM video signal to produce the picture on the face of the cathode-ray tube (CRT, also called the picture tube), and the FM audio signal to produce sound at the loudspeaker.

The picture on the CRT is accomplished by a scanning process through synchronization with the camera in the TV studio and the home receiver. A monochrome CRT consists of an electron gun and grids for accelerating and focusing the emitted electron beam, which strikes a phosphorescent chemical coating on the inside face of the tube. When the coating is struck by the electron beam, light is emitted. A yoke assembly containing the horizontal deflection coils and vertical deflection coils fits around the neck of the CRT, which produces the electron beam scanning. Simultaneously, the beam moves from left to right at 15,750 Hz and vertically at about 60 Hz. There are 525 lines on the CRT, of which only 485 are visible since during retrace (the time the electron beam moves back from right to left) the beam is

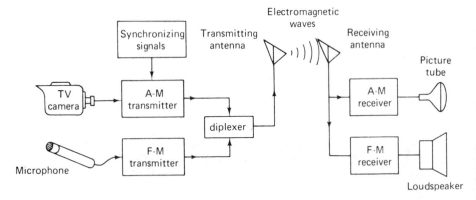

Figure 10-1 Basic television system.

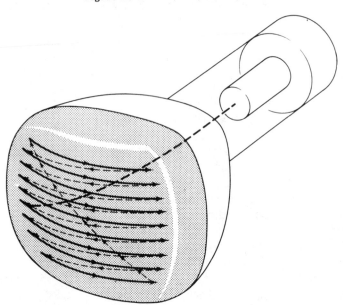

Figure 10-2 Scanning of cathode ray tube.

cut off. Through a process called *interlaced scanning*, the even lines are
scanned first to produce a field every 1/30 s and then the odd lines are
scanned to produce a second field every 1/30 s. A frame or complete
picture consists of two fields and occurs 30 times per second, but to the eye,
the illusion is 60 times a second and there is no perceptible flicker in the
picture. The blank whitish screen without picture information on the face
of the CRT is referred to as the *raster*.

A picture image (or scene) is produced when a positive signal voltage is
placed on the cathode of the CRT. This positive voltage reduces or even in
some cases, cuts off the electron beam, which produces dark areas on the

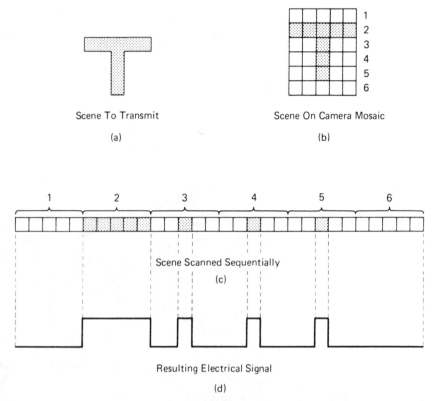

Figure 10-3 Voltage levels to produce TV scene: (a) scene to transmit;
(b) scene on camera mosaic; (c) resulting electrical signal.

face of the CRT. In producing the letter "T" as illustrated, the electron
beam is reduced at different times during the scanning process.

10-2 TELEVISION SIGNALS

A typical, complete TV channel is allocated a bandwidth of 6 MHz. The
maximum modulating signal allowed for the video signal is 4 MHz. However,
for AM detection only the upper sideband is needed; therefore, a large portion of the lower sideband is filtered out before transmission. Enough of the
lower sideband is retained to preserve the picture carrier frequency and
without affecting the amplitude and phase of the lower frequencies of the
upper sideband. This type of transmission is referred to as *vestigial-sideband*
operation. The picture carrier is situated 1.25 MHz above the lowest channel
frequency (for channel 2 this frequency is 55.35 MHz). The sound carrier
is situated 5.75 above the lowest channel frequency, or 4.5 MHz above the
picture carrier (for channel 2 this frequency is 59.75 MHz). The maximum
deviation for the TV FM sound signal is ±25 kHz. Frequency allocation for

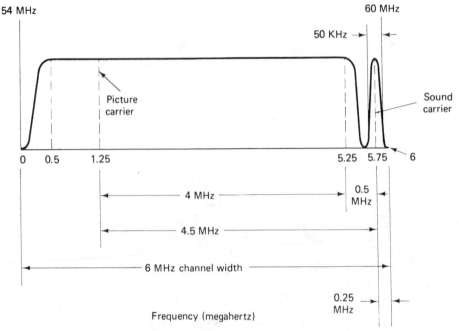

Figure 10-4 Typical TV channel allocation (channel 2).

the 12 VHF channels is from 54 to 88 MHz and 174 to 216 MHz and for the 70 VHF channels is from 470 to 890 MHz.

Horizontal synchronization (sync) pulses occur after each line of the video signal. They are used to synchronize the horizontal deflection circuits in the receiver with the transmitter and to provide blanking during the electron beam retrace. These pulses also establish the black and white level of the picture. In color TV transmission, an eight-cycle color burst of about 3.58 MHz is placed on the back porch of the horizontal sync pulse to activate the color circuits in a receiver.

A composite video waveform has the horizontal sync pulses (with video information) and the vertical blanking pulse. The equalizing pulses keep the horizontal circuits in synchronization during the vertical retrace blanking pulse time. The vertical sync pulse, which synchronizes the vertical circuits in the receiver with the transmitter, is placed between the equalizing pulses. It is serrated to keep the horizontal circuits in synchronization during the vertical sync time.

10-3 BASIC TELEVISION TRANSMITTER (STATION)

A television station basically consists of three sections: the studios, the monitor and control room, and the transmitter. These may all be located in a common facility or the transmitter may be at a remote site because of the large area needed to erect and support the antenna.

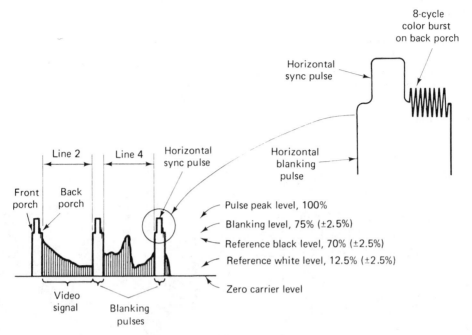

Figure 10-5 Horizontal sync pulses.

The video signal begins with the TV camera in the studio and is fed to the camera control unit. Picture quality can be adjusted by the camera control unit. It is then fed to the video switcher, a unit that selects various cameras or video tape machine to be broadcast. A sync generator provides synchronizing pulses to the camera control and adds them to the outgoing video signal. A video monitor is used for each camera and video tape machine

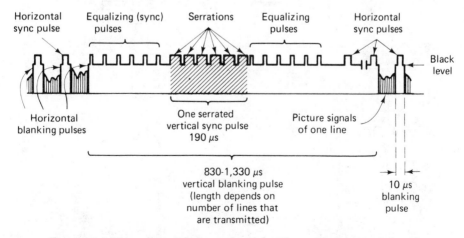

Figure 10-6 Composite video waveform showing vertical retrace interval.

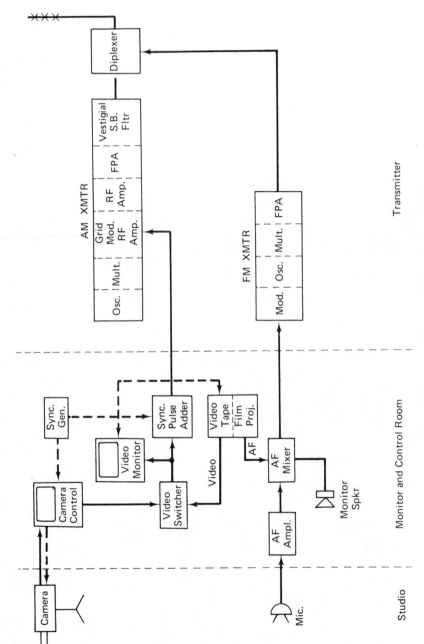

Figure 10-7 Basic TV transmitter.

as well as the master monitor that indicates the video signal being broadcast. The video signal is sent to a low-level AM transmitter, where it modulates the RF carrier, which in turn is amplified, and a portion of the lower sideband is filtered out before the AM signal is sent to the diplexer and antenna.

The audio signal starts at the microphone, is amplified, and is passed on to the audio mixer, where various audio signals can be mixed together or selected for transmission. A monitor speaker is also connected to the audio mixer. The audio signal now modulates an FM transmitter and the FM signal is multiplied and amplified before it is sent to the diplexer and antenna.

The diplexer prevents any AM signal from being coupled back into the FM transmitter, and vice versa.

10-4 BASIC TELEVISION RECEIVER

The circuits for monochrome and color receivers are essentially the same (see Section 10-14). Starting with the antenna, both the picture and sound RF carrier signals are intercepted and fed to the tuner, which consists of an RF amplifier, local oscillator, and mixer. The intermediate frequencies (IF) from the output of the mixer stage, 45.75 MHz for the picture IF carrier and 41.25 MHz for the sound IF carrier, are amplified by the video IF amplifier stages and passed on to the video detector. At this point the video signal (picture information and blanking pulses) is sent to the video amplifier and

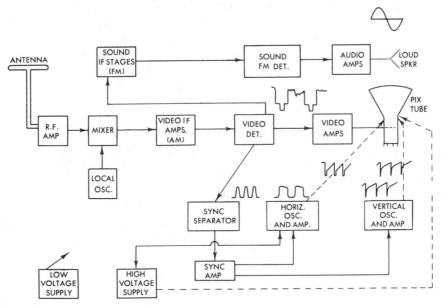

Figure 10-8 Basic TV receiver.

then applied to the cathode of the picture tube. Also, at this point a 4.5-MHz trap filters out a new 4.5-MHz IF sound signal from the video frequencies. This IF sound signal is amplified, and the audio is detected and amplified and sent to the loudspeaker.

The video signal is also picked off at the video detector or video amplifier and sent to the sync separator and sync amplifier. Here the proper horizontal sync pulses are applied to the horizontal oscillator (15,750 Hz) and the proper vertical sync pulses are applied to the vertical oscillator (60 Hz) to maintain correct synchronization with the TV station being received. The horizontal output and the vertical output are applied to the deflection coils in the yoke to accomplish the electron beam scanning on the face of the CRT. A portion of the horizontal output is applied to the high-voltage section, where a high step-up transformer, called a *flyback transformer*, and circuitry produce the necessary dc voltage to attract the electron beam to the face of the CRT. An automatic gain control (AGC) stage that controls the bias on the RF and IF amplifiers also receives its input from the video amplifier.

The low-voltage supply is applied to all other stages in the TV receiver.

10-5 TV TUNER

The *TV tuner*, also referred to as the *front end* of a TV receiver, is used to select a desired station, provide amplification, convert the RF signal to the IF signal, and prevent any unwanted radiations from the antenna. Two sections are found within the tuner, the VHF section for channels 2 to 13 and the UHF section for channels 14 to 83. Each section has its own local oscillator and shares the common RF amplifier and mixer.

Various TV channels can be selected by changing the resonant frequency of the RF amplifier, mixer, and oscillator, simultaneously. Small coils are

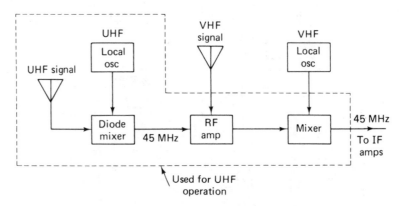

Figure 10-9 Block diagram of typical TV receiver tuner.

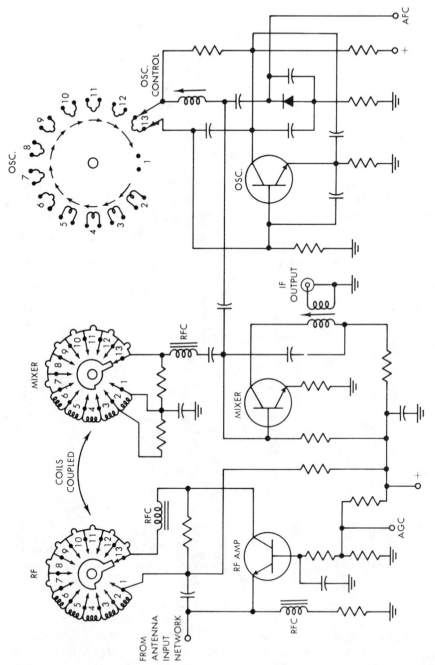

Figure 10-10 Typical VHF tuner section.

mounted on multicontact wafer or turret switches and are ganged together so that selecting a desired channel changes all three coils for each circuit at the same time. A manual fine tuning control is associated with each section to vary the oscillator frequency slightly for exact tuning of the best picture and sound. Fine tuning can be done either inductively or capacitively.

The UHF tuner is similar to the VHF tuner, but since it uses the same RF amplifier and mixer, it is somewhat smaller. Also, the high frequencies employed necessitate shorter wiring leads, because of stray capacitance and inductance. Essentially, the UHF signal is immediately stepped down to the IF signal and then goes to the VHF RF amplifier and mixer. This frequency conversion in the UHF section is usually accomplished through the non linearity of a diode.

The output of the tuner is the receiver's IF signal of 45 MHz, including the 45.7 MHz for the picture signal and 41.5 MHz for the sound signal which is fed via coaxial cable to the first IF amplifier stage.

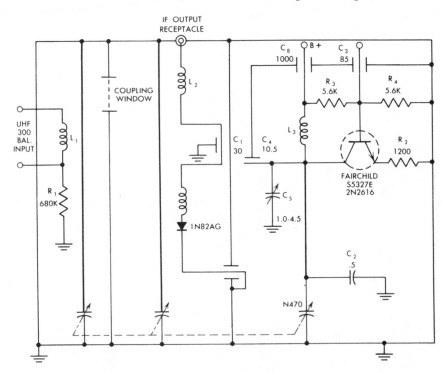

Figure 10-11 Typical UHF tuner section.

Some tuners use voltage-variable capacitance diodes (see Section 2-5) along with pushbuttons for selecting various TV channels. Other newer type of tuners use phase-locked-loop (PLL) synthesizers for channel selection.

10-6 TV IF AMPLIFIER SECTION AND VIDEO DETECTOR

The purpose of the *TV IF amplifier section* is to improve the receiver's selectivity and overall amplification. This section may be called the video IF even though the sound signal is processed also because it is part of the composite waveform. The process is referred to as an *intercarrier system.* The FM sound also has its own IF section.

The IF signal from the mixer output of the tuner is fed to the IF amplifier section, where it initially encounters wave traps. Wave traps are used to attenuate or eliminate adjacent channel carrier frequencies that may interfere with the picture and/or sound of the desired channel being received.

Since the bandwidth of the IF signal must be from 4 to 6 MHz, so as not to affect the video information, the IF stages are stagger tuned to provide an overall flat response. This stagger tuning is accomplished by adjusting the tank circuits in the collector circuit of each stage (LA_{20}, LA_{28}, and LA_{39}). Usually, Q_1 is adjusted below the center IF carrier signal, Q_2 at the IF carrier signal, and Q_3 above the IF carrier signal.

Sound IF detection may occur at the last video IF stage. The mixing of the picture carrier and sound carrier through the nonlinearity of a diode produces a new sound IF signal (45.75 MHz – 41.25 MHz = 4.5 MHz). This new sound IF signal is amplified, the audio detected (see Sections 9-18 and 9-19), and the audio signal amplified and sent to the loudspeaker.

A 41.25-MHz wave trap precedes the video detector diode so as to eliminate any sound IF from interfering with the picture information. *Video detection* is similar to AM radio detection (see Section 9-11), where the lower-frequency picture information and blanking pulses are removed from the RF modulated signal. This demodulated signal is sent to the video amplifier section.

10-7 VIDEO AMPLIFIER SECTION

The main function of the *video amplifier section* is to increase the amplitude of the video signal (detected picture information and blanking pulses), which is applied to the cathode of the CRT. This signal affects the intensity of the electron beam, producing white, black, or gray values on the face of the CRT, thus creating a TV picture. A high positive signal causes black areas, a low negative signal causes white areas, while the signal amplitude between these extremes creates the gray areas. The contrast control in the emitter circuit of Q_2 controls the transistor's gain and determines the picture contrast or black level, while the brightness control just before the CRT sets the CRT's cathode reference voltage and determines the brightness or overall white level of the picture. Peaking coils are used to extend the bandwidth of

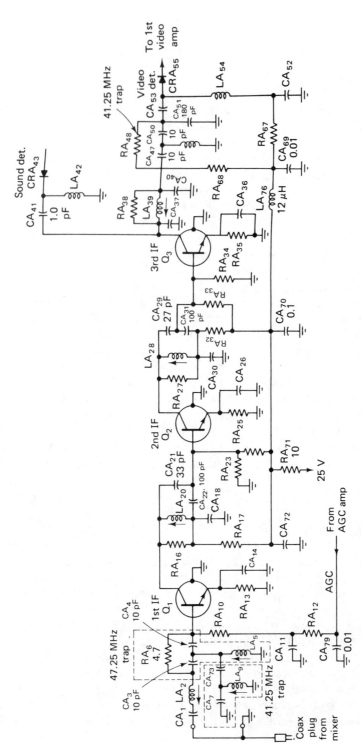

Figure 10-12 Typical IF amplifier section and video detector.

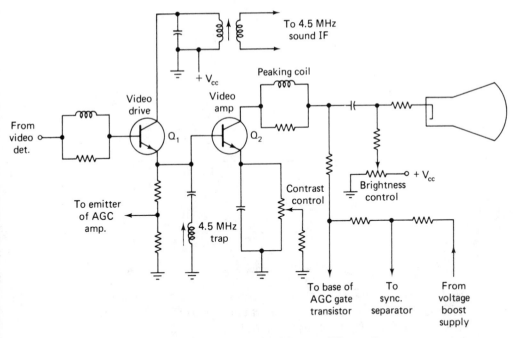

Figure 10-13 Typical video amplifier section.

the video amplifier section up to 4 MHz to produce a suitable picture. A resistor in parallel with the peaking coil reduces or prevents any extraneous oscillations from affecting the picture.

Other important functions are also picked off at the video amplifier section. The new sound IF signal may be taken from this section by the use of a tank circuit tuned to 4.5 MHz. A 4.5-MHz wavetrap in the base circuit of Q_2 prevents the new sound IF signal from interfering with the video information applied to the CRT. The detected video signal is applied to an automatic gain control (AGC) circuit that biases the RF and IF amplifiers in preceeding stages. A portion of this signal is also fed to the sync separator circuit, where the horizontal and vertical sync pulses are derived for proper receiver synchronization with the TV channel selected.

A voltage greater than the low power supply voltage ($+V_{cc}$) is needed for proper circuit operation. This *boosted voltage*, as it is called, is developed by the horizontal output section of the TV receiver.

10-8 AUTOMATIC GAIN CONTROL (AGC) CIRCUIT

The *automatic gain control* (AGC) *circuit* in TV receivers has the same function as AGC or AVC in radio receivers, and that is to maintain a relatively constant output from the video detector even though the transmitted TV

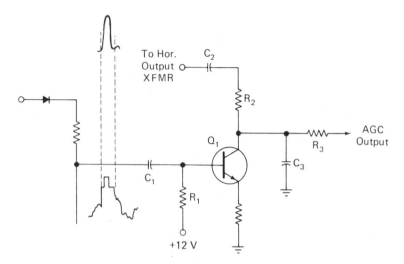

Figure 10-14 Basic AGC circuit.

signal has varying amplitude or strength. With a properly operating AGC circuit, the contrast level is maintained as different channels are selected, without the need to readjust the contrast control.

The AGC circuit is fed a portion of the video signal from the video detector or video amplifier. Here it is converted into a dc bias voltage that is applied back to the RF and IF amplifiers to control their gain. Since the video signal is constantly changing, the amount of dc bias is determined by the amplitude of the blanking pulse. A low video signal changes the dc bias and the RF and IF amplifiers produce more gain. A higher-level video signal again changes the dc bias, which reduces the gain of the RF and IF amplifiers.

In a simple AGC circuit, the blanking pulse occurring at 15,750 Hz is filtered by an RC network to produce dc bias. Noise pulses in the signal may affect the AGC voltage; therefore, a keyed AGC circuit is very often used. In a typical keyed AGC circuit, the video signal is applied to the base of an NPN transistor. The base is already forward biased by the +12 V; however, there is not constant conduction through the transistor, because the normal $+V_{cc}$ is not connected to the collector. A positive pulse from the horizontal deflection section is applied to the collector to provide conduction and it is keyed to the same time the blanking pulse is present at the base. The AGC voltage is developed across C_3.

10-9 SYNC SEPARATOR

The purpose of the *sync separator* is to remove the sync pulses from the horizontal blanking pulses and vertical blanking pulses and apply them to the horizontal and vertical sweep oscillators to provide proper synchronization

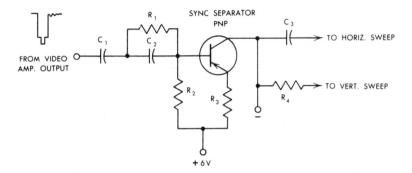

Figure 10-15 Basic sync separator circuit.

with the TV station being received. These sync pulses must be free of video signal and noise to prevent false synchronization of the sweep oscillators by spurious signals. Sync separators are sometimes called clipper circuits, since in effect, they clip the sync pulses from the video signal so that these pulses can be properly shaped to trigger the horizontal and vertical oscillators.

A basic transistor sync separator circuit normally operates with reverse bias and there is no conduction through the transistor. This reverse bias is sufficient so that the transistor remains cut off for the video signal and blanking pulse input signal. However, the sync pulse riding on the blanking pulse is able to turn on the transistor, which develops a sync pulse only at the collector circuit. The output of the sync separator is fed to RC filters to produce the necessary waveforms to trigger the horizontal and vertical oscillator. A low-pass filter or integrator is used for the 60-Hz vertical oscillator and a high-pass filter or differentiator is used for the 15,750-Hz horizontal oscillator.

10-10 HORIZONTAL DEFLECTION AND HIGH-VOLTAGE SECTION

A typical horizontal section consists of a horizontal oscillator, horizontal output amplifier, flyback transformer, and a high-voltage rectifier. The horizontal oscillator produces a linear sawtooth waveform at 15,750 Hz, which controls the left-to-right scan of the electron beam in the CRT. A variable-inductance hold control adjusts the frequency of the oscillator slightly to lock in the picture and allow the horizontal sync pulses to control the scanning operation.

Horizontal sync pulses from the sync separator are applied to an automatic frequency control (AFC) circuit using a diode phase detector. A portion of the output signal from a winding on the flyback transformer is fed to the AFC circuit as a reference voltage. A dc bias is developed when there is a difference between these two signals. In turn, this bias adjusts the horizontal oscillator to synchronize it with the incoming sync pulses. The oscillator

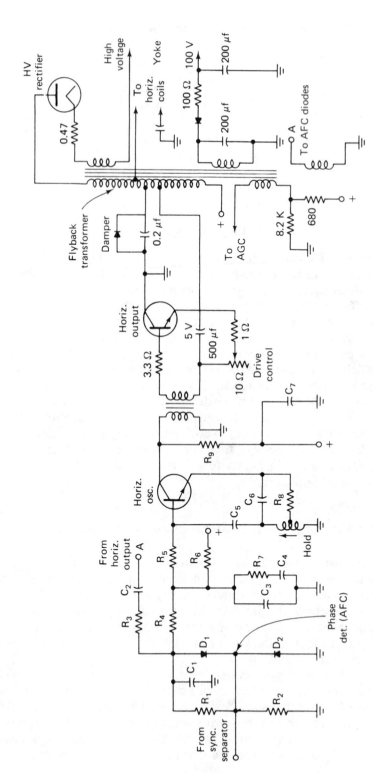

Figure 10-16 Typical horizontal deflection and high-voltage section.

is now "locked in" with the station being received to prevent heavy slanting streaks ("horizontal tear") on the screen.

The horizontal output amplifier increases the amplitude of the sawtooth wave and applies it to the flyback transformer, where several functions take place. First, the sawtooth wave is applied to the horizontal deflection coils in the yoke assembly to create the scanning. A step-up winding is used by the high-voltage rectifier to produce the high dc anode voltage (10 to 40 kV) required by the CRT. Another winding is used in a rectifier circuit to create the boosted voltage (100 V) used by circuits requiring more than the normal power supply voltage $(+V_{cc})$ of 9 to 24 V.

The damper diode is used to short circuit unwanted oscillations during flyback time when the magnetic fields are collapsing in the horizontal deflection coils. This allows the next scanning pulse to start without any interference from induced EMF from the flyback transformer.

10-11 VERTICAL DEFLECTION SECTION

A typical *vertical deflection section* consists of a multivibrator oscillator $(Q_{302}$ and $Q_{306})$ operating at 60 Hz, which produces the scanning needed to move the electron beam in the CRT from the top to the bottom of the screen. The required feedback for multivibrator operation is from the emitter of Q_{306} to the base of Q_{302}.

Sync pulses from the sync separator are detected by the integrator $(R_{302}$ and $C_{300})$, which produce the vertical sync pulses. These pulses synchronize the multivibrator with the incoming station signal to prevent the picture from rolling vertically.

The switching action of the transistors allows capacitors C_{306} and C_{308} to charge and discharge, thus producing a sawtooth waveform. This sawtooth waveform is amplified by Q_{306} and is applied to the vertical deflection coils in the yoke assembly on the neck of the CRT. Control R_{328} is the vertical hold control and adjusts the multivibrator feedback to "lock in" the picture. Control R_{318} adjusts the amplitude of the sawtooth waveform and subsequently the picture height. Control R_{314} adjusts the linearity of the sawtooth waveform to prevent the picture from appearing "stretched out" or "squeezed in."

During the vertical retrace period, Q_{306} is cut off, causing the magnetic field of L_{300} to rapidly collapse, which allows C_{312} to discharge. This returns the electron beam in the CRT from the bottom to the top of the screen and at the same time applies a blanking pulse to the CRT which cuts off the electron beam. After the vertical retrace period, the electron beam turns on and the sawtooth waveform moves it downward to complete another field.

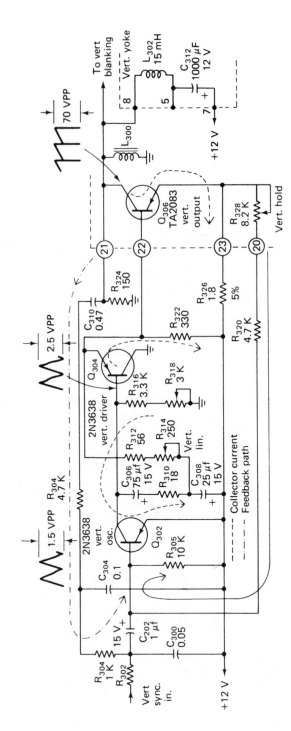

Figure 10-17 Typical vertical deflection section.

10-12 INFORMATION ON COLOR TELEVISION

Color television circuits use basically the same circuits as monochrome television, with the addition of more highly complex circuits to produce the color signals for transmission and reception.

The three primary colors (hues) used for television are red, blue, and green. These colors in their pure or saturated form can be combined in correct proportions to produce other desired colors. For example, if green and red are added, with no blue, the result is yellow. Blue and green added without any red produces cyan (blue-green). Red and blue combined, with no green, results in magenta (light purple). All three colors added produce white. By proper combinations of hues, intensities (amount of saturation), the brightness (the amount of whiteness), any desired color can be reproduced, as well as white, grays, and black.

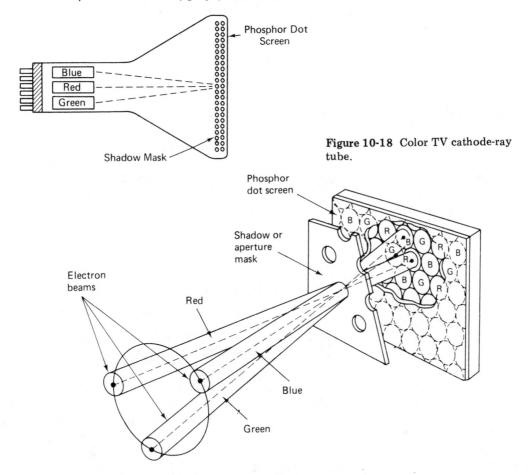

Figure 10-18 Color TV cathode-ray tube.

Figure 10-19 Color TV CRT beam arrangement.

Color receiver CRTs are made up of triads of red, blue, and green phosphor dots on the screen. Three electron guns within the tube, labeled red, blue, and green, are used to illuminate the respective color phosphors. The problem is to get the proper electron beam to strike its respective color phosphor dots. The three beams pass simultaneously through a single hole in the shadow mask which is located just behind the phosphorus screen. The shadow mask allows the proper beam to strike its respective color dot and prevents the beams from "spilling over" onto adjacent color dots. A typical shadow mask may contain over 200,000 holes. Special modifications of the horizontal and vertical deflection sections are adjusted to make the three beams "converge" on their proper color dots throughout the face of the tube.

Basically, color television transmission consists of viewing a color scene with a color camera that combines the proper proportions of the three primary colors into electrical signals that are then modulated on the composite RF television channel and transmitted into the atmosphere. At the color receiver, the color signals are demodulated from the composite RF television signal and applied to special circuits to produce the same color scene on the face of the color CRT.

10-13 BASIC COLOR TV TRANSMITTER

Initially, a color scene is viewed by a color camera, where special mirrors, called *dichroic mirrors*, separate white light into the three primary colors. The color signals produced are then combined in the proportion 59% of the green signal, 30% of the red signal, and 11% of the blue signal to make up the luminance (brightness) or Y signal (Y signal = 0.59G + 0.30R + 0.11B = 1.00). The Y signal is similar to the video signal for monochrome television and allows color transmission to be viewed on black-and-white receivers, the term being called *compatibility*. A chrominance or C signal which produces the color picture in the receiver is made up of two other signals. A matrix combines the three color signals in the proportions 60% red, 32% blue, and 28% green to produce the I (in-phase) signal, also known as the B-Y, blue-chrominance, or orange-cyan signal. The matrix also combines the three color signals in proportions of 54% green, 31% blue, and 21% red to produce the Q (quadrature out of phase by 90°) signal, also called the R-Y, red-chrominance, or blue green-magenta signal. These two signals (I and Q) each contain some of the hue and saturation information.

To transmit a color picture (chrominance information) without increasing the bandwidth of a TV channel, it is necessary to use a subcarrier. The subcarrier of 3.57945 MHz (commonly referred to as 3.58 MHz) allows the sideband energy of the C signal to interleave within the sideband energy of the Y signal. Both the I and Q signals are filtered to remove high-order frequencies and then modulate the 3.58-MHz subcarrier. During the modula-

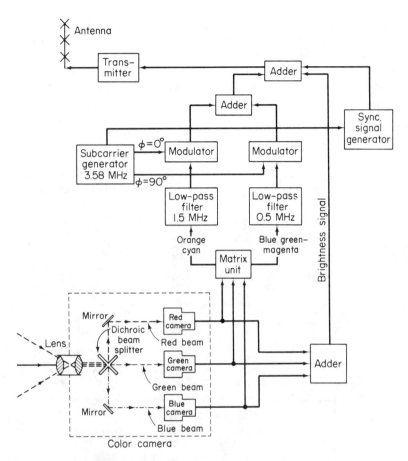

Figure 10-20 Basic color TV transmitter.

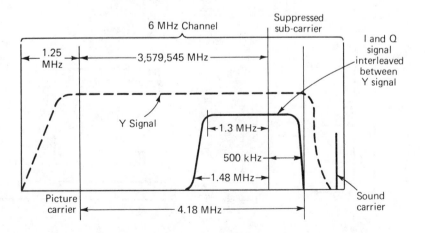

Figure 10-21 Typical composite color TV channel allocation.

tion process the Q signal is shifted 90° out of phase to be properly detected at the receiver and the subcarrier is suppressed so as not to create interference with the composite color signal.

The C signal and Y signal are then added together to form the composite color video signal, which modulates the standard TV transmitter for broadcast.

10-14 BASIC COLOR TV RECEIVER

A *color TV receiver* includes all the circuits of a monochrome receiver as indicated by the dashed lines (see Section 10-4). However, the tuner usually

Figure 10-22 Basic color TV receiver.

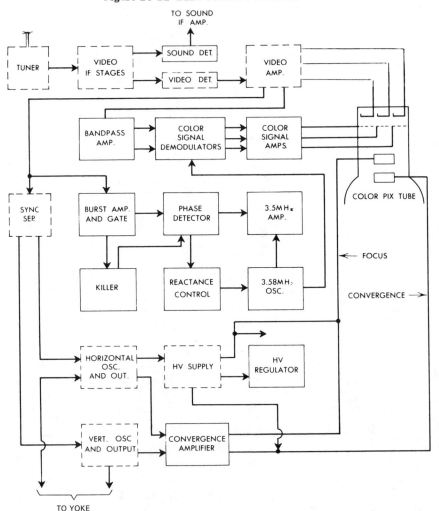

has an automatic fine tuning (AFT) circuit and the RF circuits have a wider bandwidth to permit unimpaired color reception. The required color circuits are shown in solid lines.

The Y or composite video signal is applied to a bandpass amplifier, where only the C signal (I and Q signals) is allowed to pass onto the color signal demodulators. A portion of the video signal containing the horizontal blanking pulse is fed to the burst amplifier and gate. The eight-cycle color burst on the back porch of the horizontal blanking pulse is transmitted by the TV station to synchronize the 3.58-MHz color oscillator within the color receiver. (see Section 10-2 and Figure 10-5). The color oscillator reinserts the 3.58-MHz subcarrier (suppressed at the transmitter) to accomplish demodulation of the color signals. When a black-and-white picture is transmitted, the color burst is absent, which activates the color killer circuit. This action cuts off the color circuits and prevents noise in the picture, called "confetti," which looks like snow but with larger spots in color.

The transmitted waveform contains the red information (R-Y) and the blue information (B-Y). These two signals are combined in the color signal demodulators of the receiver to also produce the green information (G-Y). The three signals are amplified and applied to control grids in the color CRT. The video (Y) signal is delayed 0.8 μs by a delay line, to allow time to process the color signals so that the Y information applied to the cathodes of the CRT coincides with the chrominance information at the grids. A mixing action within the CRT between the two signals produces the correct color picture on the screen.

Color CRTs require a higher anode voltage (up to 30 kV) than monochrome CRTs and a special high-voltage (HV) regulator is also used. Convergence circuits being fed from the horizontal and vertical sections are also needed to align the electron beams to their proper color dots throughout the entire face of the screen.

10-15 OVERVIEW OF TV RECEIVER TROUBLESHOOTING

Television receivers, especially color sets, are highly complex and require a certain amount of study and experience for efficient troubleshooting. Special test instruments are also useful and in some cases absolutely necessary for proper servicing. The multimeter (measures volts, ohms, and amperes) should have a high input resistance and the oscilloscope should have a bandwidth greater than 4 MHz for servicing color sets since the subcarrier is 3.58 MHz. Special high-voltage probes are used to measure the CRT anode voltage, and other test equipment might include sweep/marker generator, color dot/bar generator, audio generator, and transistor/solid-state tester.

Many circuits in the receiver are interconnected and share the same voltage sources. A malfunction in one section may cause a seemingly unrelated problem in another section. However, initially, an attempt should be made to localize and isolate the faulty section from the indicated trouble symptoms. All manual controls (fine tuning, horizontal hold, vertical hold volume control, brightness, contrast, tint, color, etc.) should be adjusted to obtain the best possible reception before troubleshooting procedures are started.

Table 10-1 is an overview of some of the problems that might be caused by the various sections.

TABLE 10-1 PROBLEMS OF TELEVISION RECEIVERS AND POSSIBLE CAUSES

Problem	Section to check first
1. Set completely inoperative	Low-voltage power supply
2. No raster with sound	Horizontal and HV section
3. A single line on screen with sound	Vertical section
4. Picture size or linearity incorrect	Horizontal section
	Vertical section
5. Vertical rolling	Vertical section
	Sync circuits
6. Horizontal tearing	Horizontal section (AFC diodes
	Sync circuits
7. Picture okay with sound defects or no sound	Sound section
8. Raster without picture and without sound	Tuner
	IF section
	Video section
9. Weak picture and weak sound	Tuner
	IF section
	Video section
	AGC circuits
10. Picture interference (snow, out of focus, poor quality)	Antenna
	Tuner
	IF and video section
	AGC circuits
11. Weak brightness	CRT (electron gun emission low
	Video section
12. Color problems	Convergence circuits
	Color circuits
	Video and IF circuits
	Tuner
	Antenna

11

Integrated Circuits

11-1 INFORMATION ON INTEGRATED CIRCUITS (ICs)

An *integrated circuit* (IC) is a group of interconnected circuits contained in a single package. A hybrid IC has separate components or circuits mounted on ceramic substrates and/or monolithic ICs, which are connected by wire bonds. A monolithic IC has all the components formed upon or within a single piece of silicon crystalline material. The monolithic IC has become the most popular type of circuit in current use. In can be found enclosed in every type of solid-state device package (see Figure 2-1) and in special IC packages (see Figure 13-3).

Electronic circuits are divided broadly into two areas: linear (analog) and digital. Linear circuits produce various levels at their outputs in response to various levels at their inputs, whereas digital circuits are switching devices and respond to on or off conditions at fixed levels. Sometimes ICs are labeled LM (linear module) or DM (digital module).

Nearly all ICs have two or three power supply connections in terms of V_{CC}, V_{EE}, V_{DD}, V_{SS}, $+V$, $-V$, or GND, which are needed for proper operation.

As IC technology progressed, more and more circuits could be produced in a smaller or given area, which gave rise to various types of integration processes.

1. SSI (small-scale integration) has up to 10 or 12 gates or circuits per single package.

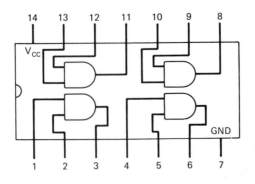

Figure 11-1 SSI (two-input quad AND gate IC).

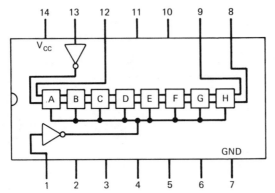

Figure 11-2 MSI (8-bit shift register IC).

2. MSI (medium-scale integration) has from 100 to 1000 gates or circuit per single package.

3. LSI (large-scale integration) has from 1000 to 10,000 gates or circuit per single package.

4. VLSI (very large scale integration) has from 10,000 to over 50,000 gates or circuits per single package.

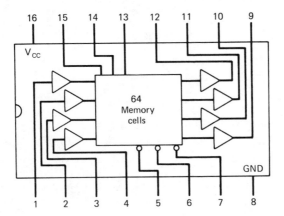

Figure 11-3 LSI (64-bit RAM).

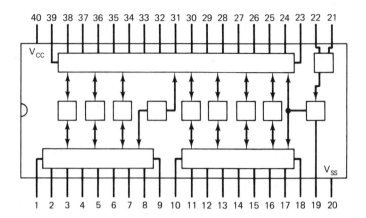

Figure 11-4 VLSI (microprocessor).

Integrated circuits are used in nearly every aspect of electronics, as indicated by the following list.

Op amps (general)	Temperature-sensing devices
Op amps (specialized)	Pressure transducers
Digital circuits	Phase-locked loops
Digital clocks	Audio circuits
Digital watches	AM radio circuits
LED displays	FM radio circuits
Memory circuits	TV RF circuits
Computer interface	TV video circuits
D/A and A/D converters	TV color circuits
Calculators	Electronic games and toys
Microprocessors	Servoamplifiers
Voltage regulators	Active-device switches
Timers	Automotive devices
Function generators	Component arrays
Optoisolators	Resistor
Control circuits	Diode and transistor

11-2 MONOLITHIC IC FABRICATION

Monolithic IC fabrication is a process in which active devices (diodes, transistors, thyristors) and passive devices (resistors, capacitors, inductors) are produced at the same time within a silicon material. Most often semicon-

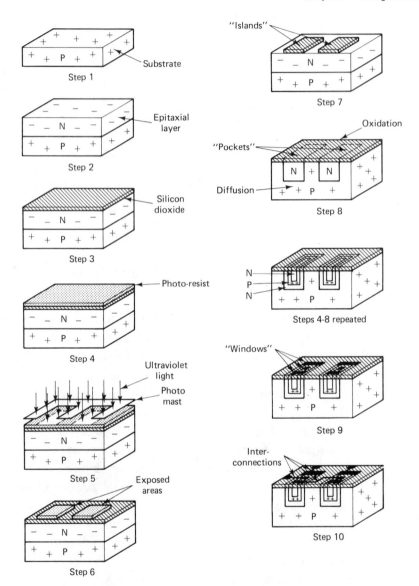

Figure 11-5 Integrated circuit fabrication.

ductor junctions and component leads are brought to a common plane surface for interconnection; this is termed the planar process. The fabrication process is the same for bipolar transistors and MOSFETs (see Figures 2-54, 2-55, 2-57, and 2-58), except that bipolar types require more steps. The MOSFET types require less area and accounts for MSI, LSI, and VLSI devices. Dimensions used in the fabrication process are on the order of mils and microns (1 mil = 0.001 inch, 1 micrometer = 0.000001 mil, 1 mil = 25.4 micrometers). Many circuits, referred to as "chips," are processed simultaneously on a "wafer" about 3 in. in diameter. The following list indicates the basic steps in IC fabrication.

Step 1: A *p*-type silicon material is produced to serve as the substrate or IC foundation.

Step 2: An *n*-type layer is grown epitaxially on the substrate. The later becomes the collector material for an NPN transistor, the emitter for a diode, or one plate of a capacitor. (*Expitaxial growth* is the process of adding a different material to a substrate while maintaining the same crystalline structure throughout the entire unit. The growing is done in special chambers set at high temperature, where the specific dopants are passed over the surface of the wafer in the form of gases.)

Step 3: A thinner layer of silicon dioxide is grown on the *n*-type layer. The silicon dioxide layer serves as an insulator between the various components and their elements.

Step 4: A photo-resist material is formed on the silicon dioxide layer. The photo-resist is in jelly form.

Step 5: Using a photo mask, the wafer is exposed to ultraviolet light (photolithographic technique). The jelly that the light strikes becomes hard. The unexposed or shaded part of the jelly remains soft.

Step 6: The unexposed photo-resist material is dissolved.

Step 7: The unprotected oxide is etched away with acid, producing "islands" of exposed silicon dioxide.

Step 8: Isolation diffusion forms the *n*-type material, not protected by the oxide layer, into *p*-type material. Diffusion is accomplished by "driving in," with high temperature, large concentrations of *p*-type impurities built up near the surface of the wafer during the diffusion process. A thin oxide layer is again produced during the diffusion process. The wafer now appears as a *p*-type material consisting of "pockets" containing *n*-type material.

Steps 4 through 8 are repeated on the pocket by etching "windows" using the respective *p*- and *n*-type diffusants to produce the base and emitter sections of transistors, collectors of diodes, or plates of capacitors and resistors.

Step 9: Windows are again etched into the different layers, so that low ohmic connections can be made to the various elements.

Step 10: Metallization is accomplished by evaporating aluminum over the entire wafer. This is then photosensitized, masked, and etched to produce the interconnection pattern or "wiring" between components. (*Metallization* is a process of forming metallic (usually aluminum) connections between the various components and to form metal contacts or bonding pads for the external circuit connections.)

Resistors are produced using photolithographic techniques by diffusion. The resistor channel is specially diffused *p*-type material. The value of resistance depends on the thickness, length, and concentration of the diffusant.

Capacitors are produced two ways; one utilizes the capacitance of a reversed-biased *pn* junction, formed within the wafer, and the other uses the oxide layer as the dielectric. One plate is formed by the metallization process, while the other plate consists of a high concentration of *n*-type material, indicated by n^+.

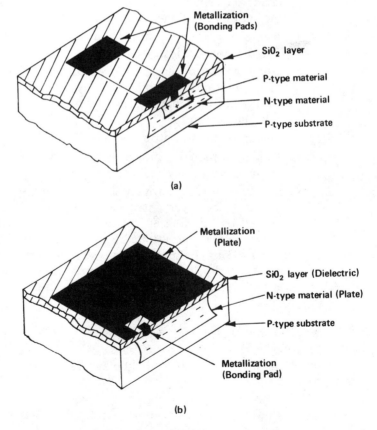

(a)

(b)

Figure 11-6 IC passive component fabrication: (a) resistor; (b) capacitor.

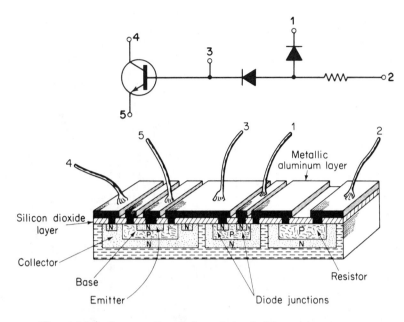

Figure 11-7 Cross-sectional view of simple IC construction.

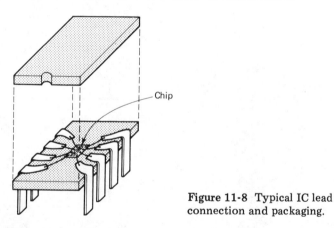

Figure 11-8 Typical IC lead connection and packaging.

After fabrication, the wafer is scribed with a diamond-tipped tool and broken into individual chips. Each chip is then mounted on a ceramic wafer and leads about 0.001 in. in diameter are attached to the metal contacts, called bonding pads. The chips are then mounted in small round cans or flat packages and the wires are connected to external pins.

11-3 FIXED IC VOLTAGE REGULATORS

Fixed IC voltage regulators can be used with power supplies or mounted on PC boards at various points in an electronic system to produce the nearly constant voltage required by specific circuits. The objective of the voltage

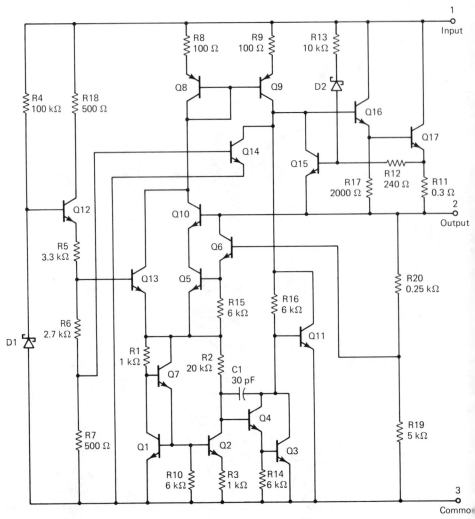

Figure 11-9 Typical voltage regulator IC. (Courtesy of Signetics Corporation, a subsidiary of U.S. Philips Corp., 811 E. Arque Ave., Sunnyvale, CA 94086.)

regulator is to provide a constant output voltage independent of input supply voltage, output load current, and temperature. A regulator usually limits its own power dissipation and its output current so that fault conditions and overload will not damage the regulator or the load. Generally, regulators contain four basic circuits: a reference source, an error detection circuit, a control circuit, and a protection circuit. Positive, negative, and dual-polarity voltages are available in IC voltage regulators and they come in a variety of output voltages, some standard values being 5, 6, 8, 12, 15, 20

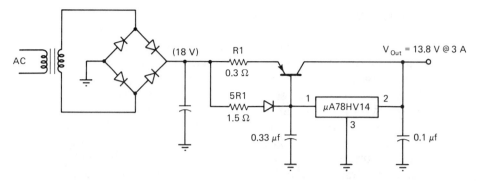

Figure 11-10 Voltage regulator IC power supply application. (Courtesy of Signetics Corporation.)

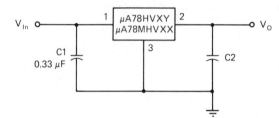

Figure 11-11 Basic fixed regulator. (Courtesy of Signetics Corporation.)

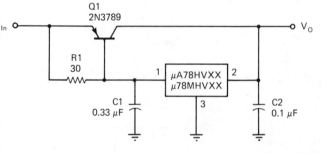

Figure 11-12 High-current regulator with pass transistor. (Courtesy of Signetics Corporation.)

and 24 V. The input voltage to such a regulator used in a power supply should be at least a few volts greater than the desired regulated voltage so that the regulator can provide satisfactory operation.

In a low-current, fixed-voltage application, an IC voltage regulator may need only a couple of external capacitors to filter out high-frequency transient noise pulses.

An IC voltage regulator can be used in a high-current application with the addition of a pass transistor circuit. The regulator establishes the regulated voltage, while allowing greater amounts of current it cannot handle to pass through the transistor. Although the pass transistor is not protected against overload, additional circuitry may provide this.

A voltage regulator can be used as a current regulator. The regulator is placed in series with the output load, with the requirement that sufficient voltage be dropped across the regulator for proper operation. For example, if the load requires a regulated current at +5 V (V_{RL}) and a +5-V regulator (V_{VR}) with a minimum input/output differential voltage of +2 V (V_D) is used, the total input voltage must be +12 V ($V_{in} = V_{RL} + V_{VR} + V_D$).

An IC voltage regulator can be used as an adjustable power supply from a minimum voltage to a maximum voltage. Here again, the input voltage must be large enough to produce the maximum output voltage plus the minimum input/output differential voltage of the regulator.

A single voltage regulator can be used with an op amp to produce a positive and a negative (dual) tracking voltage regulator. The regulator main-

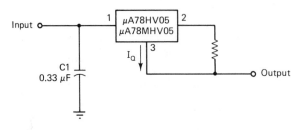

$$\text{Output current} = \left(\frac{5.0\ V}{R1} \right) + I_Q$$

Figure 11-13 Basic current regulator. (Courtesy of Signetics Corporation.)

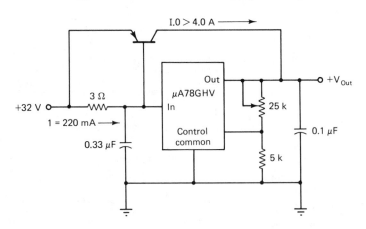

Note : External series pass device is not short circuit protected.

Figure 11-14 Positive 5 to 30 V adjustable regulator with $I_{out} > 5.0$ A. (Courtesy of Signetics Corporation.)

tains one polarity output, while the op amp connected as a voltage comparator regulates the other polarity output. Any change in the regulator output will cause the change in the op amp output—hence the term "dual tracking."

Positive and negative voltage regulators can be connected to produce a ± dual tracking voltage output. Some dual tracking voltage regulators are also available in a single package.

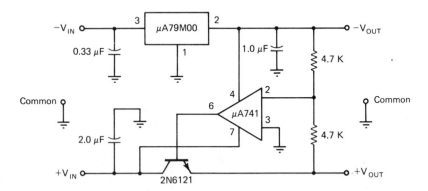

Figure 11-15 Positive and negative tracking voltage regulator with 741 op amp. (Courtesy of Signetics Corporation.)

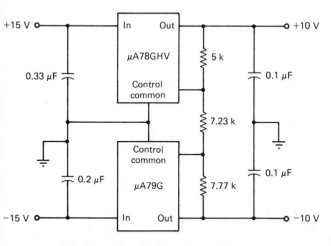

If load is not ground referenced, connect reverse biased diodes from outputs to ground.

Figure 11-16 ±10 V at 1 A dual-tracking regulator. (Courtesy of Signetics Corporation.)

Using IC voltage regulators greatly simplifies the design of regulated power supplies; however, for satisfactory results, the manufacturer's specifications and recommendations should be followed.

11-5 LM380 IC AUDIO AMPLIFIER

The LM380 IC audio amplifier is one of many low-power audio amplifier ICs intended for consumer applications (low-cost phonographs, radios, car radios, etc.) Some of its pertinent electrical characteristics are:

Power output (rms)	2.5–4 W
Supply voltage range	8–22 V
Gain	50 (average)
Output voltage swing (8-Ω load)	14 V p-p
Bandwidth	100 kHz

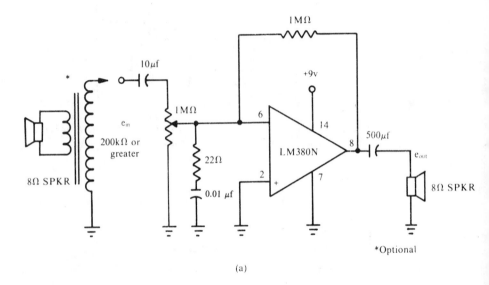

(a)

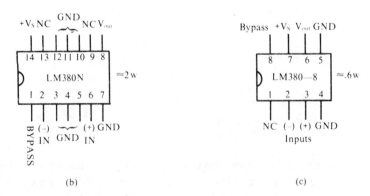

(b) (c)

Figure 11-17 LM380 audio amplifier: (a) schematic diagram; (b) 14-pin DIP; (c) 8-pin DIP.

286

Quiescent supply current	7–25 mA
Quiescent output voltage	1/2 of supply voltage
Input resistance	150 kΩ

Internal circuitry consists of a PNP emitter-follower input stage driving a PNP differential pair. This is followed by a common-emitter voltage gain amplifier with a current-source load that drives a quasi-complementary pair emitter-follower output stage. The output stage is protected with both short-circuit current limiting and thermal shutdown circuitry.

The LM380 tends to pick up noise very easily and may even oscillate at frequencies higher than the audio range. Component layout may be critical in some applications and all wires or leads should be kept as short as possible. A 22-Ω resistor in series with a 0.01-μF capacitor from pin 6 to ground helps to reduce noise without too much degradation of the input signal. A 1-MΩ feedback resistor from pin 8 to pin 6 also reduces noise.

A loudspeaker with an output transformer can be used as a microphone input for such applications as an intercom.

The LM380 IC can be used with a dual power supply by connecting the +V supply to the +V_S pin and the –V supply to the ground pin. With this arrangement quiescent output voltage should be about zero.

11-6 LM379 IC STEREO AUDIO AMPLIFIER

The LM379 IC stereo audio amplifier is typical of many similar circuits that are available. It consists of two separate audio amplifiers encased in a single 16-pin DIP. Some of the pertinent electrical characteristics for the LM379 are:

Power output/channel	6-7 W
Supply voltage range	10–35 V
Gain	50 (typically)
Output voltage swing (8-Ω load)	16 V p-p at V_S = 25 V
Bandwidth	100 kHz
Quiescent output voltage	1/2 of supply voltage
Input impedance	3 MΩ

Pin 1 is the positive supply voltage connection and pins, 2, 4, 13, and 15 are used for ground. The LM379 has on-chip frequency compensation, output current limiting, thermal shutdown protection, and a fast turn-on and turn-off, without "pops" or pulses of active gain.

Similar to a standard op amp, the LM379 can be connected in the inverting mode (input signal to – input) or the noninverting mode (input

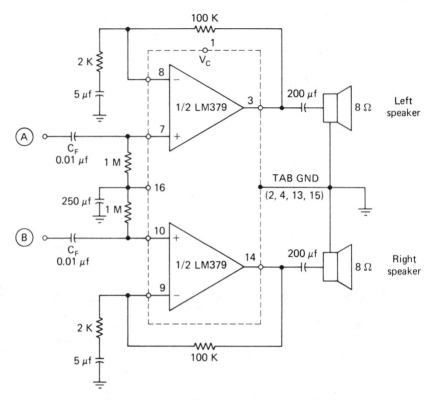

Figure 11-18 LM379 stereo audio amplifier.

signal to the + input). The noninverting connection is better suited for applications utilizing high impedance tone and volume controls. The gain can also be set the same way as with op amps.

Heat sinking is employed with the LM379 to derive maximum output power and the package is specially designed so that it can be mounted directly to a heat sink or metal chassis by means of a bolt and nut.

Some of the LM379 IC applications include:

Multichannel audio systems	Bridge output stages
Tape recorders and players	AM-FM radio receivers
Movie projectors	Intercoms
Automotive systems	Servoamplifiers
Stereo phonographs	Instrument systems

11-7 555 PRECISION TIMER IC

The 555 precision timer IC is a unique functional building block that is very stable and highly dependable for use in timing circuits. The 555 is capable of producing accurate time delays or oscillation, which is controlled by external resistors and a capacitor. Other terminals on the 555 can be used for controlled triggering, resetting, and modulation. Power supply requirements for the 555 range from 5 to 18 V and it has a relatively high output current with a power dissipation of 600 mW.

The 555 timer consists of a resistor-divider network, two voltage comparators, a bistable flip-flop, a discharge transistor, and a totem-pole output stage. All three resistors are equal and are used to set the comparator levels. Comparator 1 (threshold) is referenced at two-thirds of $+V_{CC}$ and comparator 2 (trigger) is set at one-third of $+V_{CC}$. The outputs of the comparators control the state of the bistable flip-flop. When the external voltage at pin 2 (trigger) is less than one-third of $+V_{CC}$, comparator 2 sets the flip-flop, which causes the output at pin 3 to go high. The discharge transistor is "off" or open at this time. When the external voltage at pin 6 (threshold) exceeds two-thirds of $+V_{CC}$, comparator 1 resets the flip-flop, which causes the output at pin 3 to go low. At this time the discharge transistor is "on," providing a path from pin 7 (discharge) to ground. Pin 4 (reset) can override

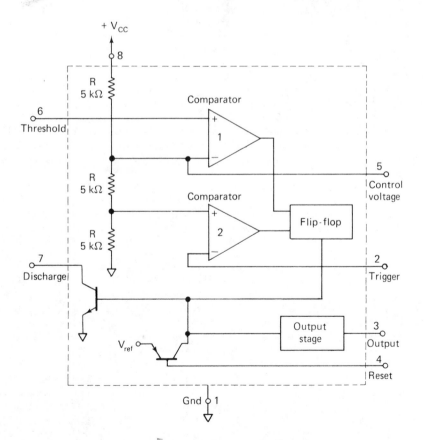

Figure 11-19 555 IC timer block diagram.

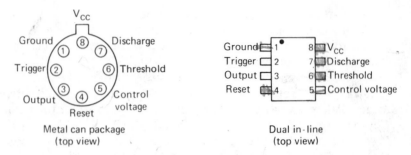

Metal can package
(top view)

Dual in-line
(top view)

Figure 11-20 555 IC timer pin connections and package types.

all other inputs and be used to initiate a new timing cycle. Pin 5 (control
voltage) is used for decoupling or modulating purpose.

Some of the applications in which the 555 timer can be used are:

Precision timing   Pulse-width modulation
Pulse generation   Pulse-position modulation
Sequential timing   Missing-pulse detector
Time-delay generation

11-8 555 MONOSTABLE MULTIVIBRATOR

A 555 timer can be used to create a monostable multivibrator with a variable
pulse-width output. Initially, the trigger (pin 2) is above one-third of $+V_{CC}$
the output (pin 3) is low (0 V), and the discharge transistor is on causing

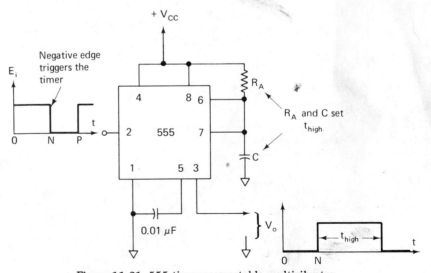

Figure 11-21 555 timer monostable multivibrator.

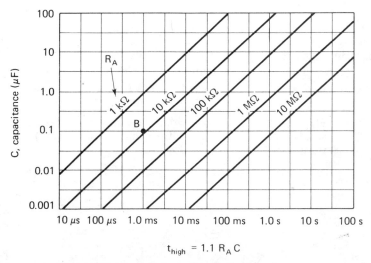

$$t_{high} = 1.1\,R_A\,C$$

Figure 11-22 Component selection graph.

pin 7 to short-out capacitor C. When the voltage at pin 2 falls below one-third of $+V_{CC}$ as a result of a negative-going pulse, the output (pin 3) goes high and the discharge transistor turns off. Timing capacitor C now starts charging toward $+V_{CC}$ through timing resistor R_A. The voltage on the capacitor increases exponentially with a time constant, $T = RC$. During this charging time the output (pin 3) remains high. When the voltage at pin 7 exceeds two-thirds of $+V_{CC}$, the output goes low and the discharge transistor turns on. The capacitor discharges through pin 7 rapidly. The timer has completed a cycle and now awaits another trigger pulse.

The pulse-width duration (time the output is high) can be found using the formula

$$T = 1.1\,R_A\,C$$

The component selection graph shown can simplify finding the components needed for a specific pulse-width or time duration. As an example, if a 10-ms pulse width is needed:

	R could equal	and	C would equal
	1 MΩ		0.009 F
or	100 kΩ		0.09 μF
or	10 kΩ		0.9 μF

11-9 555 ASTABLE MULTIVIBRATOR

The 555 timer can be connected as an astable multivibrator (pulse generator). In this case, pin 2 (trigger) is connected to pin 6 (threshold) and resistor R_B is needed between pins 6 and 7 (discharge). At power up, the capacitor

is discharged, holding pins 2 and 6 low, which causes the output (pin 3) to
go high. The discharge transistor is off at this time. The capacitor now
begins to charge toward $+V_{CC}$ through R_A and R_B. When the voltage on the

Figure 11-23 555 timer astable multivibrator: (a) connections for
astable operation; (b) capacitor voltage waveform; (c) output voltage
waveform.

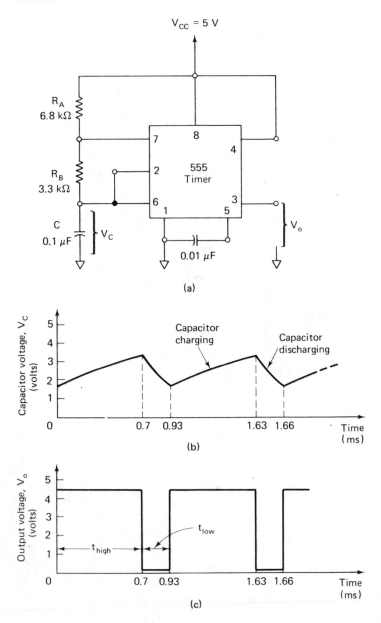

capacitor reaches two-thirds of $+V_{CC}$, the threshold comparator resets the flip-flop, causing the output (pin 3) to go low and the discharge transistor to turn on. The timing capacitor now discharges through R_B. When the voltage across the capacitor drops below one-third of $+V_{CC}$, the trigger comparator sets the flip-flop, causing the output to go high and the discharge transistor to turn off. With this action the 555 timer produces an oscillator whose frequency is independent of power supply voltage. The frequency can be found using the formula

$$f = \frac{1.49}{(R_A + 2R_B)C}$$

The component selection graph shown can simplify finding the components for a special frequency. For example, if 1 kHz is desired: ˙

$R_A + 2R_B$ could equal	and	C would be
100 kΩ		0.0149 µF
or 10 kΩ		0.149 µF
or 1 kΩ		1.49 µF

Since the timing capacitor charges through R_A and R_B but discharges only through R_B, the output waveform is not symetrical. A symetrical waveform can be produced if $R_A = R_B$ and a diode is placed from pin 7 (connected to the anode) to pin 6 (connected to the cathode).

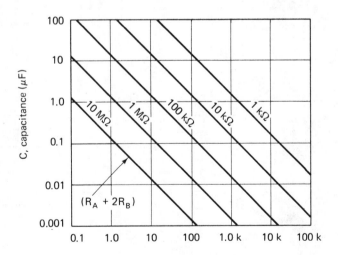

$$f = \frac{1}{T} = \frac{1.44}{(R_A + 2R_B)C}$$

Figure 11-24 Component selection graph.

The 555 timer can be used to detect a missing pulse in a train of pulses. The pulse train is applied to E_i and continuously resets the timing cycle by turning transistor Q on, which allows the timing capacitor C to discharge. The time delay of the 555 is set slightly longer than the normal time between pulses; therefore, before C has a chance to charge up to two-thirds of $+V_{CC}$, the next input pulse discharges it. The threshold comparator does not change states and the flip-flop is not reset, which keeps the output (pin 3) high.

When a pulse in the train is missing, the capacitor is allowed to charge to two-thirds of $+V_{CC}$ and the threshold comparator changes states. The

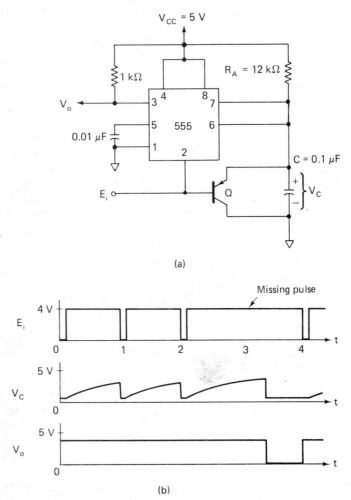

Figure 11-25 555 timer missing pulse detector: (a) circuit; (b) voltage waveforms.

flip-flop is reset and the output goes low, indicating a missing pulse. The next input pulse resets the circuit and the output goes high again.

The negative-going output from pin 3 could be used to set a flip-flop to indicate the missing pulse.

11-11 8038 PRECISION WAVEFORM GENERATOR/VCO IC

The 8038 precision waveform generator IC operates similar to a timer IC, except that it functions strictly as a voltage-controlled oscillator (VCO). There are three simultaneous outputs, consisting of square, triangle, and sine waveforms. The frequency or repetition rate can be selected by an external RC network over a range from less than 0.001 Hz to more than 1 MHz, which remains highly stable over a wide temperature and supply voltage range. The frequency is independent of supply voltage and depends only on the external components R_A and R_B connected to pins 4 and 5 and C connected to pin 10. Resistors R_A and R_B determine the symmetry and duty cycle of the output waveforms. The frequency can be found by the formula

$$f = \frac{1}{\frac{5}{3} R_A C \left(1 + \frac{R_B}{2R_A - R_B} \right)}$$

or if $R = R_B$, then

$$f = \frac{0.3}{R_A C}$$

or if a single timing resistor is used, then

$$f = \frac{0.15}{RC}$$

When the voltage across C rises to two-thirds of $+V_{CC}$, a comparator sets the flip-flop, which allows the capacitor to discharge. When the voltage across the capacitor falls to one-third of $+V_{CC}$, the other comparator resets the flip-flop and again allows the capacitor to charge up. This continuous action produces oscillations. The changing output of the flip-flop is fed to a buffer stage and produces the square-wave output. The capacitor is charged and discharged by constant-current sources within the IC; therefore, a linear triangle wave is developed and fed to a buffer output stage. A portion of the triangle wave is applied to a sine converter, where, in effect, the peaks are rounded off to produce a sine-wave output. This sine wave is not pure, but may be adjusted by external resistors to achieve minimum distortion.

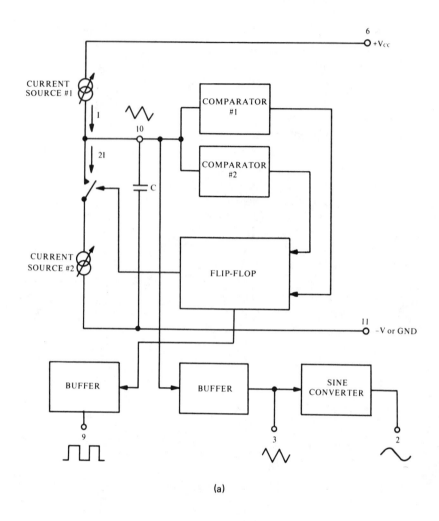

(a)

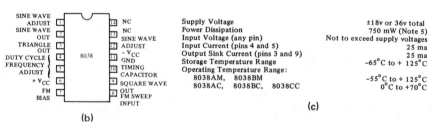

SINE WAVE ADJUST	1	14	NC
SINE WAVE OUT	2	13	NC
TRIANGLE OUT	3	12	SINE WAVE ADJUST
DUTY CYCLE FREQUENCY ADJUST	4	11	$-V_{CC}$
	5	10	GND TIMING CAPACITOR
$+V_{CC}$	6	9	SQUARE WAVE OUT
FM BIAS	7	8	FM SWEEP INPUT

8038

(b)

Supply Voltage	±18v or 36v total
Power Dissipation	750 mW (Note 5)
Input Voltage (any pin)	Not to exceed supply voltages
Input Current (pins 4 and 5)	25 ma
Output Sink Current (pins 3 and 9)	25 ma
Storage Temperature Range	$-65°C$ to $+125°C$
Operating Temperature Range:	
8038AM, 8038BM	$-55°C$ to $+125°C$
8038AC, 8038BC, 8038CC	$0°C$ to $+70°C$

(c)

Figure 11-26 8038 precision waveform generator/VCO: (a) block diagram; (b) pin connections; (c) maximum ratings. (Courtesy of Intersil Corporation, 10900 N. Tantau Ave., Cupertino, CA 95014.)

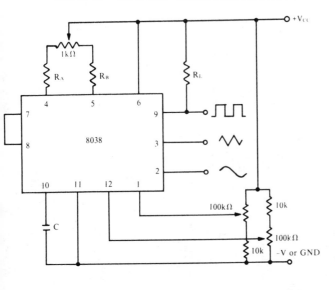

Figure 11-27 Connection to achieve minimum sine-wave distortion. (Courtesy of Intersil Corporation.)

11-12 8038 BASIC VARIABLE-FUNCTION GENERATOR

A basic variable function generator can easily be constructed using the 8038 IC and a few external components. The circuit shown covers the audio range from 0.14 to 15 kHz in five steps. Switch S_1 selects the proper capacitor for the desired range. Potentiometer R_1 is adjusted to produce the desired

Figure 11-28 Basic variable-function generator.

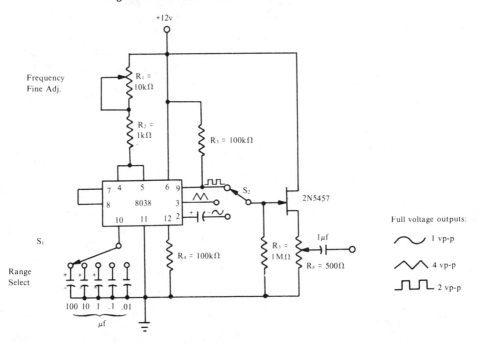

frequency. Switch S_2 selects the desired output waveform: square wave, triangle wave, or sine wave. The output of the 8038 is fed to a JFET source follower that serves as a buffer circuit for the generator. Potentiometer R_6 is adjustable to vary the amplitude of the output signal. This potentiometer can be any value, depending on the impedance matching requirements of the driven circuit. Full voltage output is 1 V p-p for the sine wave, 4 V p-p for the triangle wave, and 2 V p-p for the square wave, if a V_{CC} of +12 V is used. A +9-V power supply would perform equally well, but the full voltage outputs would be proportionately less. The 1-μF capacitor at the output blocks any dc voltage from the circuit to be driven. Reducing the values of R_1, R_2, and C will increase the output frequency; however, wiring consideration is essential, because of the increase in stray inductance and capacitance.

11-13 PHASE-LOCKED LOOP (PLL) IC

The phase-locked loop (PLL) is an electronic feedback control system. A basic PLL system consists of a phase comparator or detector, a low-pass filter, and a voltage-controlled oscillator (VCO). The phase comparator has two inputs, one for the input frequency and the other from the VCO. If the two frequencies are identical, the output signal of the phase comparator is zero. When the two frequencies are not the same, a signal is developed at the output of the comparator and passed on to the low-pass filter. The voltage level out of the low-pass filter is proportional to the difference between the two frequencies. This control signal level is fed back to the VCO input, which changes the VCO frequency in an attempt to make it exactly match the input frequency. When the VCO frequency equals the input frequency, the PLL has achieved "lock" and the control signal voltage will be zero for as long as the input frequency remains constant. There are three possible states of operation for the PLL.

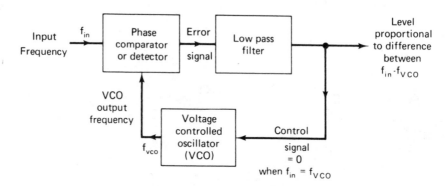

Figure 11-29 PLL block diagram.

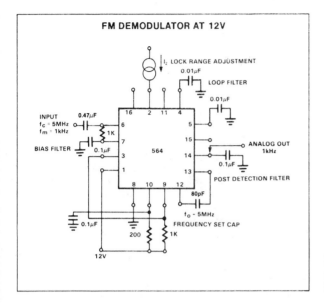

Figure 11-30 PLL IC used for FM demodulation. (Courtesy of Signetics Corporation.)

1. Free-running, where the input frequency is not within range for the VCO frequency (determined by external timing resistors and capacitor) to "lock in." The PLL runs at the nominal VCO frequency and this condition is usually avoided, since nothing is accomplished.

2. Capture, when the two frequencies are close enough to develop a control signal voltage and the VCO frequency changes until the locked condition is reached.

3. Locked or tracking, when the VCO can remain locked or tracking over a wider input frequency range variation than was necessary to achieve capture.

Many electronic systems use PLL ICs, where a frequency feedback control circuit is needed. One example of PLL use is with FM demodulation.

11-14 AM RADIO RECEIVER SUBSYSTEM IC

Communication systems utilize ICs to a great extent in the form of individual stages and/or subsystems that require a minimum of external components for proper operation. One such subsystem is the Signetics TCA440 AM radio receiver circuit, which produces a RF stage with AGC, a balanced mixer, a separate oscillator, and an IF amplifier with AGC in a single 16-pin DIP. It has low current consumption and internal stabilization, which is well suited for battery-operated portables and car and home radios.

An IC such as this features operation up to 30 MHz, a balance circuit,

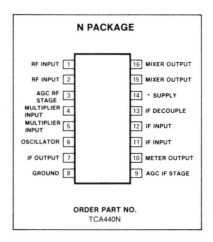

Figure 11-31 Pin configuration for 16-pin DIP AM radio receiver subsystem. (Courtesy of Signetics Corporation.)

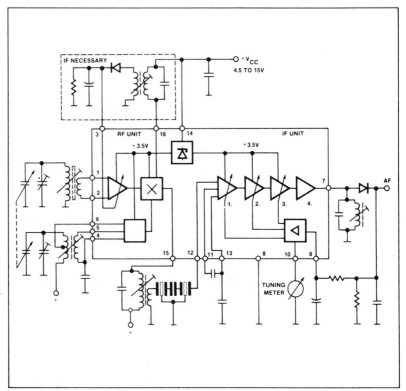

Figure 11-32 AM radio receiver subsystem IC block diagram. (Courtesy of Signetics Corporation.)

separately controllable prestage, multiplicative push-pull mixer with separate oscillator, high signal-handling capability even with a 4.5-V supply voltage, 100-dB feedback control range in five stages, direct connection for a tuning meter, and a minimum of external components. The external components consist of RF and IF transformers, ganged tuning capacitor, trimmer capac-

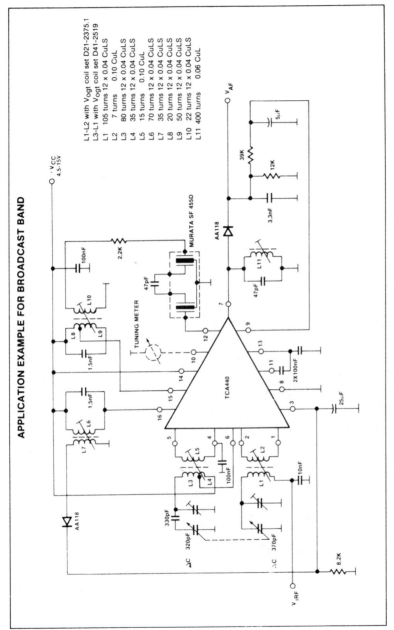

Figure 11-33 AM radio receiver IC showing external components. (Courtesy of Signetics Corporation.)

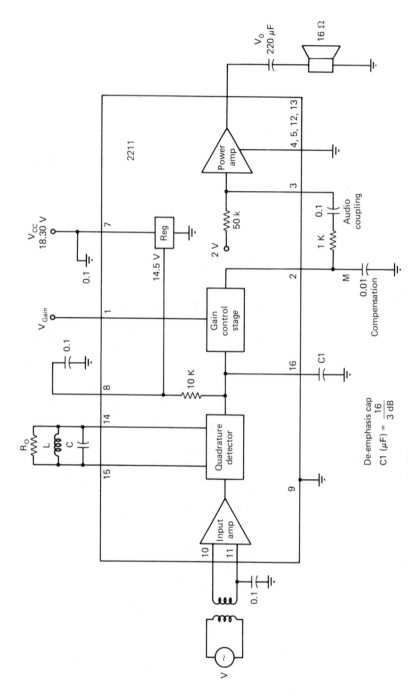

Figure 11-34 TV/FM sound channel IC block diagram. (Courtesy of Signetics Corporation.)

De-emphasis cap
$$C1 \ (\mu F) = \frac{16}{3 \ dB}$$

itors, diodes (one for AM detector) and other capacitors and resistors for
current limiting, by-passing and filtering.

The input signal from the antenna is applied at V_{iRF} and the audio
output at V_{AF} is applied to an audio amplifier stage to drive a speaker.

11-15 2-WATT TV/FM SOUND CHANNEL IC

Consumer electronics is a branch of the industry that covers television
receivers, stereo receivers, automotive sound systems, eight-track recorders
and players, cassette recorders and players, stereo hi-fi systems, video re-

Figure 11-35 TV/FM sound channel IC application.   (Courtesy of
Signetics Corporation.)

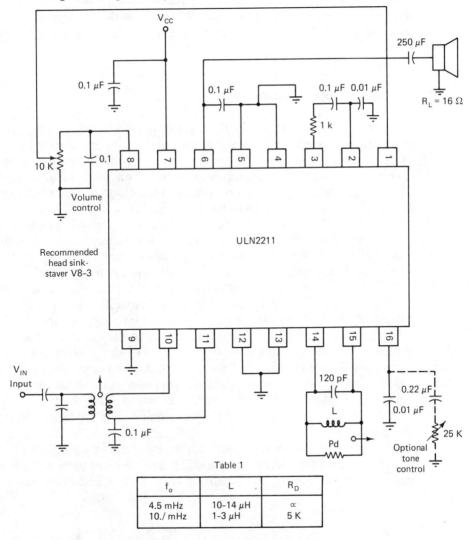

f_o	L	R_D
4.5 mHz	10–14 μH	$\propto$
10./ mHz	1–3 μH	5 K

corders, digital clocks, games, electronic musical instruments, and home/auto intrusion alarm systems. Home computers might be added to this list since so many microprocessing systems are finding their way into urban and suburban neighborhoods. Home entertainment systems now enjoy the use of ICs, which replace discrete devices and simplify the old methods of doing things. Many subsystems in IC form are used in this type of electronic equipment. One such subsystem is a 2-W TV/FM sound channel (see also Section 11-14).

The ULN2211 IC contains a limiting amplifier, an FM quadrature detector, a gain control stage, and a 2-W audio output amplifier. It is capable of detecting and amplifying any FM signal having a 0.1 to 20 MHz carrier frequency. Its primary use is a complete TV sound channel requiring few external components and only one tuning adjustment for the 4.5-MHz IF tank circuit. There are provisions for volume control and tone control, with special features of automatic thermal shutdown, overcurrent limiting, no crossover distortion, and operation from a single voltage supply from 18 to 30 V.

11-16 TESTING AND HANDLING ICs

Usually, ICs are tested within the circuit or system, which they are used by observing dc voltages and/or waveforms at the pin connections as indicated by a schematic diagram. Many ICs are too complex to test individually. However, some, containing circuits such as op amps, digital logic gates, flip-flops, counters, and timers, can be tested out of the circuit on special inexpensive checkers (see Sections 12-21, 13-41, and 13-42).

Once an IC is determined to be faulty, there comes the task of removing it from the PC board, which is sometimes a difficult and time-consuming job. Extreme care must be used when unsoldering an IC so as not to damage other circuit components or the PC board. Special desoldering tips are available that contact all pins simultaneously as the IC is rocked back and forth to free it from the board. A solder gobbler consisting of a nylon- or Teflon-coated nozzle and a rubber or plastic vacuum bulb can be used with a regular soldering iron. The iron is placed on a pin connection until the solder becomes molten. The iron is removed and the bulb of the gobbler is squeezed and placed on the connection, where it is released to draw the solder into the nozzle. Each connection is processed this way until all are fairly clean and then with a little added heat the IC is removed from the board. A similar method is to remove the solder with a hair-bristle soldering brush with a touch-and-wipe technique.

Three types of soldering techniques can be used with ICs on PC boards. In the straight-through method, the IC is held in place with its leads protruding through the holes. Molten solder is then allowed to flow over the lead

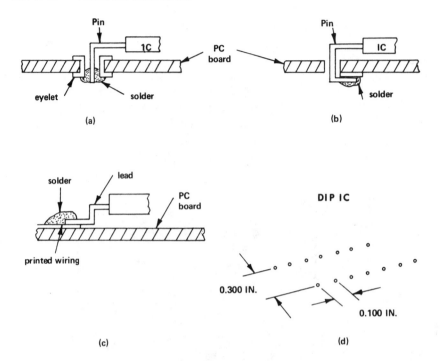

Figure 11-36 IC soldering techniques: (a) straight-through method; (b) bending method; (c) surface connection; (d) DIP mounting method.

and PC board circuit connection. The leads protruding through the holes may be bent slightly to contact the PC board connection (which also holds the IC in place during soldering) in the bending method. For the surface method of soldering, each IC lead has two reverse 90° bends. It is placed on the PC board and soldered. The bending process should be done with care using a special bending jig or long-nose pliers. A soldering iron of 50 W or less is usually acceptable for IC work. In any case, do not leave the iron on any connection longer than required for a satisfactory bond. The unnecessary additional heat could destroy the IC.

Since the pin connections are in close proximity, the connections should be cleaned with alcohol and a small hair-bristle brush after the soldering operation. This removes any residue that may short-out the pins of the IC, causing faulty circuit operation or possibly destroying the IC.

Special pin-straightening tools can be used to align IC pins. If such a tool is unavailable, long-nose pliers can be used to form the pins correctly; however, extreme care should be taken with this process.

Care is also required when working with circuits using IC sockets, so as not to bend the pins. When installing an IC into a socket, make sure that it is facing the direction corresponding to the circuit. Line up the pins of the

IC with the holes in the socket. A small tool may be needed to slightly bend some pins to match them. Then gently, with a rocking motion, push the IC into the socket.

Commercially made IC pullers are available for removing ICs from their sockets. A tool can be made from a pair of long tweezers by bending the tips 1/8 inch inward 90° (facing each other). The tips are placed under the IC at each end and with a rocking motion the IC is lifted free of the socket. If neither of these items is available, a small tool can be used to alternately pry up each end of the IC until it comes free of the socket.

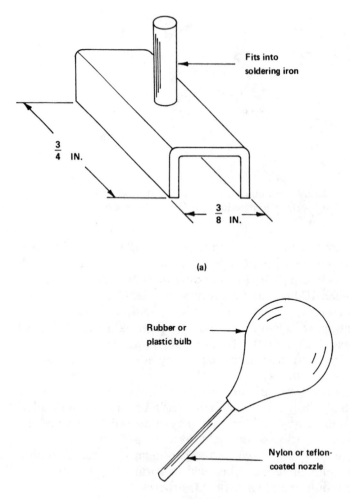

Figure 11-37 IC desoldering tools: (a) desoldering tip; (b) simple solder gobbler.

12

Operational Amplifiers

The *operational amplifier* (op amp) is a standardized circuit that can be wired externally various ways (with a minimum of components) to satisfy a wide variety of applications. Op-amp configurations include single input–single output and differential input–differential output, but the most commonly used is differential input–single output. Basically, it consists of a high-input-impedance differential amplifier, a high-gain voltage amplifier and a low-impedance output amplifier; therefore, its three main features are a very high input impedance, a very high open-loop gain, and a very low output impedance. Most op amps are manufactured in IC form. The schematic symbol is in the shape of a triangle.

In many cases an op amp requires a positive and negative voltage supply. The polarity relationship of the two input voltage determines the output polarity. If the inverting input is more positive than the noninverting input, the output will be negative. Similarly, if the inverting input is more negative than the noninverting input, the output will be positive. When the two inputs are at the same polarity and voltage level, the output will be 0 V.

Typically, an op amp may have an open-loop mode gain of 200,000. The slightest voltage difference at the inputs will cause the output to attempt to swing to the maximum power supply level. However, because of internal voltage drops, the output voltage will be about 90% of the supply voltages and is referred to as saturation voltage ($+V_{sat}$ or $-V_{sat}$). When a passive component such as a resistor is placed from the output to the inverting input (establishing a feedback path), the gain is reduced. This is referred to as the

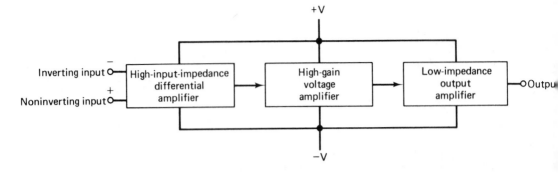

Figure 12-1 Block diagram of op amp.

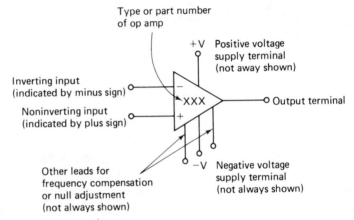

Figure 12-2 Standard op amp schematic symbol.

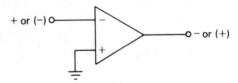

1. If − input is more positive than
   + input, output will be negative.

2. If − input is more negative than
   + input, output will be positive.

Figure 12-3 Input/output polarity
relationship.

closed-loop mode. The gain of the op amp can be accurately controlled b
the ratio of feedback resistance (R_F) to input resistance (R_i) as given by th
formula

$$A_v = -\frac{R_F}{R_i}$$

If R_F and R_i have the same value, $A_v = 1$ (unity gain), or if the feedbac
path is 0 Ω, $A_v = 1$.

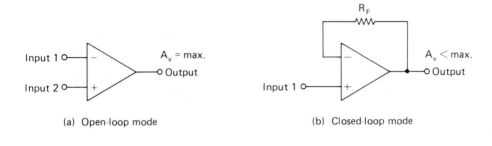

(a) Open-loop mode   (b) Closed-loop mode

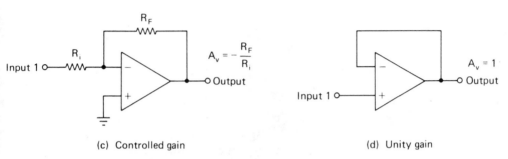

(c) Controlled gain   (d) Unity gain

Figure 12-4 Op-amp gain: (a) open-loop mode; (b) closed-loop mode; (c) controlled gain; (d) unity gain.

12-2 OP-AMP CHARACTERISTICS

Input impedance can be from 1 to 100 MΩ.

Output impedance ranges from 25 to a few thousand ohms.

An *output-offset voltage* occurs because of an input bias current. Both input currents should be equal to obtain zero output voltage; however, this

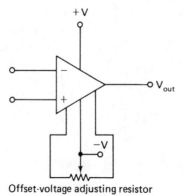

Offset-voltage adjusting resistor   **Figure 12-5** Offset nulling.

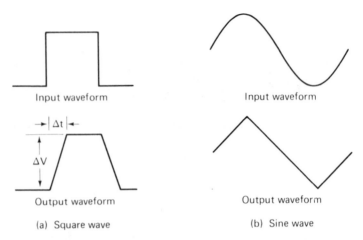

Figure 12-6 Example of slew-rate limiting on waveforms: (a) square wave; (b) sine wave.

is impossible and one input will have slightly more current than the other. This causes the output to be above or below zero. *Input-offset current* developed from the use of input-offset *voltage* can balance the inputs and the output will go to zero. This operation is called *offset nulling* and some op amps provide connections for this purpose. Some circuits using op amps are so critical that offset nulling is necessary because any error voltages at the inputs will be amplified by the gain of the op amp.

With high-frequency signals, the op amp may oscillate, so a small capacitor is added either internally or externally for *frequency compensation* which decreases the gain as frequency increases and prevents oscillations.

The *slew rate* is given thus:

$$\text{slew rate} = \frac{\text{maximum change in output voltage}}{\text{change in time}} = \frac{\Delta V_{out(max)}}{\Delta t}$$

Capacitance limits the slewing ability of the op amp and the output voltage will be delayed from that of the input voltage.

The *frequency response* of an op-amp circuit is inversely proportional to the gain. A gain of 100 might have a bandwidth of 10 kHz, while the same op-amp circuit with a gain of 10 might have a bandwidth of 100 kHz. The *unity gain frequency* is the point at which the gain equals 1. The *gain-band width product* = gain × bandwidth = unity-gain frequency. For example,

$$\text{GBP} = 100 \times 10 \text{ kHz} = 1,000,000 \text{ Hz (1 MHz)}$$

or

$$\text{GBP} = 10 \times 100 \text{ kHz} = 1,000,000 \text{ Hz (1 MHz)}$$

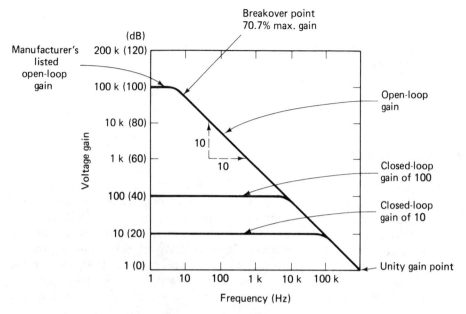

Figure 12-7 Voltage gain versus frequency.

Therefore, the *bandwidth* (BW) of a circuit with a specific gain can be found thus:

$$\text{bandwidth} = \frac{\text{unity-gain frequency}}{\text{gain}}$$

For example:

$$\text{BW} = \frac{1{,}000{,}000}{10} = 100 \text{ kHz}$$

The common-mode rejection ratio (CMRR) is the ability of the op amp to amplify the differential signal (A_D, voltage difference between inputs)

Figure 12-8 Common-mode rejection.

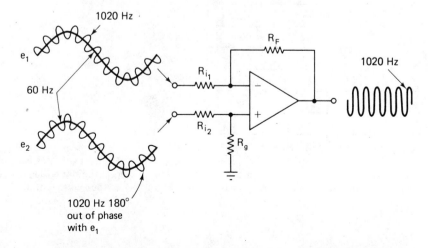

while rejecting the common-mode signal (A_{cm}), such as 60-Hz hum on both inputs and can be expressed as

$$CMRR = \frac{A_D}{A_{cm}}$$

12-3 VOLTAGE COMPARATOR

A *voltage comparator* compares the voltage on one input to the voltage on the other input. In this simplest circuit configuration, using the open-loop mode, any minute difference between the two inputs will drive the op-amp output into saturation. The direction in which the output voltage goes into saturation is determined by the polarity of the input signals. When the voltage on the inverting input is more positive than the voltage on the noninverting input, the output will swing to negative saturation ($-V_{sat}$). Similarly, when the voltage on the inverting input is more negative than

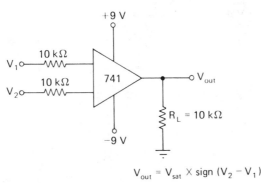

$$V_{out} = V_{sat} \times \text{sign}(V_2 - V_1)$$

(a) Schematic diagram

Input voltage		Output voltage
V_1	V_2	$\pm V_{sat}$
+1	+2	+8
+2	+1	−8
0	0	0
+1	−1	−8
−1	+1	+8
−1	−2	−8
−2	−1	+8

(b) Input/output voltage table

Figure 12-9 Voltage comparator: (a) schematic diagram; (b) input/output voltage table.

the voltage on the noninverting input, the output will swing to positive saturation ($+V_{sat}$).

The table shows that with a +1 V on the inverting input and +2 V on the noninverting input, the former is 1 V negative with respect to the latter and the output will go to $+V_{sat}$. If these input voltages are reversed, the inverting input is 1 V positive with respect to the noninverting input and the output will go to $-V_{sat}$. When the input voltages have the same amplitude and polarity, the output will be zero. Negative voltages applied to the inputs have the same effect on the output of the op amp. The important point to remember is that the polarity relationship of the inverting input to the noninverting input will cause the output to be $180°$ out of phase.

12-4 INVERTING AMPLIFIER

An *inverting amplifier* accepts a small input voltage (V_{in}) or current at its inverting input and produces a larger voltage or current $180°$ out of phase at the output (V_{out}). The voltage gain of the circuit is found by the formula

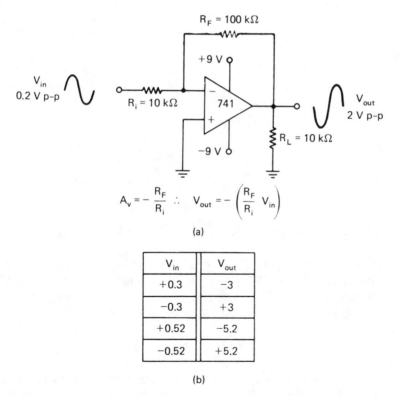

(a)

V_{in}	V_{out}
+0.3	−3
−0.3	+3
+0.52	−5.2
−0.52	+5.2

(b)

Figure 12-10 Inverting amplifier: (a) schematic diagram; (b) dc voltage table.

$$A_v = \frac{V_{out}}{V_{in}}$$

This circuit is in the closed-loop mode and the out-of-phase voltage at the output is fed back to the inverting input, where it tends to cancel the original input voltage. This type of feedback is referred to as *degenerative* or *negative feedback*. The feedback voltage will greatly reduce the effects of the input voltage and keep the inverting input at nearly 0 V. Of course, the feedback voltage cannot completely cancel the input voltage, and there is a slight difference of a few microvolts at the inverting input. The extremely high gain of the op amp produces the voltage amplification at the output. The amount of gain of the circuit is controlled by R_i and R_F and can be calculated by the formula

$$A_v = -\frac{R_F}{R_i} = \frac{100 \text{ k}\Omega}{10 \text{ k}\Omega} = -10$$

The minus sign only indicates the phase inversion for an inverting amplifier and is disregarded in the calculations. The output voltage can be found by the formula

$$V_{out} = A_v V_{in} = \frac{R_F}{R_i}(V_{in})$$

The inverting input at the op amp is referred to as a *virtual ground* and will tend to follow the potential at the noninverting input. The input resistance (or impedance) of the circuit is determined by R_i.

Formulas for finding current:

$$I_{in} = \frac{V_{in}}{R_i}, \qquad I_F = \frac{V_{out}}{R_F}, \qquad I_L = \frac{V_{out}}{R_L}$$

$$I_{out} = I_F + I_L, \qquad I_F = I_{in}$$

12-5 NONINVERTING AMPLIFIER

A *noninverting amplifier* accepts a small input voltage (V_{in}) at its noninverting input and produces a larger voltage in phase at the output (V_{out}). The circuit gain is determined the same as with the inverting amplifier, except that it has an increased factor of 1 and is given as

$$A_V = \frac{R_F}{R_i} + 1 = \frac{100 \text{ k}\Omega}{10 \text{ k}\Omega} + 1 = 10 + 1 = 11$$

Since the potential at the inverting input tends to be the same as the noninverting input, $V_{in} = V_A$ and the gain of the circuit can be expressed as

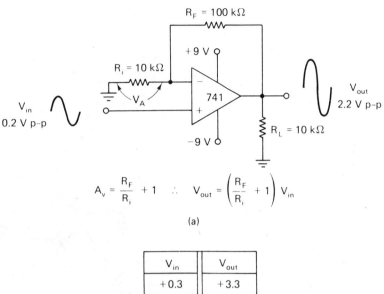

$$A_v = \frac{R_F}{R_i} + 1 \qquad \therefore \qquad V_{out} = \left(\frac{R_F}{R_i} + 1\right) V_{in}$$

(a)

V_{in}	V_{out}
+0.3	+3.3
-0.3	-3.3
+0.52	+5.72
-0.52	-5.72

(b)

Figure 12-11 Noninverting amplifier: (a) schematic diagram; (b) dc voltage table.

$$A_V = \frac{V_{out}}{V_{in}} \quad \text{or} \quad \frac{V_{out}}{V_A}$$

The output voltage can be found by the formula

$$V_{out} = \left(\frac{R_F}{R_i} + 1\right) V_{in}$$

The major advantage of the noninverting amplifier over the inverting amplifier is a higher input impedance, since R_i is used to determine the gain of the circuit and in many cases is required to be a low value.

12-6 VOLTAGE FOLLOWERS

The *voltage follower* is a circuit with a gain of 1, with the output voltage following the input voltage. It can be used for isolation or buffering and matching high impedance to low impedance.

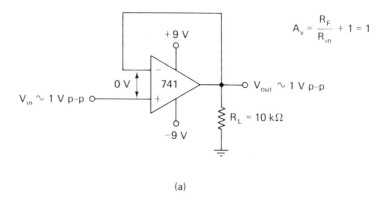

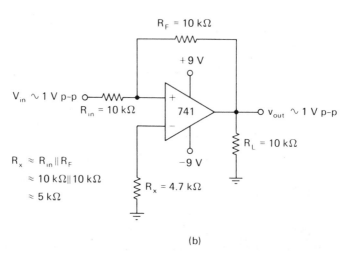

(b)

Figure 12-12 Voltage followers: (a) noninverting voltage follower; (b) inverting voltage follower.

In a *noninverting* voltage follower, the output is connected directly to the inverting input, with the input voltage at the noninverting input. The feedback resistance equals zero; therefore, according to the stage gain for a noninverting amplifier,

$$A = \frac{R_F}{R_{in}} + 1 = \frac{0}{R_{in}} + 1 = 1$$

This circuit has an extremely high input impedance and a low output impedance, which is ideal for buffering or isolation between circuits.

If a particular application required a voltage follower with phase inversion, an *inverting* amplifier with a gain of 1 could be used. In this case R_{in} equals R_F and the gain formula states:

$$A_v = -\frac{R_F}{R_{in}} = \frac{10\ k\Omega}{10\ k\Omega} = -1$$

The limitations of this circuit would be a greatly reduced input impedance, since R_{in} equals the input impedance. Resistor R_X at the noninverting input is used to reduce input bias currents and is equal to the value of R_{in} in parallel with R_F:

$$R_X = R_{in} \parallel R_F$$

12-7 VOLTAGE SUMMING AMPLIFIER

A *voltage summing amplifier* is an inverting amplifier with two or more inputs connected together. It can be noninverting or inverting; however, the inverting type is easier to design and construct. If the output needs to

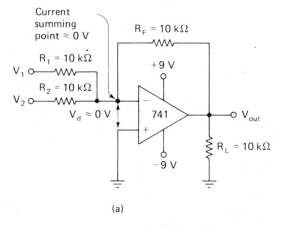

(a)

Input voltage		Output voltage
V_1	V_2	Algebraic sum
+1	+1	−2
+1	−1	0
+2	+1	−3
−1	+1	0
−1	+2	−1
−2	+1	+1

(b)

$$V_{out} = -\frac{R_F}{R_1}V_1 + \frac{R_F}{R_2}V_2 + \ldots + \frac{R_F}{R_n}V_n$$

When $R_1 = R_2 = R_F = \ldots R_n$

$$V_{out} = -(V_1 + V_2 + \ldots + V_n)$$

Figure 12-13 Voltage summing amplifier: (a) schematic diagram; (b) input/output voltage table.

be inverted, an inverting voltage follower can be used after the summing amplifier.

The output voltage of the inverting summing amplifier is inverted and equals the algebraic sum of each input voltage times the ratio of its appropriate input resistor to the feedback resistor, and can be expressed

$$V_{\text{out}} = -\frac{R_F}{R_1} V_1 + \frac{R_F}{R_2} V_2 + \cdots + \frac{R_F}{R_N} V_N$$

where R_N is the total number of inputs. If all of the external resistors equal one another ($R_F = R_1 = R_2 = \cdots = R_N$), the output voltage can be found simply by algebraically adding up the input voltages and then inverting the sign, which can be expressed as

$$V_{\text{out}} = -(V_1 + V_2 + \cdots + V_N)$$

The virtual ground at the inverting input of the op amp is now called the *current summing point* (≈ 0 V), which produces the algebraic summing effect of the circuit. All currents flowing in the inputs must algebraically equal the current flowing through R_F.

A summing amplifier with gain will have all input resistors equal the same value and the feedback resistor (R_F) will be greater than any single input resistor.

A scaling adder will have different value input resistors, so that some input voltages are amplified more than others.

12-8 VOLTAGE DIFFERENCE AMPLIFIER

The *voltage difference amplifier* is similar to the voltage comparator, but the circuit utilizes the closed-loop mode, which results in a controlled and predictable output voltage. Both inputs are used to sense a difference of potential between them, which appears at the output. The output voltage is the inverted algebraic difference between the two input voltages and can be found from the formula

$$V_{\text{out}} = -\frac{R_F}{R_1} V_1 + \left(\frac{R_g}{R_2 + R_g}\right)\left(\frac{R_1 + R_F}{R_1}\right) V_2$$

If the ratio of R_F to R_1 equals the ratio of R_g to R_2, which is normally the case, the output voltage can easily be found by the formula

$$V_{\text{out}} = -\frac{R_F}{R_1} (V_2 - V_1)$$

If all the external resistors are equal, the voltage-difference amplifier functions as an analog mathematical circuit and is often called a *voltage sub-*

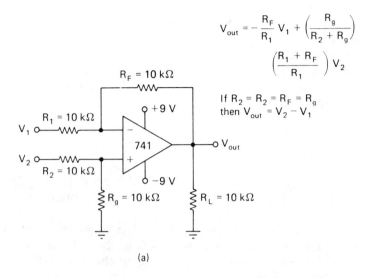

$$V_{out} = -\frac{R_F}{R_1} V_1 + \left(\frac{R_g}{R_2 + R_g}\right)$$

$$\left(\frac{R_1 + R_F}{R_1}\right) V_2$$

If $R_2 = R_2 = R_F = R_g$
then $V_{out} = V_2 - V_1$

$R_F = 10 \text{ k}\Omega$

$R_1 = 10 \text{ k}\Omega$

V_1

$+9 \text{ V}$

741

-9 V

$R_2 = 10 \text{ k}\Omega$

V_2

V_{out}

$R_g = 10 \text{ k}\Omega$ $R_L = 10 \text{ k}\Omega$

(a)

Input/voltage		Output/voltage
V_1	V_2	Algebraic difference
+2	+4	+2
+4	+2	−2
+4	−2	−6
−2	+4	+6
−4	+2	+6
+2	−4	−6
−4	−2	+2
−2	−4	−2

(b)

Figure 12-14 Voltage difference amplifier: (a) schematic diagram; (b) input/output voltage table.

tractor. In this case the output voltage formula is simply $V_{out} = -(V_2 - V_1)$. The first minus sign in the formulas indicates the output polarity (similar to the comparator). If the voltage at the inverting input is more positive than the voltage at the noninverting input, the output voltage will be negative, and vice versa.

The voltage difference amplifier has the ability to sense a small differential voltage buried within larger signal voltages. Unfortunately, the input impedance is very low and this type of circuit may require voltage followers for buffering or isolation.

12-9 INTEGRATOR

An *integrator circuit* continuously adds up a quantity being measured over a period of time. The output waveform is proportional to the time interval of the input signal.

A basic op-amp integrator consists of an input resistor to the inverting input and a capacitor in the feedback path from output to the inverting input. When a +dc voltage is placed at the input, the output will develop a negative-going linear ramp voltage. Even when the input voltage is at 0 V, the input bias current will cause the capacitor to charge until the output reaches saturation and the circuit is unusable. A practical op-amp integrator will have a large-value feedback resistor in parallel with the feedback capacitor to prevent saturation and provide reduced noise, less offset drift, and better stability.

As long as V_{in} is constant, the output voltage can be found using the formula

$$V_{out} = - \frac{1}{R_{in}C_F} \int_0^t V_{in}\,dt$$

Figure 12-15 Practical op-amp integrator.

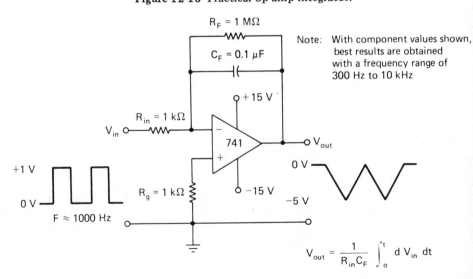

Note: With component values shown, best results are obtained with a frequency range of 300 Hz to 10 kHz

$$V_{out} = \frac{1}{R_{in}C_F} \int_o^t d\,V_{in}\,dt$$

The integral sign $\int_0^t$ indicates the period or limits of integration to be calculated where V_{in} is constant and dt is the time or period of integration. The minus sign indicates only that the output is 180° out of phase with the input.

Since capacitive reactance (X_c) varies with frequency, less signal is fed back for lower frequencies, while at high frequencies more signal is fed back. Because of the change in X_c with frequency, the integrator circuit behaves like a low-pass filter.

12-10 DIFFERENTIATOR

The *differentiator* is the reverse concept of that of the integrator (see Section 12-9), where the output voltage is proportional to the rate of change of the input voltage. The basic op-amp differentiator consists of a capacitor to the inverting input and a feedback resistor from the output to the inverting input. If a triangle voltage waveform is applied to the input, the output will develop a square-wave voltage, which can be found by the formula

$$V_{out} = -2R_F C_{in} \frac{dV_{in}}{dt}$$

When a square wave is applied to the differentiator input, a sharp leading-edge pulse will be developed at the output, which can be used to trigger or drive other circuits.

Since the capacitive reactance (X_c) of the input signal decreases with an increase in signal frequency, the differentiator behaves like a basic high-pass filter. This action results because when X_c decreases, the gain of the circuit increases and is expressed as

$$A_v = - \frac{R_F}{X_c}$$

Figure 12-16 Practical op-amp differentiator.

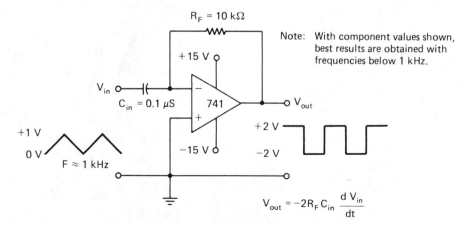

A *low-pass filter* has a constant output voltage from dc up to a specific *cutoff frequency* (f_c). This cutoff frequency occurs at the half-power point, or 70.7% of the maximum output voltage, and is referred to as the cutoff frequency, the 0.707 frequency, the −3-dB frequency, the corner frequency, or the breakpoint frequency. Frequencies above the f_c are attenuated (decreased). The range of frequencies below f_c are called the pass band, and the frequencies above f_c are known as the stop band.

An ideal cutoff is indicated by the dashed lines on the frequency-response curve; however, filters are usually not this efficient and tend to roll off or even peak and then roll off. Op-amp filters can be designed to have different roll-off characteristics, resulting in various slopes. A slope of −20 dB/decade means that as the frequency increases by ten (×10) from f_c, the output voltage will decrease 20 dB. The more decibel loss per decade results in a steeper slope, which is most desirable. With the −40-dB/decade low-pass filter shown, capacitor C_2 shunts current away from the input for frequencies above f_c. The X_c of capacitor C_1 is low for frequencies in this stop-band range and more negative feedback is applied at the input, which reduces the gain of the circuit.

The cutoff frequency for the −40-dB/decade low-pass filter can be found by the formula

$$f_c \approx \frac{1}{2\pi \sqrt{R_1 R_2 C_1 C_2}}$$

Figure 12-17 Low-pass-filter frequency-response curve.

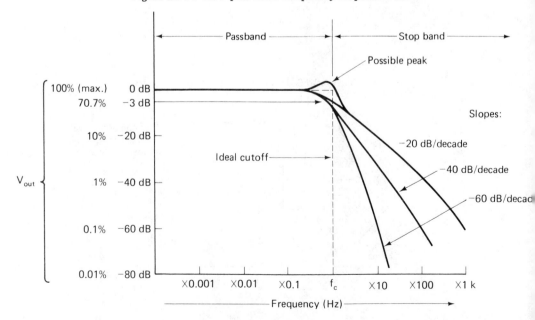

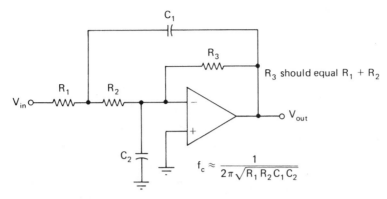

R_3 should equal $R_1 + R_2$

$$f_c \approx \frac{1}{2\pi\sqrt{R_1 R_2 C_1 C_2}}$$

Figure 12-18 -40 dB/decade low-pass filter.

12-12 ACTIVE HIGH-PASS FILTER

A *high-pass filter* performs the opposite function of that of a low-pass filter (see Section 12-11). The high-pass filter attenuates all frequencies below a specific cutoff frequency (f_c), while passing all frequencies above f_c. The f_c for a high-pass filter occurs at 70.7% of the maximum output voltage.

In the -40-dB/decade high-pass filter shown, C_1 should equal C_2 and R_2 should be twice as large as R_1. Resistor R_3 should equal R_2 and is used

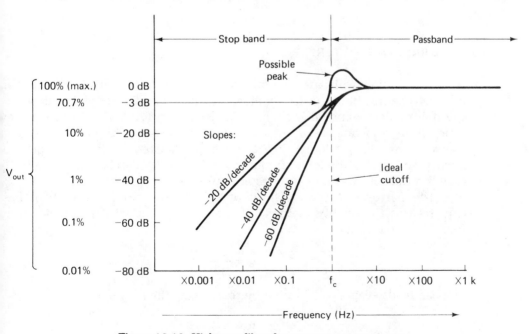

Figure 12-19 High-pass-filter frequency-response curve.

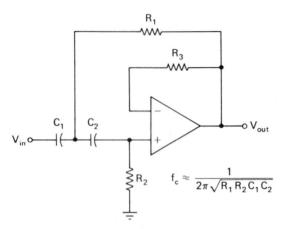

$$f_c \approx \frac{1}{2\pi\sqrt{R_1 R_2 C_1 C_2}}$$

Figure 12-20 −40 dB/decade high-pass filter.

for dc offset. With V_{in} connected to the noninverting input, C_1, C_2, and R_2 form a voltage divider. When V_{in} is below f_c, the X_c of C_1 and C_2 is large and drops most of V_{in}. The voltage drop across R_2 is low and V_{out} is low When V_{in} increases above f_c, the X_c of C_1 and C_2 decreases, allowing more V_{in} to be dropped across R_2; hence V_{out} is larger.

The cutoff frequency for the −40-dB/decade high-pass filter can be found by the formula

$$f_c \approx \frac{1}{2\pi\sqrt{R_1 R_2 C_1 C_2}}$$

12-13 ACTIVE BANDPASS FILTER

A *bandpass filter* will pass a certain group of frequencies while rejecting all others. The maximum output voltage of this type of filter will peak at one specific frequency known as the resonant frequency (f_r). When the frequency varies from resonance, the output voltage decreases. The point above and below f_r that V_{out} falls to 70.7% can be designated f_H for the upper frequency and f_L for the lower frequency. The frequencies between f_L and f_H establish the bandwidth of the circuit (BW = f_H − f_L).

The narrower the bandwidth of a filter, the more selectivity it has expressed as the quality factor Q of the circuit. The Q of a circuit can be found by the formula

$$Q = \frac{f_r}{\text{BW}}$$

The bandwidth can also be found by rearranging the formula:

$$\text{BW} = \frac{f_r}{Q}$$

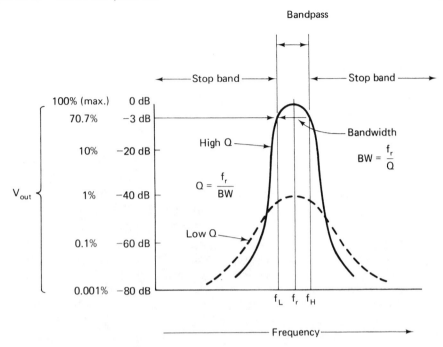

Figure 12-21 Bandpass filter frequency-response curve.

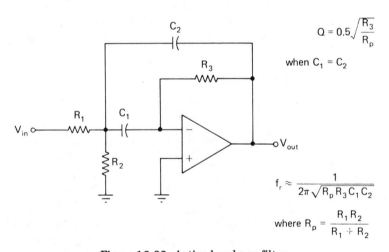

Figure 12-22 Active bandpass filter.

A high-Q filter has a narrow bandwidth and tends to have a large V_{out}, whereas a low-Q filter has a wider bandwidth and tends to have a smaller V_{out}.

An active bandpass filter is a combintion of low-pass and high-pass circuits with a resonant frequency that can be found by the formula

$$f_r \approx \frac{1}{2\pi \sqrt{R_p R_3 C_1 C_2}}, \qquad \text{where } R_p = \frac{R_1 R_2}{R_1 + R_2}$$

The Q of the filter can be found with the formula

$$Q = 0.5\sqrt{\frac{R_3}{R_p}}, \qquad \text{when } C_1 = C_2$$

12-14 ACTIVE NOTCH (BANDREJECT) FILTER

A *notch filter*, sometimes called a *bandreject filter*, functions opposite to that of a bandpass filter (see Section 12-13). The output voltage will remain fairly constant until the applied frequency approaches f_r, at which point it is attenuated. As the applied frequency increases, the output voltage will again increase to its previous level. The stop band (bandwidth) occurs for amplitudes less than 70.7% of V_{out} and the same Q and BW formulas used with the bandpass filter also apply to the notch filter.

One type of notch filter has V_{in} applied to both inputs. Resistors R_2 and R_3 form a voltage divider, with R_3 typically 50 times greater than R_2. Components R_1, R_4, C_1, and C_2 form a frequency-selective feedback network. For frequencies below f_r, the X_c of the capacitors is very high and there is little feedback; thus the output is maximum. When the frequency of V_{in} approaches f_r, the reactance and resistance form the appropriate relationship and phase angle to produce feedback which decreases the output.

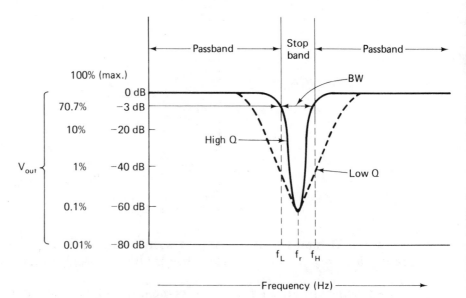

Figure 12-23 Notch filter frequency-response curve.

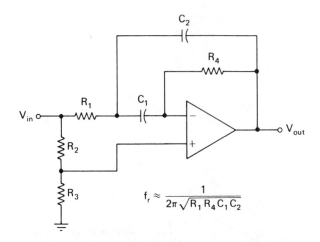

$$f_r \approx \frac{1}{2\pi \sqrt{R_1 R_4 C_1 C_2}}$$

Figure 12-24 Active notch filter.

As the frequency of V_{in} increases above f_r, the X_C of the capacitors decrease and the feedback factor approaches 1, or the gain of a voltage follower, and V_{out} increases to its previous level.

The f_r of the notch filter can be found from the formula

$$f_r = \frac{1}{2\pi \sqrt{R_1 R_4 C_1 C_2}}$$

The Q of the filter can be found by the formula

$$Q = 0.5\sqrt{\frac{R_4}{R_1}}$$

12-15 SQUARE-WAVE GENERATOR

The *square-wave generator* is also referred to as a *free-running* or *astable multivibrator*. Its output is constantly changing states (high and low) without any input signal. There are two feedback paths for this circuit. One goes from the output to the noninverting input and contains R_2 and R_3, which establishes the reference voltage (V_{ref}) at that input. The other feedback path, which goes from the output to the inverting input, contains R_1 and C and determines the fundamental operating frequency of the generator. If R_3 is selected so that it is 86% of the value of R_2, the approximate frequency can be found by the formula

$$f_{out} = \frac{1}{2R_1 C}$$

When power is first applied to the circuit, the inverting input will be 0 V; therefore, V_{out} will be $+V_{sat}$ and the noninverting input will be at the positive threshold ($+V_T$) reference voltage. The capacitor begins to charge

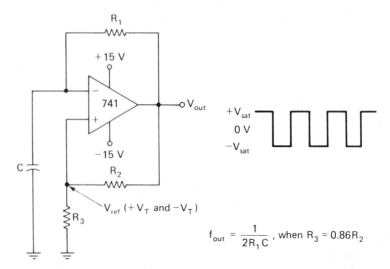

Figure 12-25 Basic square-wave generator.

toward $+V_{sat}$ through R_1. When the charge of C increases to a fraction greater than $+V_T$ on the noninverting input, the comparator action of the op amp causes the op-amp output to switch states and go to $-V_{sat}$. The V_{ref} at the noninverting input is now at the negative threshold $(-V_T)$. The capacitor now changes its charging direction and begins to charge toward $-V_{sat}$. The instant the charge on the capacitor goes below $-V_T$, the op amp switches back to the original state and V_{out} goes to $+V_{sat}$. A cycle has now been completed and the process is repeated. The threshold voltages can be determined by the formula

$$\pm V_T = 0.46(\pm V_{sat}), \quad \text{when } R_3 = 0.86R_2$$

12-16 SAWTOOTH-WAVE GENERATORS

A *sawtooth-wave generator*, sometimes referred to as a *ramp-voltage generator*, is similar to an integrator circuit (see Section 12-9). The basic action of the circuit is that a minus voltage at the input will cause the capacitor to charge up linearly toward $+V_{sat}$. At some point, the PUT, acting as a switch, momentarily closes to discharge the capacitor and the cycle is repeated, thus producing a sawtooth waveform at the output.

Voltages V_{ref} and V_p partly determine the frequency of the generator. Resistors R_1 and R_2 are used to develop V_{ref}. Diodes D_1 and D_2 help to stabilize the voltage across R_2 when it is adjusted to vary V_{ref}. The firing point (V_p) on the gate of the PUT is set by R_4. When the voltage across the PUT (V_{AK}) exceeds V_p, the PUT fires (turns on) and a new cycle is initiated.

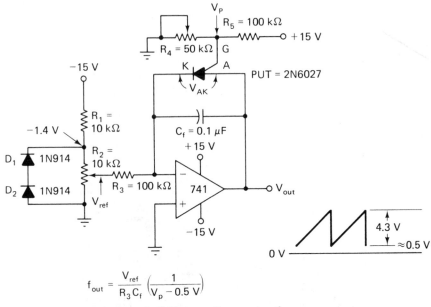

$$f_{out} = \frac{V_{ref}}{R_3 C_f} \left(\frac{1}{V_p - 0.5\ V} \right)$$

Figure 12-26 Self-generating sawtooth-wave generator.

The output frequency can be approximated with the formula

$$f_{out} \approx \frac{V_{ref}}{R_3 C_f} \left(\frac{1}{V_p - 0.5\ V} \right)$$

Since the input voltage plays a part in f_{out}, a circuit such as this is sometimes referred to as a voltage-to-frequency converter or a voltage-controlled oscillator.

12-17 TRIANGLE-WAVE GENERATOR

The *triangle-wave generator* usually requires at least two op amps to operate. A basic circuit consists of a square-wave generator (see Section 12-15) connected to a ramp generator (see Section 12-9).

When the output of the square-wave generator goes positive, the output of the ramp generator goes negative. Similarly, when the output of the square-wave generator goes negative, the output of the ramp generator goes positive.

The frequency of the triangle-wave output is the same as the frequency of the square-wave generator, which can be determined from Section 12-15. However, it is desirable to have the RC time constant of R_4 and C_2 be twice as large as the time constant of R_1 and C_1 to prevent distortion of the triangle waveform. The amplitude of the square wave will be nearly $\pm V_{sat}$,

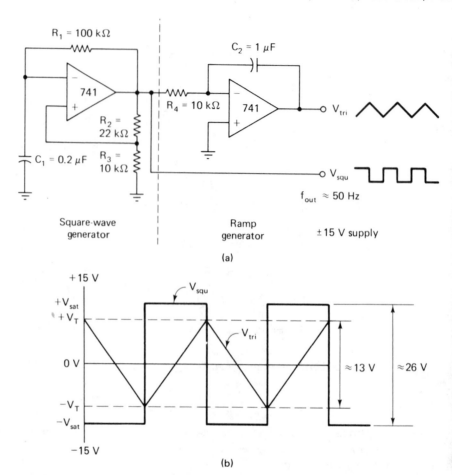

Figure 12-27 Simple triangle-wave generator: (a) combination of two basic circuits; (b) output waveforms.

while the amplitude of the triangle wave can be determined from the integrator circuit in Section 12-9. A signal generator, such as this, which produces two or more different output waveforms, is sometimes referred to as a *function generator.*

12-18 SINE-WAVE OSCILLATOR

A *sine-wave oscillator* can be produced using two op amps, one a bandpass filter and the other a voltage comparator. Since a square wave contains a sine wave and its odd harmonics, filtering a square wave will produce a sine wave. The comparator is fed with a sine wave from the bandpass filter to obtain a square-wave output. The square wave is fed back to the input of

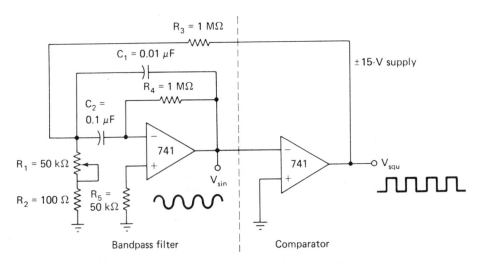

Figure 12-28 Two-op-amp sine-wave oscillator.

the bandpass filter to cause oscillation and produce the fundamental sine wave.

The output frequency can be determined by the formula

$$f_{\text{out}} = \frac{1}{2\pi \sqrt{R_p R_4 C_1 C_2}}, \quad \text{where } R_p = \frac{R_1 R_3}{R_1 + R_3}$$

Resistor R_2 can be considered part of R_1, but is negligible, because of its extremely small value. It is used only to keep from shorting the feedback to ground.

The frequency range of the oscillator with the component values shown is about 7 to 1.6 kHz and can be adjusted by R_1.

Two outputs are provided by this circuit, a sine wave (V_{sin}) from the bandpass filter and a square wave (V_{squ}) from the comparator.

12-19 QUADRATURE OSCILLATOR

A *quadrature oscillator* produces two sine-wave outputs 90° out of phase. The sine output comes from op amp 1 and the cosine output comes from op amp 2. Basically, the circuit consists of two integrators with positive feedback via R_3. Since the phase shift of an integrator is 90°, the cosine output is 90° out of phase with the sine output.

Resistor R_1 is usually slightly less in value than R_3, to ensure that the circuit oscillates. If R_1 is too small in value, the outputs will be clipped and resemble square waves. A potentiometer may be used to adjust for minimum distortion of the output voltages. The output voltages may also

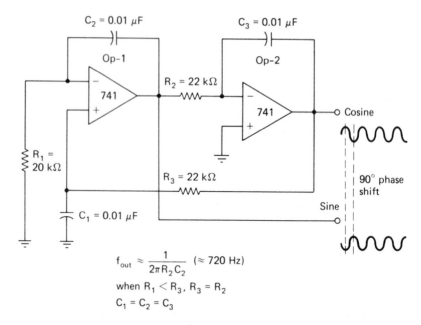

$$f_{out} \approx \frac{1}{2\pi R_2 C_2} \quad (\approx 720 \text{ Hz})$$

when $R_1 < R_3$, $R_3 = R_2$

$C_1 = C_2 = C_3$

Figure 12-29 Basic quadrature oscillator.

reach op-amp saturation. If this situation prevails, two zener diodes may be connected face to face across C_3 to limit the output.

When $R_2 = R_3$ with $R_1 < R_3$ and $C_1 = C_2 = C_3$, the output frequency can easily be found by the expression

$$f_{out} = \frac{1}{2\pi R_2 C_3}$$

12-20 FUNCTION GENERATOR

A *function generator* with three outputs—sine wave, square wave, and triangle wave—can be constructed from basic op-amp circuits. A sine wave/square wave generator (see Section 12-18) is used to establish f_{out}. The sine-wave output is taken at the output of op amp 1. The comparator is fed to a noninverting voltage follower to prevent loading down the basic oscillator and affecting its frequency. The square-wave output is taken from the output of op amp 3 (see Section 12-6). This square wave is connected to an integrator to produce the triangle-wave output at the output of op amp 4 (see Section 12-9).

This generator can be built from a single 14-pin DIP 324 (quad op amp) IC. This pin identification is shown to facilitate construction. A ±15-V

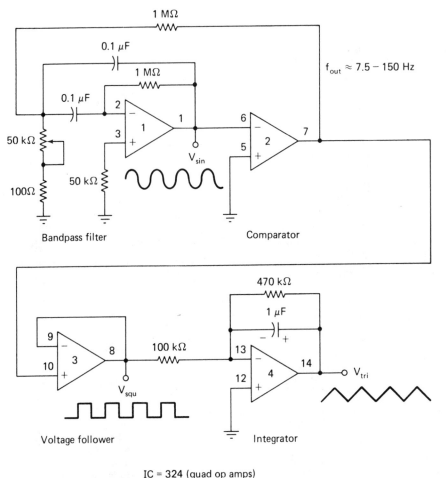

Figure 12-30 Basic low-frequency trifunction generator.

power supply is used, which results in amplitudes of the following output voltages:

square wave = 26 V p-p

sine wave = 16 V p-p

triangle wave = 0.3–6 V p-p (depending on f_{out})

Adjusting the 50-kΩ potentiometer will vary f_{out} from about 7.5 to 150 Hz. Decreasing the value of the two capacitors will increase f_{out}, and if their values are equal, it simplifies calculating f_{out}.

A voltmeter is used to test the dc voltages of an op-amp circuit. The supply voltage should be checked right at the $+V$ and $-V$ pins of the IC. Improper readings may indicate that the IC is bad; however, the power supply could be defective or another circuit is upsetting the power to all circuits.

One method of testing an op amp is to short the inputs together, which causes the differential input voltage to be zero; therefore, the output voltage should also drop to zero. If this does not occur, the op amp is defective. There are a few precautions to be observed with this technique:

1. Make sure the pins being shorted are the inputs, because shorting other pins might destroy the IC.
2. Open the output circuit if the op amp is coupled directly to another circuit. The resulting test could damage following circuits.
3. This test should be used only on circuits using a dual-polarity power supply.

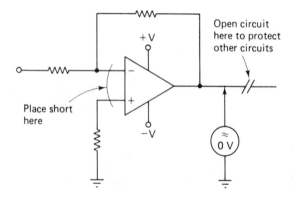

Figure 12-31 Testing for zero-volts output when inputs are shorted.

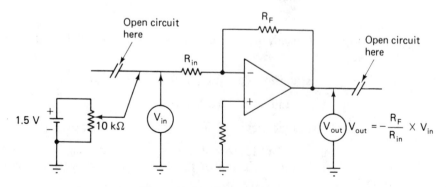

Figure 12-32 Testing stage dc gain.

The dc gain of an op-amp amplifier can be tested by opening the input and output circuits and then placing a small dc voltage at the input. The dc input voltage could be a 1.5-V battery with a 10-kΩ potentiometer across it. The gain formula can be applied as the output is monitored with a voltmeter.

An oscilloscope is very useful in tracing a signal through op-amp circuitry. The oscilloscope is placed from input to output of each op amp until the defective circuit is located.

An automatic LED GO/NO GO op-amp checker is easily constructed and can be used to test new ICs or ones that are removed from a circuit. The checker will require an IC socket and two 9-V batteries to make it portable. The circuit consists of a low-frequency square-wave generator and two reverse-connected LEDs at the output of the device under test (D.U.T.).

When the test is conducted, the output of the op amp should swing alternately negative and positive, which will cause LED_1 and LED_2, respectively, to light.

If one or both of the LEDs fail to light, the op amp has an internal problem, which may be an open circuit.

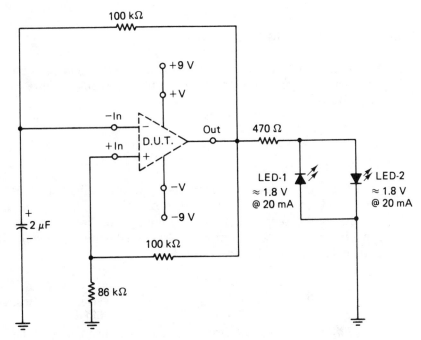

Figure 12-33 Automatic LED GO/NO GO op-amp checker.

13

Digital Circuits

13-1 INFORMATION ON DIGITAL CIRCUITS

All digital logic circuits, whether in a simple circuit or complex computer arrangement, are made up of basic building-block circuits called logic gates and flip-flops.

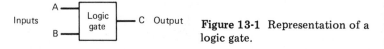

Figure 13-1 Representation of a logic gate.

Logic gates have two or more inputs and a single output. The output is controlled by the various conditions at the inputs. Inputs and output conditions are represented by the binary numbers 0 and 1. A gate can be considered on if its output is 1. Logic gates may turn on, with all inputs: at 0, at 1, or a combination of both.

Figure 13-2 Representation of a flip-flop.

Flip-flops are multivibrators that can be in one of two states: on or off. The flip-flop may have one, two, or three inputs, and usually two outputs.

It may take one or two inputs to turn a flip-flop on or off. When a flip-flop is on, its Q output is 1 and the $\overline{Q}$ output is 0. When the flip-flop is off, the opposite is true: $Q = 0$ and $\overline{Q} = 1$.

All inputs and outputs of digital circuits should be in one or the other of two states: 1 or 0. These states may also be given other names:

1	*0*
High	Low
ON	OFF
Closed	Open
Positive	Negative (or ground)
Yes	No
True	False

A particular circuit may consist of discrete logic gates and flip-flops wired together, or fabricated in an integrated circuit (IC), or a combination of both. The current trend in the electronics industry is toward IC arrangements. Two types of ICs are transistor–transistor logic (TTL or T^2L) and complementary metal-oxide semiconductor (CMOS).

The TTL IC consists of bipolar transistors and usually requires a dc supply voltage of +5 V. The CMOS IC uses MOSFETs and usually requires a dc supply voltage in the range +3 to +15 V. CMOS ICs have a lower power consumption than TTL ICs.

Digital ICs are generally found in four types of packages. In the standard transistor TO-92 and TO-5 cases, there may be up to eight leads protruding from the bottom. When the flat side or tab is held upward, the pins are numbered counterclockwise. With the dual In-line Package (DIP) and the flat pack, the notch or dot is placed to the bottom left and the pins are numbered counterclockwise from the lower left.

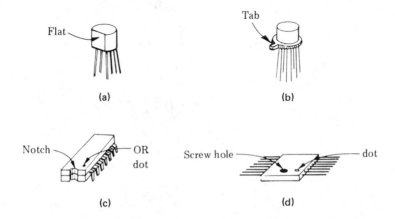

Figure 13-3 IC packages: (a) cylindrical ceramic case; (b) cylindrical metal case; (c) dual-in-line plastic case; (d) 16-pin flat-pack case.

13-2 AND GATE

The *AND function* requires that two or more conditions be met before desired result is obtained. In a basic electrical circuit containing a battery two switches in series, labeled A and B, and a lamp C, both switches must b closed for the lamp to light. A basic logical AND gate has two inputs, labele A and B, with an output C. Both inputs must be a 1, simultaneously, for th output also to be a 1. The Boolean expression $A \cdot B = C$ means "A and equals C." The mathematical symbol for multiplication ($\cdot$) stands for th

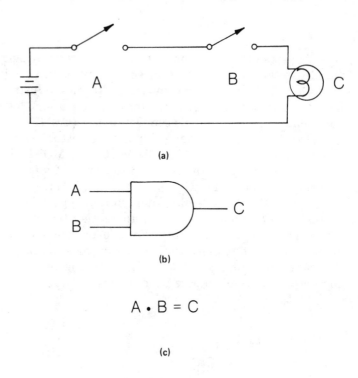

(a)

(b)

$$A \cdot B = C$$

(c)

INPUTS		OUTPUT
A	B	C
0	0	0
0	1	0
1	0	0
1	1	1

(d)

Figure 13-4 AND gate: (a) electrical circuit; (b) logic symbol; (c) Boolean expression; (d) truth table.

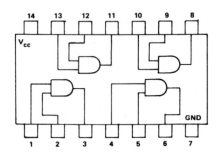

Figure 13-5 7408 quad two-input AND gate IC.

AND operation in Boolean algebra. A truth table indicates the conditions that the inputs must be to get an output of 1.

A 7408 quad two-input AND gate IC has four separate AND gates within the package. Pin 14, marked V_{cc}, is connected to +5 V and pin 7, marked GND, is connected to ground.

To summarize, the output of an AND gate will be high when all of its inputs are high.

13-3 OR GATE

The *OR function* requires that any one of two or more conditions be met before a desired result is obtained. In a basic electrical circuit containing a battery, two switches in parallel, labeled A and B, and a lamp C, closing either switch will cause the lamp to light. A basic logical OR gate has two

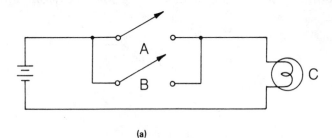

(a)

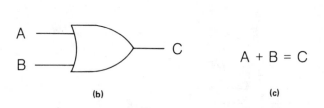

INPUTS		OUTPUT
A	B	C
0	0	0
0	1	1
1	0	1
1	1	1

$A + B = C$

(b)   (c)   (d)

Figure 13-6 OR gate: (a) electrical circuit; (b) logic symbol; (c) Boolean expression; (d) truth table.

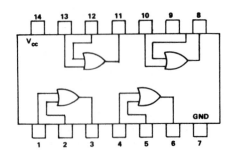

Figure 13-7 7432 quad two-input OR gate IC.

inputs, labeled A and B, with an output C. Either input must be a 1 for the output to be a 1. The Boolean expression $A + B = C$ means "A or B equals C." The mathematical symbol for addition (+) stands for the OR operation in Boolean algebra. A truth table indicates the conditions that the inputs must be to get an output of 1.

A 7432 quad two-input OR gate IC has four separate OR gates within the package. The power supply connections are indicated at pins 7 and 14.

To summarize, the output of an OR gate will be high when any one of its inputs is high.

13-4 EXCLUSIVE-OR GATE/EXCLUSIVE-NOR GATE

The OR gate of Section 13-3 is termed an *inclusive-OR gate*, since any or all of its inputs with a 1 will produce a 1 at the output. The *exclusive-OR gate* will only produce a 1 at its output, when only one input is 1 and all other

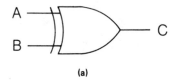

(a)

$$A\bar{B} + \bar{A}B = C$$

(b)

INPUTS		OUTPUT
A	B	C
0	0	0
0	1	1
1	0	1
1	1	0

Figure 13-8 Exclusive-OR gate: (a) logic symbol; (b) truth table.

(c)

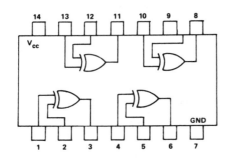

Figure 13-9 7486 quad two-input Exclusive-OR gate.

inputs are 0. When more than one input is 1, the output will be 0. The truth table shows this relationship. Its Boolean expression reads "*A* and *B* NOT or *A* NOT and *B* equal *C*." The equation can also be written $A \oplus B = C$, where the circle around the + indicates exclusive. The exclusive-OR function is recognized by an extra line drawn across the inputs on the logic symbol.

A 7486 quad two-input exclusive-OR gate IC has four separate exclusive-OR gates within the package. The power supply connections are indicated at pins 7 and 14.

To summarize, the output of an exclusive-OR gate will be high when any one input is high and all other inputs are low.

Adding an inverter after an exclusive-OR gate produces an *exclusive-NOR gate*. The output will be 1 when both inputs are 0 or both inputs are 1. This logic function serves as an equality detector. The output will be 1 when input *A* equals input *B*.

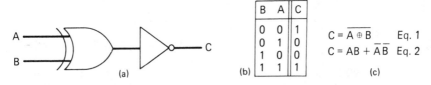

Figure 13-10 Exclusive-NOR gate: (a) logic symbol; (b) truth table; (c) Boolean expressions.

13-5 BUFFER/DRIVER AND INVERTER

A *buffer/driver* represented by a triangle has a single input and output. The output follows the input where 0 in equals 0 out and 1 in equals 1 out. This noninverting gate is used to drive other gates and output-indicating devices or to serve as a buffer, isolating different circuits.

The *inverter* also has a single input, but has a small bubble at the apex of the triangle which indicates negation. Sometimes referred to as a NOT gate, the output is the complementary (or opposite) of the input. When the input is 0, the output will be 1 and when the input is 1, the output will be 0.

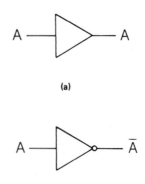

(a)

(b)

INPUT	OUTPUT
A	$\overline{A}$
0	1
1	0

(c)

Figure 13-11 Buffer/driver and inverter: (a) driver; (b) inverter; (c) inverter truth table.

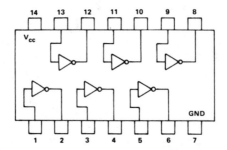

Figure 13-12 7404 hex inverter IC.

If A is the input, $\overline{A}$ is the output. The bar above the A is the Boolean expression for negation and is read "A NOT." Some inverters may be shown with the bubble at the input; however, the logic function remains the same, and this indicates only the signal condition for a desired result.

A 7404 hex inverter IC has six separate inverters within the package. The power supply connections are indicated at pins 7 and 14.

To summarize, the output of a buffer/driver is the same as its input, while the output of an inverter is opposite to its input.

13-6 NAND GATE

An AND gate followed by an inverter produces a "NOT-AND," which is shortened to *NAND*. The inverter triangle is removed and only the bubble appears at the output of a NAND gate logic symbol. The Boolean expression for the AND gate is $A \cdot B = C$. After going through an inverter, it would be

342

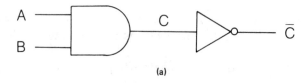

(a)

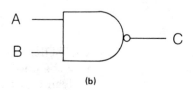

(b)

$$\overline{A} + \overline{B} = C$$

(c)

INPUTS		OUTPUT
A	B	C
0	0	1
0	1	1
1	0	1
1	1	0

(d)

Figure 13-13 NAND gate: (a) AND gate with inverter; (b) logic symbol; (c) Boolean expression; (d) truth table.

$\overline{AB} = C$, and would read "A and B NOT equal C." A Boolean algebra conversion known as DeMorgan's theorem shows that $\overline{AB}$ can be written as $\overline{A} + \overline{B}$, and reads "NOT A or NOT B." Therefore, the Boolean expression $\overline{A} + \overline{B} = C$ reads that "NOT A or NOT B" will produce an output at C for a NAND gate. Inspecting the truth table shows that when any NAND gate input is 0, the output will be 1. The NAND gate can also be considered a negated input OR gate. The NAND function is the most used of all the other functions, since it can be used to build or create the other functions.

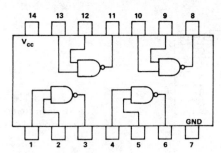

Figure 13-14 7400 quad two-input NAND gate IC.

343

A 7400 quad two-Input NAND gate IC has four separate NAND gate
within the package. The power supply connections are indicated at pins
and 14.

To summarize, the output of the NAND gate will go high when any on
of its inputs go low.

13-7 NOR GATE

An OR gate followed by an inverter produces a "NOT-OR," which is shortene
to *NOR*. The inverter triangle is removed and only the bubble appears at th
output of the NOR gate logic symbol. The Boolean expression for an OI
gate is $A + B = C$. After going through an inverter, it would be $\overline{A + B} = C$
and would read "A or B NOT equals C." A Boolean algebra conversio
known as DeMorgan's theorem shows that $\overline{A + B}$ can be written as $\overline{A} \cdot \overline{B}$ an
reads "NOT A and NOT B." Therefore, the Boolean expression $\overline{A} \cdot \overline{B} = 6$
reads that "NOT A and NOT B" will produce an output of C for a NOI

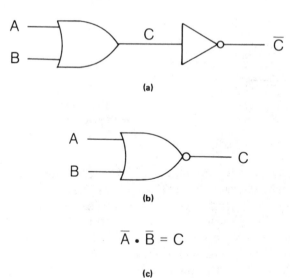

(a)

(b)

$$\overline{A} \cdot \overline{B} = C$$

(c)

INPUTS		OUTPUT
A	B	C
0	0	1
0	1	0
1	0	0
1	1	0

(d)

Figure 13-15 NOR gate: (a) NOR
gate with inverter; (b) logic symbe
(c) Boolean expression; (d) truth
table.

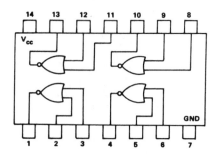

Figure 13-16 7402 quad two-input NOR gate IC.

gate. Inspecting the truth table shows that when all inputs are 0, the output will be 1. The NOR gate can also be considered a negated input AND gate.

A 7402 quad two-input NOR gate IC has four separate NOR gates within the package. The power supply connections are indicated at pins 7 and 14.

To summarize, the output of the NOR gate will go high when all of its inputs go low.

13-8 EQUIVALENT LOGIC GATES

Combining the various logic gates can produce other basic logic functions. A negated input OR gate becomes a NAND gate. A negated input AND gate becomes a NOR gate. Cross-coupling the inputs through inverters to AND gates, which in turn feed an OR gate, produces an exclusive-OR gate.

Inverters can be produced from NAND and NOR gates. If one input to a NAND gate is permanently placed at a logic 1, the output will be the complement of the other input. If both inputs to a NAND gate are connected together, the output will be opposite the input. Similarly, if one input to a NOR gate is permanently placed at a logic 0, the output will be the comple-

(a)

(b)

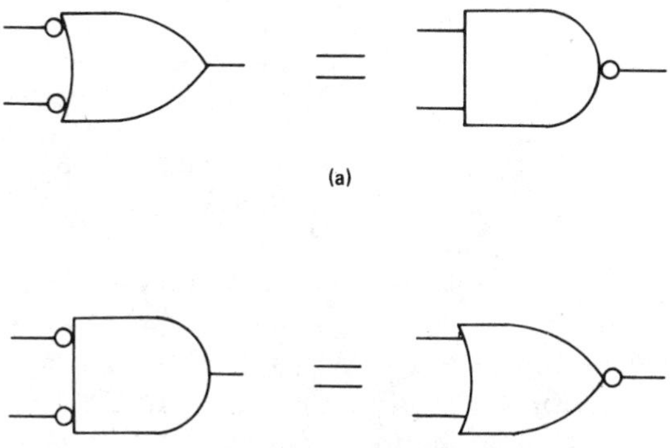

Figure 13-17 Equivalent gates: (a) negated input OR becomes NAND gate; (b) negated input AND becomes OR gate.

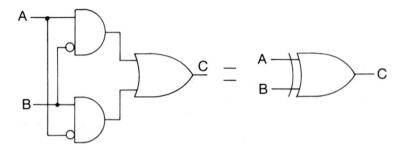

Figure 13-18 Inverters, AND gates, and OR gate becomes exclusive-OR gate.

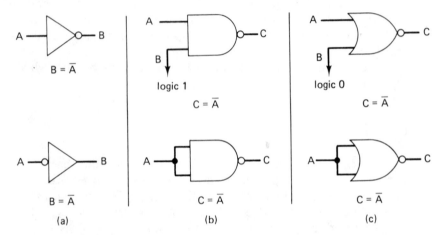

Figure 13-19 Producing inverters: (a) two ways of showing inverter; (b) NAND gate inverter; (c) NOR gate inverter.

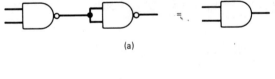

Figure 13-20 Producing AND and OR gates: (a) NAND gate with NAND inverter equals AND gate; (b) NOR gate with NOR inverter equals OR gate.

ment of the other input. Also, if both inputs are connected together, the output will be opposite the input.

An AND function can be produced from NAND gates by connecting a regular NAND gate to a NAND-inverter gate. Similarly, an OR function can be produced from NOR gates by connecting a regular NOR gate to a NOR-inverter gate.

Logic gates are used in combinations to control various functions and perform certain operations. A basic AND-to-OR gate network will produce a 1 at the output if A and B are 1, or if C and D are 1. The Boolean expression would read "output = $AB + CD$." A basic OR-to-AND gate network will produce a 1 at the output if A or B is 1 and C or D is 1. The Boolean expression would read "output = $(A + B)(C + D)$."

A 7451 expandable dual two-wide two-input AND-OR invert gate IC has two separate combination logic networks within the package. To develop a Boolean expression for one of these networks, let pin 2 = input A, pin 3 = input B, pin 4 = input C, pin 5 = input D, and pin 6 = the output. Inputs A and B are ANDed (AB). Inputs C and D are ANDed (CD). These inputs

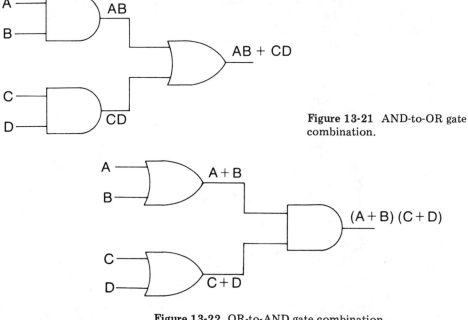

Figure 13-21 AND-to-OR gate combination.

Figure 13-22 OR-to-AND gate combination.

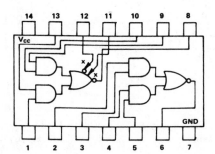

Figure 13-23 7451 expandable dual two-wide two-input AND–OR invert gate IC.

347

are then ORed $(AB + CD)$ and finally, inverted $(\overline{AB + CD})$ to produce an output. Using DeMorgan's theory of conversion, $\overline{AB + CD}$ becomes $(\overline{A} + \overline{B})$ $(\overline{C} + \overline{D})$. It can now be seen that if A or B is low and C or D is low, the AND gates feed two 0's to the NOR gate, which will produce a 1 output.

13-10 *RS* FLIP-FLOP

The basic flip-flop used for storing data in the form of a 1 or 0 can be constructed from two NAND gates or two NOR gates. This *RS* (reset-set) *flip-flop* has two inputs and two complementary outputs. The NAND-gate *RS* flip-flop changes state when 0 pulses are placed at its inputs. Both inputs are normally 1 and should not be 0 at the same time. A 0 pulse on the S input will cause the Q output to go high, while the $\overline{Q}$ will go low. The flip-flop is set and considered storing a 1 at its Q output. A 0 pulse on the R input will cause the Q output to go low and the $\overline{Q}$ output to go high. The flip-flop is now reset, storing a 0 at the Q output.

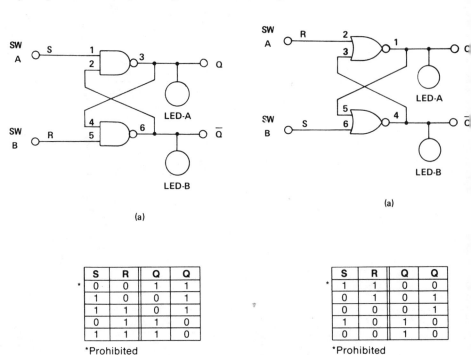

(a)

S	R	Q	Q
*0	0	1	1
1	0	0	1
1	1	0	1
0	1	1	0
1	1	1	0

*Prohibited

(b)

S	R	Q	Q
*1	1	0	0
0	1	0	1
0	0	0	1
1	0	1	0
0	0	1	0

*Prohibited

(b)

Figure 13-24 NAND gate *RS* flip-flop: (a) logic diagram; (b) truth table.

Figure 13-25 NOR gate *RS* flip-flop: (a) logic diagram; (b) truth table.

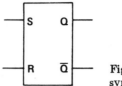

Figure 13-26 *RS* flip-flop logic symbol.

The NOR gate *RS* flip-flop operates the same way, except that its inputs are normally low and a 1 to the inputs is needed to change states. Because the outputs are cross-coupled back to the inputs, these flip-flops will latch up (or remember) the last command given by the pulse at the input.

An *RS* flip-flop can be made from NAND gates, NOR gates, or specially designed circuits; however, very often, its function is shown as a standard logic symbol.

13-11 CLOCKED *RS* FLIP-FLOP

It is desirable in many instances to control flip-flops in a synchronous operation. This can be accomplished by a NAND gate latch with the addition of two more control gates and a third input. The third input is commonly

Figure 13-27 Clocked *RS* flip-flop: (a) logic diagram; (b) logic symbol; (c) truth table.

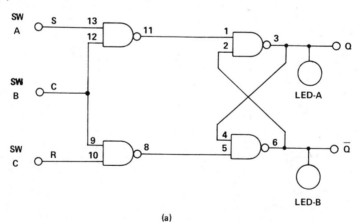

(a)

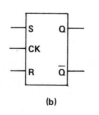

(b)

inputs		state of flip-flop after clock pulse	
S	R	Q	$\overline{Q}$
0	0	unchanged	
1	0	1	0
0	1	0	1
1	1	indeterminate	

(c)

referred to as a *clock* or *trigger input* (abbreviated CK, C, or T). With the addition of the clock input, two input pulses are needed to change the state of the clocked *RS* flip-flop. A NAND gate clocked *RS* flip-flop responds to high input pulses. In order to set it on, where $Q = 1$, the S input and CK input must be high simultaneously. If the S input goes low but the CK input remains high or is pulsed, the flip-flop will not change states, but remain on. Pulses simultaneously applied to the R input and CK input will reset (turn off) the flip-flop, where $Q = 0$.

A clocked *RS* flip-flop can be constructed of NOR gates, but will respond to 0 input pulses.

13-12 *T*-TYPE FLIP-FLOP

A *T-type* (toggle) *flip-flop* has a single input. Each time this input is pulsed, the complementary outputs change state. When the NAND gates are used to construct a *T*-type flip-flop, the Q and $\overline{Q}$ outputs are cross-coupled back into the input control gates. In the reset state, a 1 is applied to the upper control gate and a 0 is applied to the lower control gate. When a 1 trigger pulse appears at the T input, this upper gate initiates a change, which causes Q to go to 1 and $\overline{Q}$ to go to 0. The flip-flop is now in the set state, and a 1 is

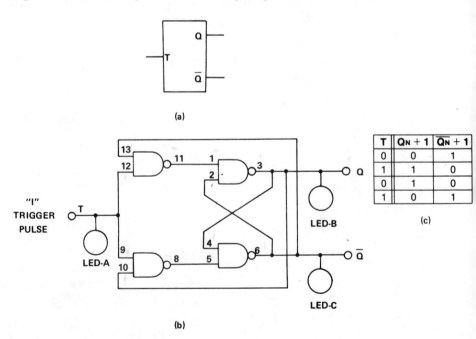

(a)

T	Q_N + 1	\overline{Q}_N + 1
0	0	1
1	1	0
0	1	0
1	0	1

(c)

(b)

Figure 13-28 *T*-type flip-flop: (a) logic symbol; (b) NAND gate logic diagram; (c) truth table.

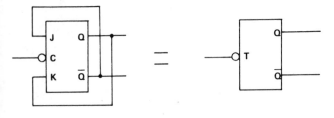

Figure 13-29 *JK* flip-flop to *T*-type flip-flop.

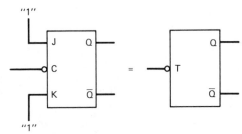

Figure 13-30 *JK* flip-flop conversion to *T*-type flip-flop.

applied to the lower control gate and a 0 to the upper control gate. The next trigger pulse causes the lower control gate to initiate a change, which resets the flip-flop, causing Q to go to 0 and $\overline{Q}$ to go to 1. The truth table shows the output relationships, where $QN+1$ is the condition of the Q output after the trigger pulse and $\overline{QN}+1$ is the condition of the $\overline{Q}$ output after the trigger pulse.

A *T*-type flip-flop can be made from a *JK* flip-flop by cross-coupling the outputs back into the inputs or simply by connecting J and K inputs to a 1. The bubble on the T input means that the flip-flop operates on a negative-going pulse.

13-13 *JK* FLIP-FLOP

The *JK flip-flop* is a combination of the clocked *RS* flip-flop and the T-type flip-flop. Two inputs pulses are required to set or reset the flip-flop. Simultaneous pulses on J input and C input will set it (on: $Q=1$, $\overline{Q}=0$). Simultaneous pulses on K input and C input will reset it (off: $Q=0$, $\overline{Q}=1$). The output relationships are indicated by the truth table, where $QN+1$ is the Q output after the clock pulse and $\overline{QN}+1$ is the $\overline{Q}$ output after the clock pulse. Most of the time the J and K inputs are preselected or steered to set or reset. When the next clock pulse appears at the C input, the flip-flop will go to the indicated state. If both J and K inputs are 1 and a clock pulse appears at the C input, the flip-flop behaves like a *T*-type and the output will change state: if set, it will reset, and if reset, it will set.

A *JK* flip-flop constructed from NAND gates would require two ICs, a 7400 quad two-input NAND gate, and a 7410 triple three-input NAND gate. Using a specifically designed IC, such as a 7470 edge-triggered *JK* flip-flop,

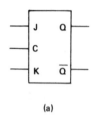

(a)

J	K	$Q_N + 1$	$\overline{Q_N} + 1$
0	0	Q_N	$\overline{Q_N}$
1	0	1	0
0	1	0	1
1	1	$\overline{Q_N}$	Q_N

(b)

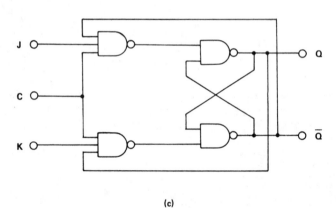

(c)

Figure 13-31 *JK* flip-flop: (a) logic symbol; (b) truth table; (c) NAND gate logic diagram.

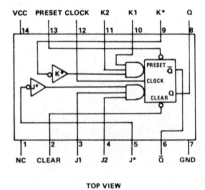

TOP VIEW

Figure 13-32 7470 edge-triggered *JK* flip-flop.

would be more efficient and give better performance. An IC like this might have multiple inputs to accommodate various circuit configurations and preset and clear inputs. A preset input with a bubble means that a negative-going pulse will turn on the flip-flop regardless of the normal input conditions. Similarly, a clear input with a bubble means that a negative-going pulse will override any other normal input conditions and turn off the flip-flop.

13-14 *JK* MASTER/SLAVE FLIP-FLOP

Flip-flops used in high-speed counters and registers are clocked (or triggered) in a synchronous manner. The output of one flip-flop which conditions the inputs of a following flip-flop may change so rapidly that an uncontrollable

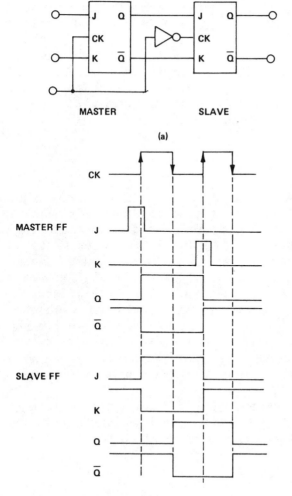

Figure 13-33 *JK* master/slave flip-flop: (a) logic symbol; (b) output waveforms.

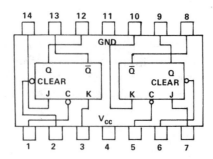

Figure 13-34 7473 dual *JK* maste
slave flip-flop IC.

situation known as the *race condition* occurs. This results in a wrong coun
or answer. The *JK master/slave flip-flop* was developed to solve this problem
There exists a master flip-flop which is set or reset by the leading edg
(positive going) of the clock pulse. The outputs of the master flip-flop ar
present at the inputs of the slave flip-flop when the trailing edge (negativ
going) of the same clock pulse sets or resets the slave flip-flop via an inverte
to the clock input.

A typical IC containing a master/slave flip-flop usually does not indicat
the individual flip-flops, such as the 7473 dual *JK* master/slave flip-flop IC
All that is shown is the normal inputs, outputs, negatively triggered cloc
input, and a clear input. The fact that it is a master/slave arrangement i
defined only in the nomenclature.

13-15 *D*-TYPE FLIP-FLOP

The *D-type flip-flop*, where the *D* stands for delay or data, has an outpu
that is a function of the *D* input that occurred one clock pulse earlier. *i
clocked *RS* flip-flop or *JK* flip-flop can be used to construct a *D*-type flip
flop. The reset input is connected through an inverter to the set inpu
which is now referred to as the *D* input. A 1 present at the *D* input during
clock pulse will cause the *Q* output to go to a 1. If the *D* input is a 0 durin
the next clock pulse, the inverter places a 1 on the reset input and the *i
output goes back to 0. The truth table indicates this relationship, wher
$QN + 1$ is the *Q* output after the clock pulse and $\overline{QN} + 1$ is the $\overline{Q}$ outpu
after the clock pulse, depending on the status of *D* input.

These types of flip-flops are referred to as *D* latches and are used fc
temporarily holding data as input registers or output registers (perhaps fc
LED displays).

A typical IC configuration is the 7474 dual *D*-type edge-triggered flip
flop. These *D*-type flip-flops are triggered by a leading-edge pulse, but othe
may be triggered by a trailing-edge pulse, where a bubble will be found at th
clock input.

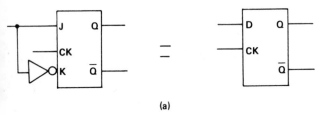

(a)

D	Qₙ + 1	Qₙ + 1
1	1	0
0	0	1

(b)

Figure 13-35 *D*-type flip-flop.

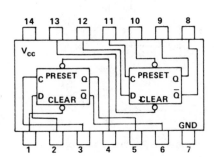

Figure 13-36 7474 dual *D*-type edge-triggered flip-flop.

13-16 ASTABLE MULTIVIBRATOR

The *astable* or *free-running multivibrator* is an oscillator that produces a square-edge output signal. In digital systems it is often referred to as a *clock*.

A basic clock consists of two inverters, a resistor, and a capacitor. The resistor provides the proper bias, while the capacitor is used to supply nec-

Figure 13-37 Astable multivibrator (clock) using inverters.

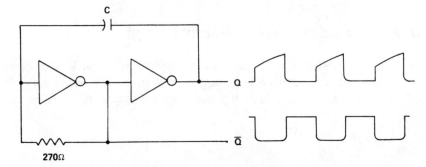

270Ω

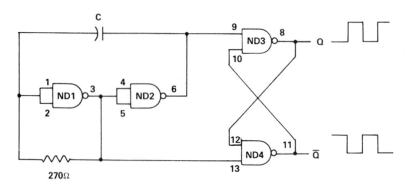

Figure 13-38 NAND gate clock.

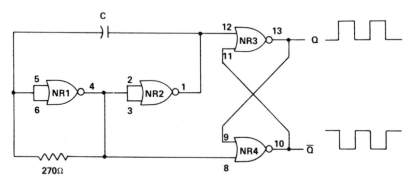

Figure 13-39 NOR gate clock.

essary feedback to sustain oscillation. The approximate frequency is found by the formula $F = 1/3RC$. There is a complementary output of Q and $\overline{Q}$. These outputs are not ideally suited for digital circuits, because of the distortion caused by the charging and discharging of the capacitor. However, the outputs from the inverters can be connected to an RS flip-flop (latch), which will produce square-wave pulses satisfactory for digital circuits.

A NAND gate clock can be constructed from a single 7400 quad two-input NAND gate IC, where two of the gates are wired as inverters. Also, a NOR gate clock can be constructed from a single 7402 quad two-input NOR gate IC in a similar manner.

13-17 MONOSTABLE ("ONE-SHOT") MULTIVIBRATOR

The *monostable* (one-shot) *multivibrator* remains in a normal stable state from which it can be triggered to change to the other state for a predetermined time, after which it returns to the initial state. The one-shot can be

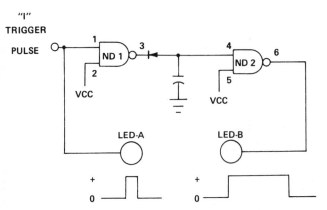

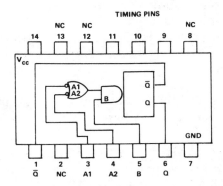

Figure 13-40 NAND gate time delay circuit.

used to delay a pulse, reshape a pulse, create a longer or shorter pulse width, detect a very narrow pulse, and eliminate false triggering due to mechanical switch contact bouncing.

Two NAND gates with a diode and capacitor can be connected to form a basic time-delay circuit. However, this circuit is not a true one-shot, being susceptible to false triggering and failing to have other desirable features.

The TTL 74121 monostable multivibrator IC is well suited for digital IC circuits and has many excellant features. The complementary output (Q and $\overline{Q}$) provides both positive- and negative-going pulses. When inputs A_1 and A_2 are low, triggering is accomplished by a positive pulse at input B. Negative triggering can be used at inputs A_1 or A_2, when input B is high. The timing range of the 74121 can be varied from 30 ns up to 40 s. When pin 9 is connected directly to V_{CC}, a pulse width of 30 ns is obtained. Connecting an external capacitor between pins 10 and 11 and an external resistor from pin 9 to V_{CC} will vary the output pulse width. For a capacitance range of 10 pF to 10 μF and a resistance range of 2 to 40 kΩ, the pulse width can be found by the formula $t_p = 0.693RC$.

Figure 13-41 74121 monostable multivibrator IC.

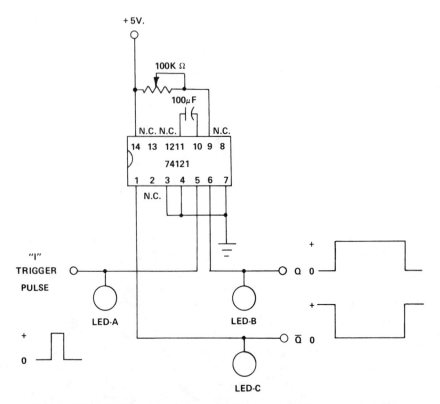

Figure 13-42 74121 one-shot multivibrator showing external timing components.

13-18 BINARY REGISTER

A *binary register* is made up of flip-flops connected together to hold binary data. The number of flip-flops determines the "word" length of data. Four flip-flops connected together form a 4-bit word. Registers are usually grouped in 4-, 8-, 16-, and 32-bit word lengths.

Registers must also be able to move data in specified operations. Some registers perform only a few operations, whereas others perform many. A sample of operations include:

1. Entering data in parallel by parallel inputs. (Preset inputs can be used for this purpose. Inputs with bubbles means lows will set the flip-flops.)
2. Entering data in serial via the least significant bit flip-flop, usually in conjunction with the clock input.
3. Retrieving data simultaneously by parallel outputs.

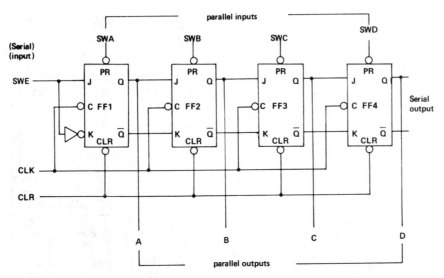

Figure 13-43 Binary register.

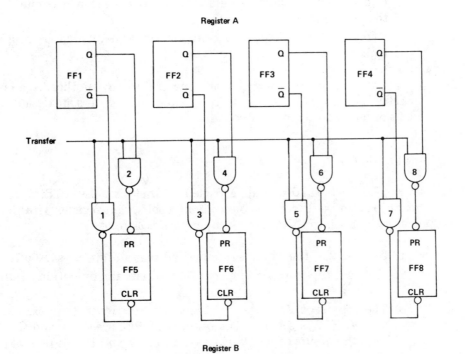

Figure 13-44 Parallel transfer of data.

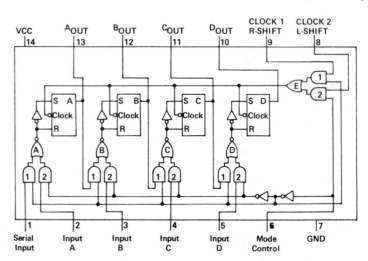

Figure 13-45 7495 4-bit right-shift/left-shift register IC.

4. Retrieving data serially, usually at the maximum-significant-bit flip flop output by shifting data through the register in conjunction with the clock input.

5. Resetting the register by a common clear line, which is brought either high or low.

Data can be transferred from one register to another by a controlled parallel method or by serial shifting. Some IC registers are designed to shift right or left with the proper control.

13-19 RING COUNTER

A *ring-counter* can distribute a single line of clock pulses to different circuits at various times. It consists basically of a shift right register with the outputs coupled back into its inputs. Its operation is as follows:

1. Assume that the clock is running and that all flip-flops are off.
2. The first flip-flop (FF1) is set on to initiate the operation. Outputs $Q = 1$ and $\overline{Q} = 0$.
3. The inputs of FF2 are $J = 1$ and $K = 0$. When the next clock pulse arrives, FF2 is set on. The outputs of FF4 are $Q = 0$ and $\overline{Q} = 1$; therefore, the inputs of FF1 are $J = 0$ and $K = 1$ and FF1 is reset at this time.
4. The outputs of FF2 are $Q = 1$ and $\overline{Q} = 0$, which feed the J and K inputs of FF3, respectively. The outputs of FF1 are $Q = 0$ and $\overline{Q} = 1$, which feed the J and K inputs of FF2 respectively. When the next clock pulse arrives, FF3 is set on and FF2 is reset.

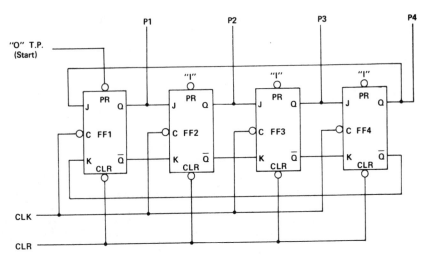

Figure 13-46 Ring counter.

5. This operation continues and FF4 feeds back the pulse to FF1 to start the cycle again.
6. The operation is stopped when the clear line is brought low.

Four positive-going pulses (P1 through P4) are available at the Q outputs of the flip-flops. A complementary output with negative-going pulses could also be taken from the $\overline{Q}$ outputs.

Figure 13-47 Ring counter output waveforms.

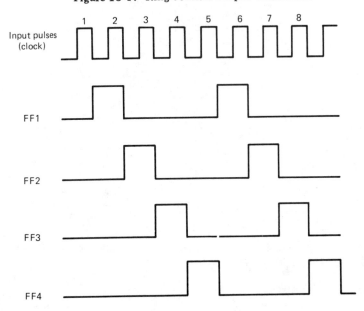

13-20 SWITCH-TAIL RING COUNTER

The *switch-tail ring counter* is very similar to the ring counter (see Section 13-19) except that the outputs of FF4 are cross-coupled back to the inputs of FF1. Because of this, the counter is self-starting when the first clock pulse appears.

This counter differs from the regular ring counter, since each flip-flop is turned on, and remains on, in succession. Each flip-flop conditions the following flip-flop. When all flip-flops are on, the outputs of FF4 are $Q = 1$ and $\overline{Q} = 0$, which steer FF1 to turn off on the next clock pulse. The rest of the flip-flops turn off in succession and the cycle begins again. The operation is stopped when the clear line is brought low.

Logic gates can be connected to the Q and $\overline{Q}$ outputs of the flip-flops to provide eight separate pulses from a 4-bit counter.

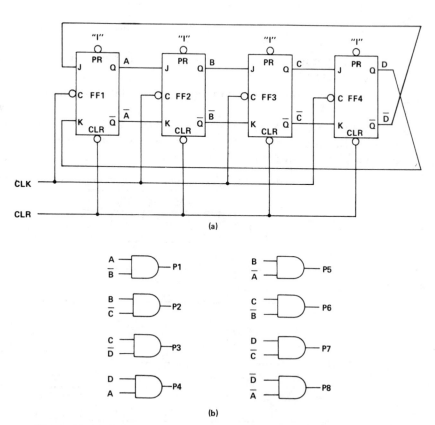

Figure 13-48 Switch-tail ring counter: (a) logic diagram; (b) decoders for 8 pulse output.

362

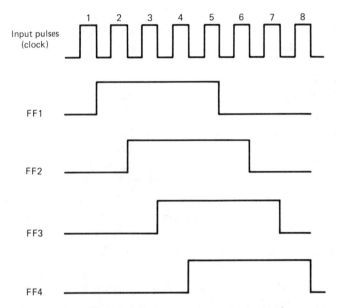

Figure 13-49 Switch-tail ring counter output waveforms.

Connecting LED or other visual indicators to the outputs of the flip-flops can produce an interesting "light effect" such as that seen in movies and television shows about computers and space exploration. The clock should be adjusted to low to give the "ripple" effect.

13-21 BINARY UP-COUNTER

The *binary up-counter* is made up of a group of flip-flops and is used to tally a series of input pulses. Visual indicators can be connected to the Q outputs to display the tally in binary form.

Figure 13-50 Binary up-counter.

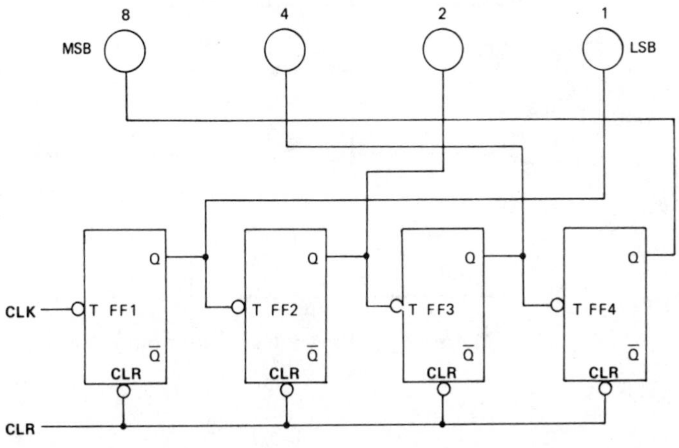

A basic binary up-counter is constructed with trailing-edge triggered T-type flip-flops. The Q output of each flip-flop is connected to the T input of the following flip-flop. The operation sequences are as follows:

1. All flip-flops are off.
2. The trailing edge of the first clock pulse turns on FF1. Its Q output goes high, so nothing occurs to the rest of the flip-flops.
3. The trailing edge of the second clock pulse turns off FF1. Its Q output goes low, which turns on FF2. The Q output of FF2 is high.
4. The trailing edge of the third clock pulse turns on FF1. Its Q output goes high and FF2 is not effected.
5. The trailing edge of the fourth clock pulse turns off FF1. Its Q output goes low, which turns off FF2. The Q output of FF2 goes low, which turns on FF3.
6. This action continues until after 15 clock input pulses, all flip-flops are on.
7. The sixteenth clock input pulse turns all the flip-flops off and the cycle begins again.

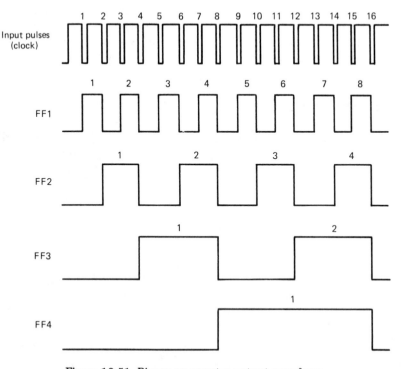

Figure 13-51 Binary up-counter output waveforms.

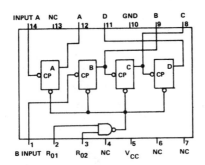

Figure 13-52 7493 4-bit binary counter IC.

Each flip-flop requires two pulses at its input to complete a cycle; therefore, the binary up-counter acts as a frequency divider and the output of each flip-flop is $FF1 = \frac{1}{2}f$, $FF2 = \frac{1}{4}f$, $FF3 = \frac{1}{8}f$, and $FF4 = \frac{1}{16}f$, where f is the input frequency.

The counting sequence can be reset at any time with a low pulse applied to the clear line.

13-22 BINARY DOWN-COUNTER

The *binary down-counter* is similar to the binary up-counter (see Section 13-21) except that the $\overline{Q}$ output of each flip-flop is connected to the T input of the following flip-flop. The operation sequences are as follows:

1. All flip-flops are off.

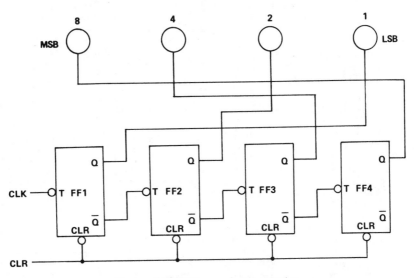

Figure 13-53 Binary down-counter.

2. The trailing edge of the first clock pulse turns on FF1. Its $\overline{Q}$ output goes low and turns on FF2. This action continues and all flip-flops turn on.

3. The trailing edge of the second clock pulse turns off FF1. Its $\overline{Q}$ output goes high and no other flip-flops are affected.

4. The trailing edge of the third clock pulse turns on FF1. Its $\overline{Q}$ output goes low and turns off FF2. The $\overline{Q}$ output of FF2 goes high and no other flip-flops are affected.

5. The trailing edge of the fourth clock pulse turns off FF1. Its $\overline{Q}$ output goes high and no other flip-flops are affected.

6. The trailing edge of the fifth clock pulse turns on FF1. Its $\overline{Q}$ output goes low and turns on FF2. The $\overline{Q}$ output of FF2 goes low and turns off FF3.

7. This counting down action continues until the sixteenth pulse has turned off all of the flip-flops. The cycle then begins again.

The counting-down sequence can be reset at any time with a low pulse applied to the clear line.

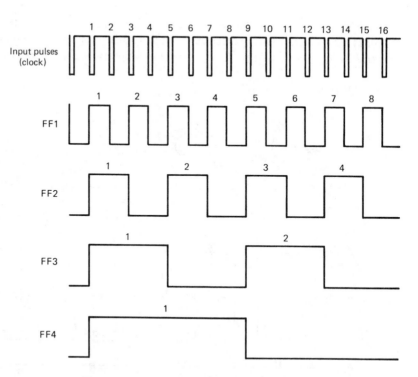

Figure 13-54 Binary down-counter output waveforms.

A 4-bit up counter will count up to 15 and then reset on the sixteenth input pulse. It is referred to as a *modulo-16 counter* or simply MOD-16 counter. Three flip-flops are needed to produce a MOD-8 counter, but if a MOD-10 counter is needed, a 4-bit counter will have to be used with some method of resetting it, when the tenth input pulse occurs. The easiest way of accomplishing this is with a NAND gate decoder connected to the common clear line of the counter. When the counter contains the binary number 10, FF2 and FF4 are on, placing a 1 at each input to the NAND gate. The output of the NAND gate goes low, resets the counter, and the count cycle begins again. The 7493 4-bit binary counter IC can be wired in this manner to produce counters from MOD-2 to MOD-15 (other external gates are needed). A MOD-16 counter is produced by connecting output A to input B. In this mode, outputs A, B, C, and D provide divisions by 2, 4, 8, and 16, respectively (see Section 13-21).

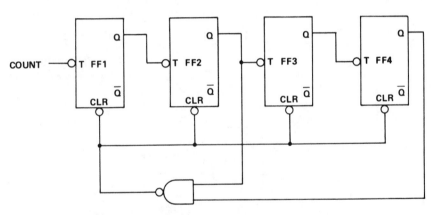

Figure 13-55 MOD-10 counter using reset control.

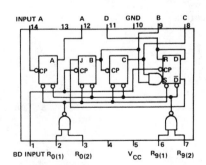

Figure 13-56 7490 decade counter.

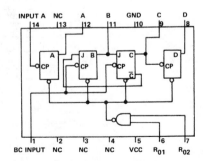

Figure 13-57 7492 divide-by-12 counter IC.

367

The 7490 decade counter IC can produce a MOD-10 counter when output A is connected to the BD input. A MOD-5 counter is available, when BC input and D output are used. This allows flip-flop A to be used independently as a MOD-2 counter.

Also available in IC form is the 7492 divide-by-12 counter, with frequency divisions of 2, 3, 6, and 12.

13-24 BINARY ADDER

The *binary adder* is used for most of the arithmetic operations of a digital system or computer. It can perform normal addition and subtraction by complementing, and adding and multiplication by successive addition, division by comaring, and subtracting and various other operations. A *half-adder* consists of an exclusive-OR gate and an AND gate. There are two inputs, A and B, and two outputs, S (sum) and C (carry). When both inputs

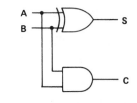

Figure 13-58 Binary half-adder using exclusive-OR gate and AND gate.

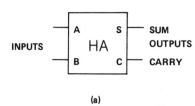

(a)

INPUTS		OUTPUTS	
A	B	S	C
0	0	0	0
0	1	1	0
1	0	1	0
1	1	0	1

(b)

$$S = A\bar{B} + \bar{A}B$$
$$C = AB$$

(c)

Figure 13-59 Binary half-adder: (a) logic symbol; (b) truth table; (c) boolean expressions.

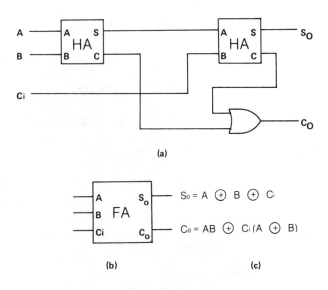

Figure 13-60 Binary full adder: (a) two half-adders with OR gate; (b) logic symbol; (c) Boolean expressions; (d) truth table.

are 0, both outputs are 0. If any input is 1, the outputs will be $S = 1$ and $C = 0$. When both inputs are 1, the outputs will be $S = 0$ and $C = 1$.

The half-adder is capable of adding only two 1-bit binary numbers; therefore, a *full adder* is more practical and easily constructed from two half-adders and an OR gate. The inputs are A, B, and C_i (carry in), and the outputs are S_o (sum out) and C_o (carry out). When all inputs are 0, the outputs are 0. If any single input is a 1, the outputs will be $S_o = 1$ and $C_o = 0$. If any two inputs are 1, the outputs are $S_o = 0$ and $C_o = 1$. When all inputs are 1, the outputs will be $S_o = 1$ and $C_o = 1$.

Binary adders can be used in serial mode, where two multiposition numbers are shifted into inputs A and B. The C_o is connected to a D-type flip-flop for any resulting carry and then fed back to input C_i. The S_o is connected to a sum register or fed back into register A.

Binary parallel addition is faster than the serial mode of addition using shift ing. A full adder is required for each bit of the system word length. Th LSB of each register is connected to the lowest-order adder (FAO). Th next-highest-order bit of each register is connected to the next-highest-orde adder. The C_o of the lowest-order adder is connected to the C_i of the next highest-order adder. An overflow indicator can be connected to the highes order adder C_o. The S_o of the adders can be connected to a third register c fed back into register A (often called the *accumulator*).

In a *binary parallel subtraction* operation the outputs of register B ar fed through a complement circuit to the *B* inputs of the adders. Input C_i c the lowest-order adder is set to a 1 and the result is subtraction.

A basic adder is the 7483 4-bit binary full adder IC. Register A is con nected to inputs A_1 through A_4, and register B is connected to inputs B

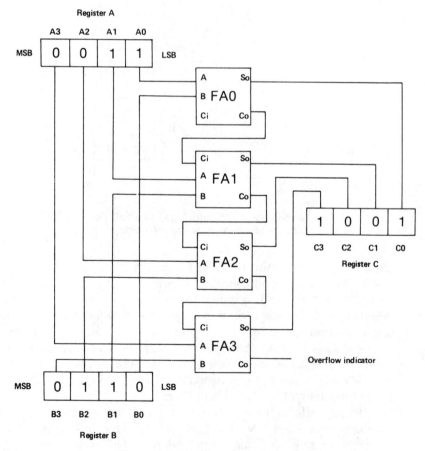

Figure 13-61 Binary parallel addition.

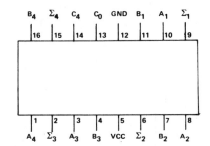

Figure 13-62 7483 4-bit binary full-adder IC.

through B_4. The outputs are Σ_1 through Σ_4. The initial C_i (normally not used) is C_{in} and the final C_o is C_4.

13-26 ENCODERS

Encoders are devices used to convert other number systems into the binary number system.

A simple decimal-to-binary encoder can be constructed using OR gates. The inputs to the OR gates are connected to switches. The switches are connected to a +5 V or logic 1. When any switch is depressed, the equivalent

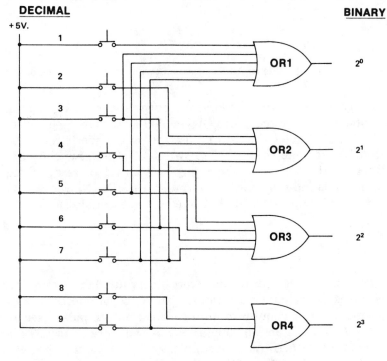

Figure 13-63 Decimal-to-binary OR gate encoder.

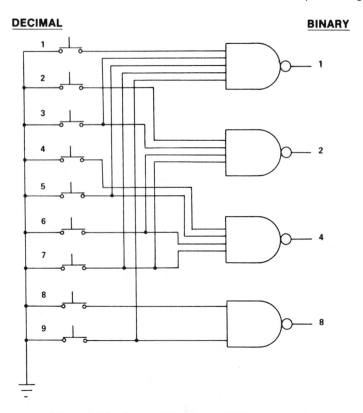

Figure 13-64 Decimal-to-binary AND gate encoder.

binary number is indicated at the output of the OR gates. As an example, i
switch 6 is depressed, gates OR_2 and OR_3 turn on. The output reads 0110.

NAND gates can also be used in an encoder circuit. The inputs to th
NAND gates are connected to switches. These switches are then connecte
to ground or logic 0. When any switch is depressed, the equivalent binar
number is indicated at the outputs of the NAND gates. As an example, i
switch 5 is depressed, the indicated output would read 0101.

13-27 DECODERS

Decoders are devices that convert binary number systems into other numbe
systems or execute operations. An AND gate decoder is used with a 4-bi
switch-tail ring counter to produce eight output pulses (see Section 13-20).

A specific action, operation, or sequence of events may be desired a
a particular time. As an example, an indication may be needed on the thir
and sixth pulses of an eight-pulse sequence. The Q and $\overline{Q}$ outputs of the flip

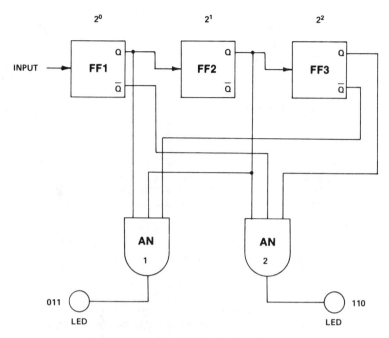

Figure 13-65 AND gate decoders.

flops in a 3-bit up-counter are connected to AND gates, which turn on LED indicators. The inputs to AN1 are coded FF1-Q, FF2-Q, and FF3-$\overline{Q}$, which will cause the LED to light when the counter contains a binary 3. The inputs to AN2 are coded FF1-$\overline{Q}$, FF2-Q, and FF3-Q, which will cause the LED to light when the counter contains a binary 6.

   In digital systems and computers, decoders are used to decode memory address locations and operation codes of a program, to select input/output devices, to display alphanumeric information, and for countless other tasks.

13-28 LED SEVEN-SEGMENT DISPLAY

The *LED seven-segment display* is used as a digital system output indicator. All of the numbers of the decimal system can be displayed and also a few alphabetical letters: A B C D E F H I J L P U.

   Two styles of LED seven-segment displays may be found. In the older version, each segment was made up of a few LEDs. When a segment was selected, all the LEDs would turn on. A newer style uses solid diffused reflective bars, which are connected with light pipes to individual LEDs.

   These displays are of two types, the common anode and the common cathode. The common-anode type requires lows to turn on the segments, while the common cathode requires highs to turn on the segments. External

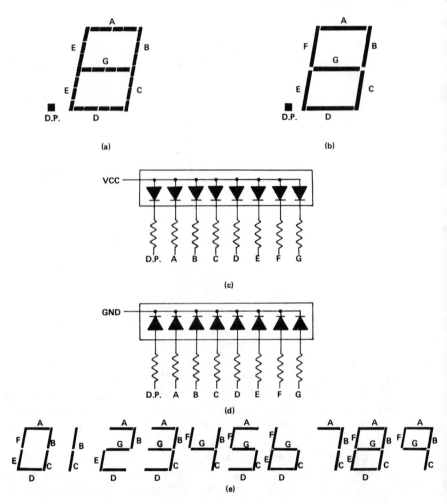

Figure 13-66 LED seven-segment display: (a) multiple-LED segment; (b) solid segment; (c) common anode; (d) common cathode; (e) decimal display.

current-limiting resistors are needed with some displays, but others have th resistors built into the IC package.

Various segments, alphabetically lettered, are selected to produce th 10 decimal digits. There may also be a decimal-point indicator with eac seven-segment display. BCD-to-seven segment decoders are used to select th proper segments.

Displays are generally red, green, orange, or yellow, with character size ranging from 0.11 to 1.0 in.

BCD-to-seven segment decoders are used to convert a 4-bit binary number input into seven outputs, which drive LED seven-segment displays that indicate decimal numbers.

There are three standard BCD-to-seven segment decoder/driver ICs: 7446, 7447, and 7448. They all have the same pin configurations. The 1-2-4-8 BCD input is applied to I_A, I_B, I_C, and I_D, respectively. The seven outputs O_a, O_b, O_c, O_d, O_e, O_f, and O_g are connected to the seven-segment display. Input LT is a lamp test, which enables all seven outputs simultaneously and will cause the display to show number eight. Input RB/I is used to extinguish the entire output when brought to ground.

The 7446 and 7447 have active-low open-collector outputs and can handle 40 mA of current. Current-limiting resistors of about 330 Ω are used with LED displays. The 7446 can withstand 30 V at the output when high, whereas the 7447 can only handle up to 15 V.

The 7448 has active-high outputs (the complement of the 7446 and 7447). Pull-up resistors ranging from 330 to 1000 Ω are usually needed with this IC. It can supply a source current of 2 mA and has a sink capability of 6.4 mA.

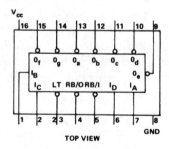

Figure 13-67 Pin configuration for the 7446, 7447, and 7448 BCD-to-seven segment decoder/driver IC.

Figure 13-68 Applications of decoder/driver with LED seven-segment display: (a) common anode; (b) common cathode.

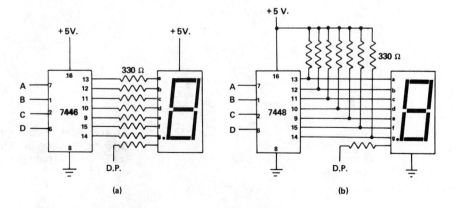

(a)　　　　　　　　　　　　(b)

Decimal numeral or symbol	BCD inputs D C B A	Seven-segment outputs a b c d e f g
0	0 0 0 0	0 0 0 0 0 0 1
1	0 0 0 1	1 0 0 1 1 1 1
2	0 0 1 0	0 0 1 0 0 1 0
3	0 0 1 1	0 0 0 0 1 1 0
4	0 1 0 0	1 0 0 1 1 0 0
5	0 1 0 1	0 1 0 0 1 0 0
6	0 1 1 0	1 1 0 0 0 0 0
7	0 1 1 1	0 0 0 1 1 1 1
8	1 0 0 0	0 0 0 0 0 0 0
9	1 0 0 1	0 0 0 1 1 0 0
⊏	1 0 1 0	1 1 1 0 0 1 0
⊐	1 0 1 1	1 1 0 0 1 1 0
⊔	1 1 0 0	1 0 1 1 1 0 0
c	1 1 0 1	0 1 1 0 1 0 0
E	1 1 1 0	1 1 1 0 0 0 0
Blank	1 1 1 1	1 1 1 1 1 1 1

Figure 13-69 Truth table for 7446 and 7447 BCD-to-seven-segment decoder/driver ICs. (7448 IC output is complemented.)

13-30 COMPARATOR

A *comparator circuit* is used to sample two bits and give an output indication of 1, when they are the same value. In other words, A and B equal C
also $\overline{A}$ and $\overline{B}$ equal C, or when A equals B, the output will be 1. This basi
circuit is also called the exclusive-NOR gate (see Section 13-4).

The 7485 4-bit magnitude comparator IC is capable of testing tw
words, A and B, that are 4 bits in length. Word A uses inputs A_0 throug
A_3 and word B uses inputs B_0 through B_3. Three outputs are available

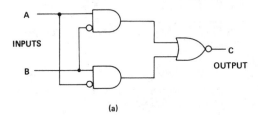

(a)

A	B	C
0	0	1
0	1	0
1	0	0
1	1	1

(b)

$$
\left.
\begin{array}{l}
AB = \overline{C} \\
\overline{A}\overline{B} = \overline{C}
\end{array}
\right\} \; C = 1 \;\text{WHEN}\; A = B
$$

(c)

Figure 13-70 Simple comparator
(a) logic diagram; (b) truth table;
(c) Boolean expression.

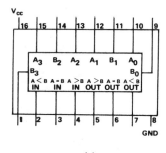

(a)

COMPARING INPUTS				CASCADING INPUTS			OUTPUTS		
A3, B3	A2, B2	A1, B1	A0, B0	A > B	A < B	A = B	A > B	A < B	A = B
A3 > B3	X	X	X	X	X	X	H	L	L
A3 < B3	X	X	X	X	X	X	L	H	L
A3 = B3	A2 > B2	X	X	X	X	X	H	L	L
A3 = B3	A2 < B2	X	X	X	X	X	L	H	L
A3 = B3	A2 = B2	A1 > B1	X	X	X	X	H	L	L
A3 = B3	A2 = B2	A1 < B1	X	X	X	X	L	H	L
A3 = B3	A2 = B2	A1 = B1	A0 > B0	X	X	X	H	L	L
A3 = B3	A2 = B2	A1 = B1	A0 < B0	X	X	X	L	H	L
A3 = B3	A2 = B2	A1 = B1	A0 = B0	H	L	L	H	L	L
A3 = B3	A2 = B2	A1 = B1	A0 = B0	L	H	L	L	H	L
A3 = B3	A2 = B2	A1 = B1	A0 = B0	L	L	H	L	L	H

NOTE: H = high level, L = low level, X = irrelevant.

(b)

Figure 13-71 7485 4-bit magnitude comparator IC: (a) pin configuration; (b) truth table.

which indicate whether A is larger than B ($A > B$), A equals B ($A = B$), or A is less than B ($A < B$). There are three cascading inputs, labeled $A < B$, $A = B$, and $A > B$, which can be used with other 7485 ICs when larger words are being compared. If only a single IC is used, the $A = B$ input is wired to a permanent high and the $A < B$ and $A > B$ inputs are wired to a permanent low.

13-31 MULTIPLEXER

A basic *multiplexer* is a swithcing device that can select one of several inputs and connect it to a single common output. The device acts as a single-pole multiposition switch that passes digital information in one direction only. With a four-input digital multiplexer, input A, B, C, or D is switched to output X, depending on the binary data present at the select control inputs. As an example, if inputs E and F are 1 and 0, respectively, input C is connected through AN3 and OR1 to output X. The output will respond directly to the data present at input C.

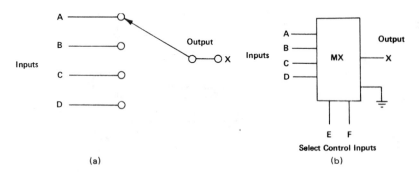

Figure 13-72 Basic multiplexers: (a) single-pole four-throw switch; (b) logic diagram of digital controlled multiplexer.

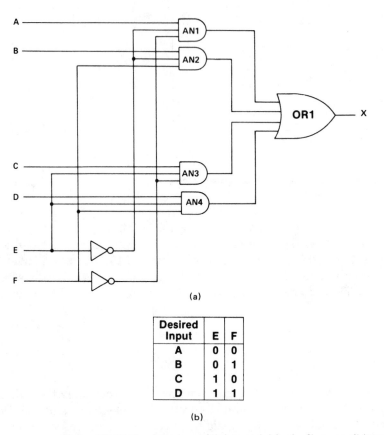

(a)

Desired Input	E	F
A	0	0
B	0	1
C	1	0
D	1	1

(b)

Figure 13-73 Logic gate multiplexer: (a) logic diagram; (b) truth table.

IC multiplexers are fabricated in many configurations. The 74153 dual 4:1 multiplexer can serve as a double-pole four-throw (DP4T) switch. This means that data can come in on a four-input pair and leave on a two-line

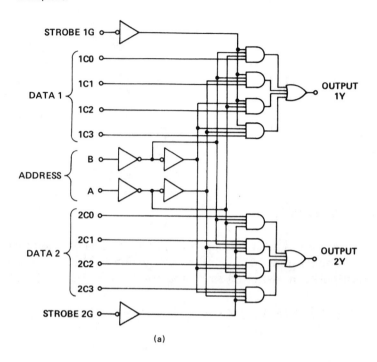

(a)

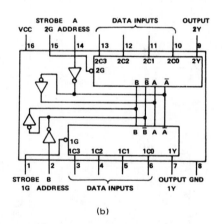

(b)

Figure 13-74 74153 dual 4:1 multiplexer IC: (a) logic diagram; (b) pin configuration.

output.   A single section can be selected by the proper strobe input to produce a single 4:1 multiplexing device.

Other IC multiplexers consist of the 74150 16 line-to-1 line, 74151 8 channel-to-1 line, and 74157 quad two-input multiplxer (allows two groups of 4 bits each to be multiplexed to four parallel outputs).

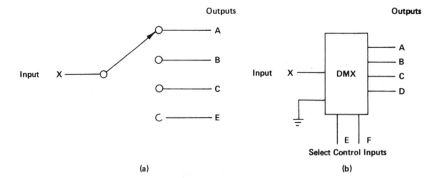

Figure 13-75 Basic demultiplexers: (a) single-pole four-throw switch; (b) logic diagram of digital-controlled demultiplexer.

13-32 DEMULTIPLEXER

A basic *demultiplexer* is a switching device that can connect a single input by selection to one of a number of outputs. The device acts as a single-pole multiposition switch that passes digital information in one direction only,

Figure 13-76 Logic gate demultiplexer: (a) logic diagram; (b) truth table.

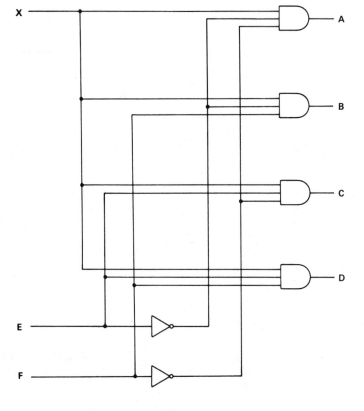

Desired Output	E	F
A	0	0
B	0	1
C	1	0
D	1	1

(b)

(a)

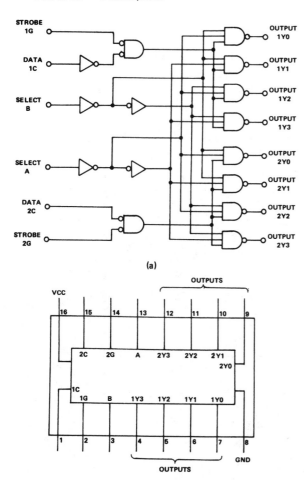

(a)

(b)

Figure 13-77 74155 dual 2:4 demultiplexer IC: (a) logic diagram; (b) pin configuration.

opposite to that of a multiplexer. With a four-output digital demultiplexer, input X is switched to output A, B, C, or D, depending on the binary data present at the select control inputs. As an example, if inputs E and F are 0 and 1, respectively, input X will be connected to output B. The output B will respond directly to the data present at input X.

IC demultiplexers are fabricated in many configurations. The 74155 dual 2:4 demultiplexer IC can serve as a double-pole four-throw (DP4T) switch. This means that data can come in on a two-line input and leave on a four-output pair. A single section can be selected by the proper strobe input to produce a single 1:4 demultiplexing device.

Other IC demultiplexers consist of the 74154 4 line-to-16 line decoder/demultiplexer and the 74139 2 line-to-4 line decoder/demultiplexer.

A *random-access memory* (RAM) is a device into which data in the form of 0's and 1's can be written (stored) and then read out again (retrieved). Any location where data can be stored can be selected (accessed) directly with

Figure 13-78 7489 64-bit random-access read/write memory IC: (a) logic diagram; (b) pin configuration; (c) truth table.

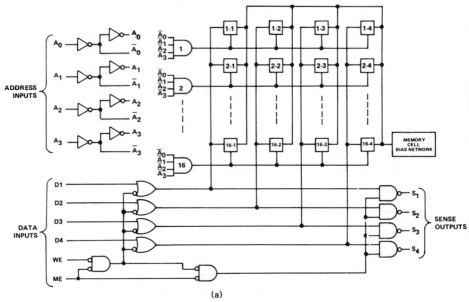

(a)

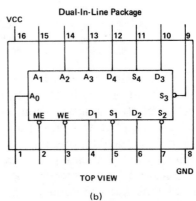

(b)

MEMORY ENABLE	WRITE ENABLE	OPERATION	OUTPUTS
0	0	Write	Logical "1" State
0	1	Read	Complement of Data Stored in Memory
1	X	Hold	Logical "1" State

(c)

the proper address number.   A semiconductor memory cell where a single bit can be stored is a flip-flop.   A basic RAM, the 7489 64-bit random access read/write memory IC, has 64 bits organized into 16 words of 4 bits each. Its sequences of steps for read/write operations follow.

To write (store data):

1. Binary data are placed at address inputs to select desired memory location.   Decoders set up the desired location for operations.
2. Binary data to be stored are placed at the data inputs.
3. The ME and WE are brought low and the data are stored at the selected location.

Figure 13-79 Increasing word length (16 × 8-bit word RAM): (a) wiring diagram; (b) truth table.

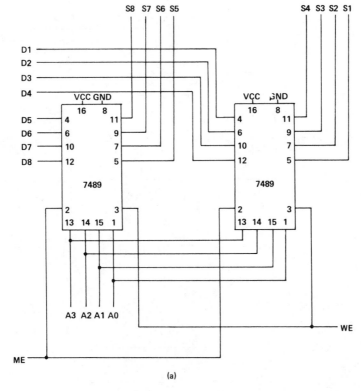

(a)

Operation	Address Locations	ME	WE	Ouputs S_1-S_8
Read	A_0-A_3	0	1	O's for stored data
Write	A_0-A_3	0	0	All 1's

(b)

To read (retrieve data):

1. Same as step 1 for storing data.
2. The ME is placed at 0 and the WE is placed at 1.
3. The sense outputs will detect the data at the selected location, but for this particular memory the data will be the complement of what is stored.

The word length of a memory can be increased by connecting two ICs together. The address inputs, ME inputs, and WE inputs are wired together.

Figure 13-80 Increasing the number of words (32 × 4-bit RAM): (a) wiring diagram; (b) truth table.

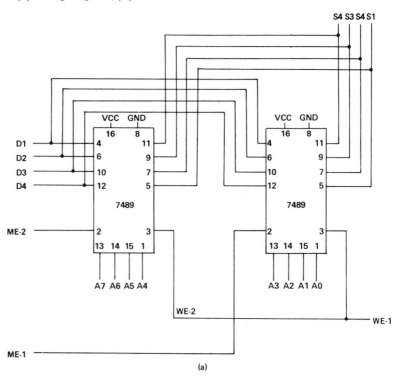

(a)

Operation	Address Locations	ME-1	ME-2	WE-1	WE-2	Outputs S_1-S_4
Read	A_0-A_3	0	1	1	1	O's for stored data
	A_4-A_7	1	0	1	1	
Write	A_0-A_3	0	1	0	0	All 1's
	A_4-A_7	1	0	0	0	

(b)

When addressed, the same location of each IC is selected; however, eight data inputs and eight sense outputs are now available, resulting in a 16 × 8-bit word RAM.

The number of words in a memory may be increased by connecting two ICs together. The data inputs, WE inputs, and sense outputs are connected together. Each IC is addressable by the address inputs, when its corresponding ME input is placed low, resulting in a 32 × 4-bit word RAM.

Semiconductor memories may range from 1-bit words of several hundred to 8- and 16-bit words of several thousand. The smaller memories may be TTL or CMOS, whereas the larger memories are usually always CMOS.

The RAM is a volatile memory in that if the power is removed from the IC, the stored data will be lost or rearranged.

13-34 READ-ONLY MEMORY (ROM)

The *read-only memory* (ROM) cannot be written into, but can be read out repeatedly from the data that are placed into it during the manufacturing process. In the fabricating process certain transistors or circuits are left open representing the desired data to be stored. The ROM may have only address

Figure 13-81 Schematic representation of a ROM.

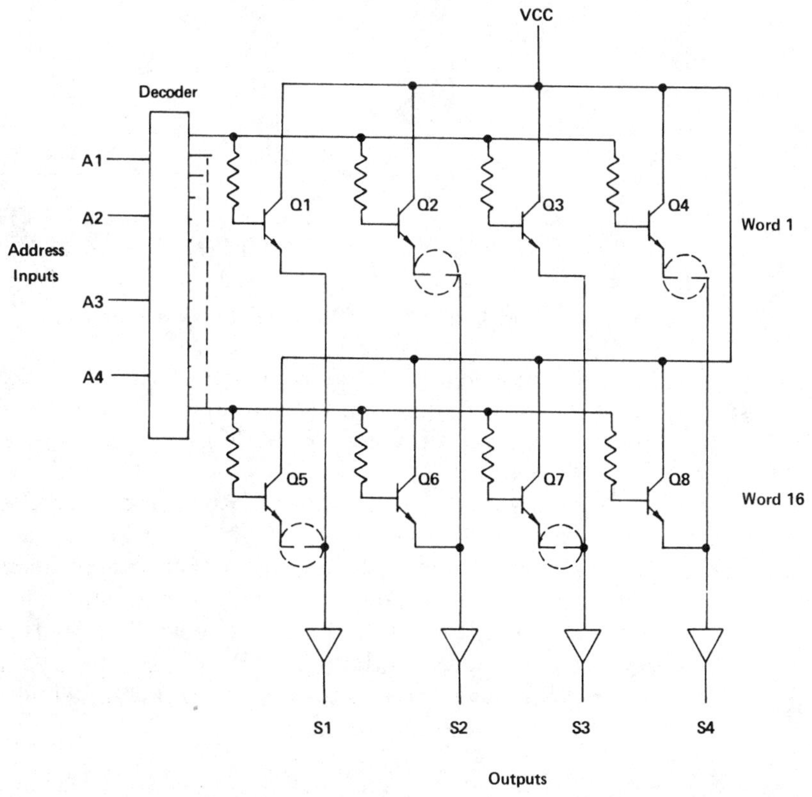

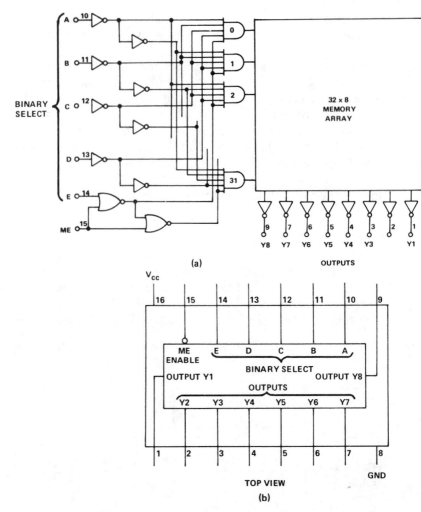

Figure 13-82 7488 256-bit read-only memory IC.

inputs, a ME (memory enable) input, and sense outputs. A desired location is selected by the binary data at the address inputs. When the ME input goes low (or high, depending on the type of IC), the data will appear at the outputs.

A programmable read-only memory (PROM) is a ROM that is field programmable by the user. The PROM has fusible links in the transistor circuit which the user is able to remove electrically (or blow these fuses) with special external power supplies. This type of PROM is programmable only once.

The erasable PROM is a MOSFET device with a floating silicon gate located between the source and drain. During the programming process, certain selected cells have these gates assume a charge, which turns on the transistor and a ROM condition is created. The erasable PROM package has a clear window over the memory cell array. When this package is exposed to ultraviolet light for a few minutes, the charge on the gates flows back into the IC substrate. The data previously stored are erased and the memory can be reprogrammed.

13-35 THREE-STATE LOGIC DEVICES

In complex digital circuit arrangements, the outputs of gates and flip-flops are connected to common lines or data transfer buses. If a particular gate is supposed to be high, while the others are low, there results a "shorting" effect which impairs circuit operation and may cause excessive current flow, possibly damaging the devices. *Three-state logic* (TSL), also called *three-state TTL* or *tri-state logic*, was developed to overcome the problems of common busing lines. A control input is added to the logic devices, which when low will allow the device to operate in the normal manner, but when high will cause the device to appear as a high-impedance output (or be disconnected from the line).

For example, when three flip-flops with TSL are connected to a common bus, only one flip-flop at a time has control of the bus. If flip-flop A is to send its data over the bus, its control input is low, while the control inputs

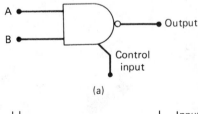

(a)

Control	Output	Input currents at A and B
LOW (enable)	Same as conventional NAND gate: output goes LOW only when A = B = HIGH.	$I_{IH} = 40\ \mu A$ $I_{IL} = 1.6\ mA$
HIGH (disable)	Totem-pole transistors Q_3 and Q_4 both OFF so that output is *high* impedance to V_{CC} and ground.	$I_{IH} = I_{IL} = 40\ \mu A$ $\Big\}$ Hi – Z

Hi – Z state

(b)

Figure 13-83 Three-state NAND gate.

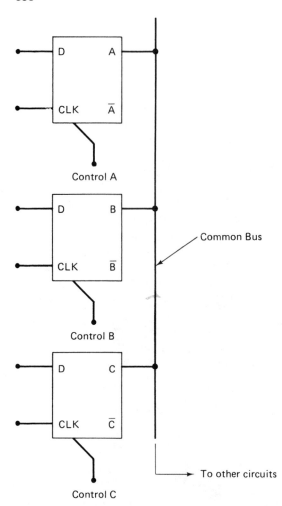

Figure 13-84 Three-state D-type flip-flops connected to a common bus.

of flip-flops B and C are high. Similarly, if flip-flop B is to send data, its input control is low, while the input controls to flip-flops A and C are high.

13-36 BINARY NUMBER SYSTEM

The *binary number system* has a base or radix of 2 with numbers that are 0 and 1. Groups of 0's and 1's can represent numbers of other systems. To convert a decimal number to a binary number, simply divide the decimal number by 2, and the remainder from each division results in the binary number. The first remainder is the least significant bit and the order increases to the last remainder, which is the maximum significant bit of the binary number. Example:

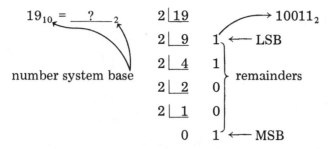

Converting from binary to decimal can be done by the positional notation method. As each bit of a binary number moves to the left, its exponent increases, such as

$$2^4 \quad 2^3 \quad 2^2 \quad 2^1 \quad 2^0$$
$$1 \quad 0 \quad 0 \quad 1 \quad 1_2$$

If the exponents are multiplied out, the number can be written as

$$16 \quad 8 \quad 4 \quad 2 \quad 1$$
$$1 \quad 0 \quad 0 \quad 1 \quad 1_2$$

Now simply add up the products of the raised powers for each position that a 1 occurs, thus:

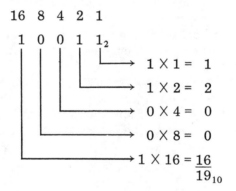

13-36.1 Binary Addition

There are four conditions of binary addition:

$$0 \quad 0 \quad 1 \quad 1$$
$$\underline{+0} \quad \underline{+1} \quad \underline{+0} \quad \underline{+1}$$
$$0 \quad 1 \quad 1 \quad 10 \text{ (with carry)}$$

The first three are easily understood, but the last one reads "1 plus 1 equals zero with a carry of 1." Remember, 10_2 is 2 in binary. When three

1's occur in a column, the result is 1 with a carry of 1 or 11_2, which is a binary 3.

13-36.2 Binary Subtraction

There are four conditions of binary subtraction:

$$
\begin{array}{cccc}
0 & 1 & 1 & 0 \\
-0 & -0 & -1 & -1 \\
\hline
0 & 1 & 0 & 1 \text{ (with borrow)}
\end{array}
$$

Here, again, the first three are understandable, but the last one would have to be able to borrow from a column to the left in order to make sense. If a borrow occurs, it reads $10_2 - 1_2 = 1_2$ (two minus one equals one). Remember, 10_2 is a binary 2.

Binary multiplication and division are the same as with decimal operations and no special problems should be encountered.

13-37 OTHER NUMBER SYSTEMS

Any decimal number can be converted to another number system by dividing the decimal number by the base (*radix*) of the desired system and the remainders become the converted number. The numbers can be converted back to decimal by the positional notation method, remembering to use the base of the number system. (Follow the same procedure as with the binary number system.)

The *octal number system* (base 8) is a multiple of the binary system since $2^3 = 8$. Conversion from octal to binary, and vice versa, is very simple. Section the binary number into groups of 3 bits and find the octal equivalent for each section. Reverse the procedure to go from octal to binary. Example:

$$
101011_2 = 101\ 011 = 53_8 \quad \text{or} \quad 72_8 = 111\ 010 = 111010_2
$$

The *hexadecimal number system* (base 16) is also a multiple of the binary system, since $2^4 = 16$. Decimal numbers from 10 through 15 are represented by alphabetic letters A through F, respectively, for the hexadecimal system.

Conversion from hexadecimal to binary, and vice versa, is similar to octal conversion except that the binary number is sectioned into groups of four bits. Example:

$$
11110110_{16} = 1111\ \ 0110 = F6_{16}
$$

and

$$95_{16} = 1001 \quad 0101 = 10010101_2$$

Binary-coded decimal (BCD) numbers are not a number system, but a code for making binary operations easier. Decimal numbers are converted to BCD numbers, and vice versa, similar to hexadecimal and binary conversion. Example:

$$69_{10} = 0110 \quad 1001 = 01101001_{BCD}$$

and

$$10010011_{BCD} = 1001 \quad 0011 \quad 93_{10}$$

The numbering systems discussed above are compared in Table 13-1.

TABLE 13-1   COMPARISON OF NUMBER SYSTEMS

Decimal	Binary	Octal	Hexadecimal	BCD
0	0000	0	0	00000000
1	0001	1	1	00000001
2	0010	2	2	00000010
3	0011	3	3	00000011
4	0100	4	4	00000100
5	0101	5	5	00000101
6	0110	6	6	00000110
7	0111	7	7	00000111
8	1000	10	8	00001000
9	1001	11	9	00001001
10	1010	12	A	00010000
11	1011	13	B	00010001
12	1100	14	C	00010010
13	1101	15	D	00010011
14	1110	16	E	00010100
15	1111	17	F	00010101

13-38 BOOLEAN ALGEBRA AXIOMS

Boolean algebra is similar to normal mathematical algebra for the first five axioms, in that the same laws or axioms apply. The order in which quantities are added or multiplied makes no difference; therefore, the order in which quantities are ANDed and ORed makes no difference. The other axioms

TABLE 13-2 BOOLEAN ALGEBRA AXIOMS

Axiom	Formula		Description
1	$a + b = b + a$	}	Commutative law
2	$ab = ba$		
3	$(a + b) + c = a + (b + c)$	}	Associative law
4	$(ab)c = a(bc)$		
5	$a(b + c) = ab + ac$		Distributative law
6	$a \cdot 0 = 0$		
7	$a \cdot 1 = a$		AND operation laws
8	$a \cdot a = a$		
9	$a \cdot \overline{a} = 0$		
10	$a + 0 = a$		
11	$a + 1 = 1$		OR operation laws
12	$a + a = a$		
13	$a + \overline{a} = 1$		
14	$a(a + b) = a$		
15	$a + ab = a$		
16	$a + \overline{a}b = a + b$		
17	$\overline{\overline{a}} = a$		DeMorgan's laws
18	$\overline{abc} = \overline{a} + \overline{b} + \overline{c}$		
19	$\overline{a + b + c} = \overline{a}\,\overline{b}\,\overline{c}$		
20	$a + bc = (a + b)(a + c)$		

pertain to two-input AND gate and OR gate operations, DeMorgan's laws and special cases for simplifying complex Boolean expressions, thereby reducing the number of components required to construct a circuit (Table 13-2).

13-39 LOGIC PULSER

A *logic pulser* is a digital test instrument for injecting logic signals into digital circuits. A simple logic pulser can be made from a 7400 quad two-input NAND gate. An LED is used to indicate when a pulse is injected. The basic circuit consists of a NAND gate latch. S_1 is the pushbutton spring-return switch; therefore, the probe output is normally low. When S_1 is activated this output goes high, to about +3.6 V. The LED is isolated from the latch by two NAND gate inverters and gives the proper indication when 1 is injected into a circuit.

Protection diodes are used in the V_{cc} and ground leads in case these leads are incorrectly connected. The diodes will be reverse biased and no current will flow to the circuit. However, when the leads are correctly connected, the output in the low condition will be about +0.7 V due to the forward-biased voltage drop of the diodes.

The components can be mounted on a small perforated board and placed into a small container, such as a plastic medicine bottle. Mini alligator

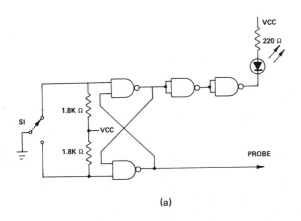

(a)

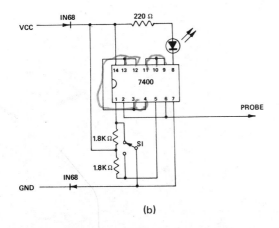

(b)

Figure 13-85 Logic pulser: (a) logic diagram; (b) wiring diagram.

clips are attached to the wire leads for connection to the power source of the circuit under test.

To use the pulser, connect the leads to the power source, then place the probe at the point of the circuit where a pulse is to be entered and press the pushbutton switch. If the circuit requires a low pulse for operation, keep the switch depressed and then release it to enter a 0.

This logic pulser can be used with digital TTL and other circuits with a V_{cc} of +5 V.

13-40 LOGIC PROBE

A *logic probe* is a digital test instrument used to determine the logic condition of a digital circuit. A simple logic probe can be made from a few standard electronic components and a 74121 monostable multivibrator. Power leads from the logic pulser are connected to the power source of the circuit under test. Observe proper connections so as not to damage the probe circuitry. When S_1 is in the normal position, the probe can test static conditions or slow-changing conditions. When the input to the probe is 0, the

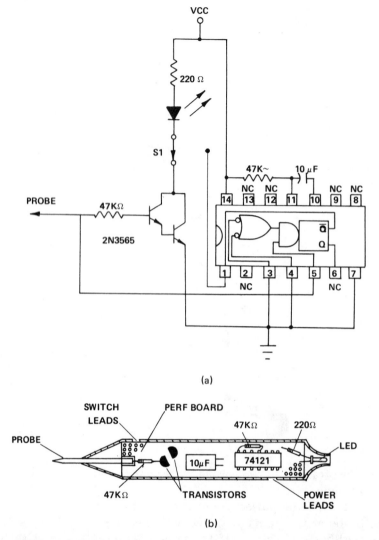

(a)

(b)

Figure 13-86 Logic probe: (a) wiring diagram; (b) component layout.

Darlington pair transistors are cut off and the LED does not glow. If a 1 is present at the probe, the transistors turn on and the LED will glow.

Some pulses in digital circuits occurs so rapidly that the LED will not respond sufficiently for the human eye to detect their presence. In this case, S_1 is positioned to use the 74121 monostable multivibrator, which serves as a pulse stretcher. The 47-kΩ resistor and the 10-μF capacitor form about a 0.5-s delay, which enables the LED to be seen. This circuit can detect an input pulse width of less than 30 ns. When viewing the train of pulses, the LED will flicker, and as the frequency of the pulses increase, the LED will become dimmer.

The components can be mounted on a perforated board and then placed into a container. The switch S_1 can be mounted externally on the case. Mini alligator clips are attached to the wire leads.

The logic probe can be and is often used in conjunction with the logic pulser of Section 13-39.

13-41 DIGITAL TTL IC TESTER

In-circuit testing of digital ICs is usually performed with logic pulsers, logic probes, and/or special test equipment. However, it can save a lot of time and aggravation if faulty ICs are detected before they are replaced in circuits.

Figure 13-87 Digital IC tester.

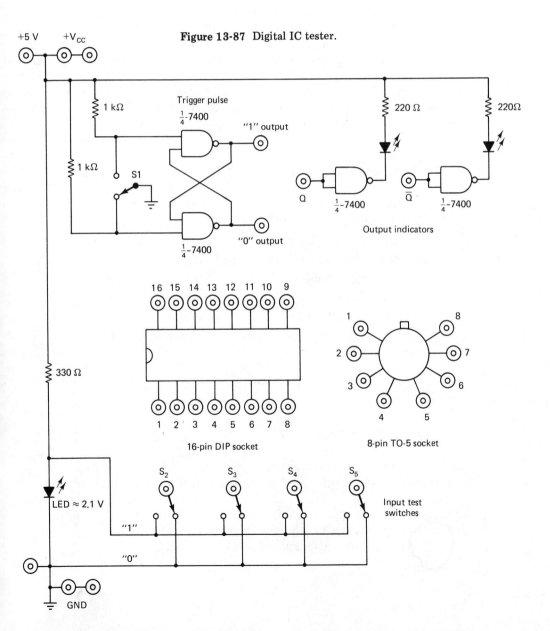

A simple *digital IC tester* can be used for this purpose and also serve as a basic digital logic trainer.

No special power supply is required for the tester since it can be connected to the power source of an existing digital system. An LED pilot light is used to indicate that the unit has power. Because the LED has good regulating properties, input test switches can be wired across it and not cause loading problems. The NAND gate latch provides a high and low bounceless pulse for triggering flip-flops, counters, and other circuits. A high or 1 placed at the inputs of the output indicators will cause the LEDs to light. A 16-pin DIP socket and an 8-pin TO-5 socket are used to test the ICs. Terminals are needed for the pins of the sockets and other indicated points of the tester. A wiring method is needed to connect the terminals for various circuit configurations. Any method may include mini pin plug and jack, solder terminal and clip lead, mini banana plug and jack, or spring-type connectors with tinned wires. Switch S_1 is a momentary pushbutton type, while switches S_2 through S_5 can be slide or SPDT toggle. The entire unit can be assembled on a small perforated board.

13-41.1 Method of Testing ICs.

1. Carefully and correctly place the IC into a proper socket.
2. Wire the proper pins for V_{cc} and GND.
3. Wire the output indicators to the IC output pins.
4. Wire the input test switches to the IC inputs.
5. Wire the proper trigger pulse to the IC input, if needed.
6. Carefully connect the tester to the power source.
7. Proceed to test the IC for proper operations.
8. After the test, remove the power source and then the IC.

13-42 TESTING DIGITAL ICs

13-42.1 Testing Single Gates

Each input to a logic gate must be tested in order for the gate to be considered good. While one input is being tested, the other inputs must be connected to a 1 or 0. A logic pulser can be used at the input under test and a logic probe can be placed at the output to observe proper operation. The digital IC tester of Section 13-41 makes testing even easier.

Table 13-3 presents a guide for testing gates:

TABLE 13-3 GUIDE FOR TESTING GATES

Type of gate	Inputs not under test connected:	Input being tested goes:	Output indication
NAND	High	Low	High
NOR	Low	Low	High
AND	High	High	High
OR	Low	High	High
Exclusive-OR	Low	High	High
Inverter	None	High	Low

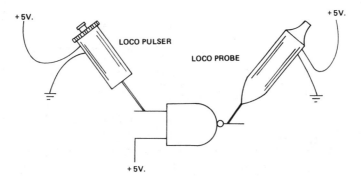

Figure 13-88 Testing single gates.

13-42.2 Testing More Complex ICs

More complex ICs can have data entered into them and then a check is made for the proper output indications. As an example, a counter could have a specific number of pulses entered with a logic pulser. A logic probe can now check the output for proper 1's and 0's.

Figure 13-89 Testing complex digital ICs.

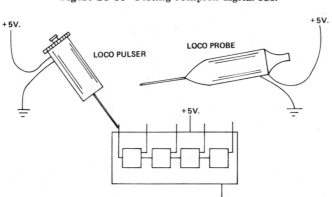

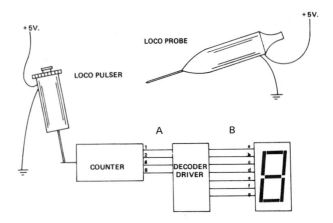

Figure 13-90 Testing a digital system.

13-42.3 Testing a Basic Digital System

Testing a digital system may involve checking the inputs and outputs of a few ICs, which are connected together. As an example, a binary counter is connected to a seven-segment decoder/driver, which in turn, is connected to a LED seven-segment display.

A specific number of pulses would be entered into the counter by a logic pulser. A logic probe would then be used to check for the proper output at point A. The output of the decoder/driver is then checked at point B. It must be determined if the LED display is common anode or common cathode to properly test the 1's and 0's for the seven outputs.

If the output at point A is incorrect, the counter is defective. An incorrect output at point B would indicate a faulty decoder/driver. If these points are correct and the display still indicates a wrong output, the display itself is probably at fault.

Microcomputers

14-1 INFORMATION ON THE BASIC COMPUTER

A *microprocessor* is a central processing unit (CPU) which controls the interpretation and execution of instructions, including an arithmetic/logic unit (ALU) that is fabricated into one integrated circuit. The microprocessor is assembled with other ICs (memory, interface circuits, etc.) to form a *microcomputer*. Understanding a basic computer system is essential to working with microprocessors and microcomputers.

The sections of a basic computer consist of:

1. The ALU that performs arithmetic functions and logic operations, such as ANDing and ORing.
2. A memory section to store programs (a chosen sequence of operations) and data (to be used in calculations).
3. The control section that produces the necessary decoded conditions and timing pulses to execute the instructions (operations) of the programs.
4. Input devices, such as keyboard, punch card reader, punch tape reader, magnetic tape, and electronic sensing devices, to enter instructions and data into the memory section.
5. Output devices, such as printer, CRT, tape recorder, and electronic control devices, to show results or changed conditions of a computer-operated function.

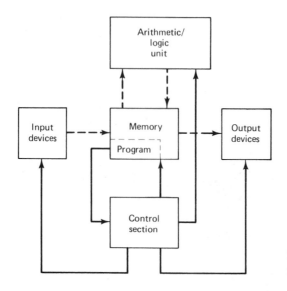

Figure 14-1 Functional block diagram of a computer. (Dashed lines indicate data flow, solid lines indicate program control.)

The computer is under the command of a human-made program at all times. A program consists of a set of instructions that is stored in the working memory that tells the computer what to do. The desired instructions in the form of *instruction codes* or *operation codes* (op code) are stored in the sequence in which they are to be performed. The data used in the operations (referred to as *operands*) are usually variable so that the instruction refers to a location where the data to be used are stored and is called an operand address. The mnemonic code is an abbreviated explanation of the op code and is easier to understand by the user.

In the sample program shown ($A + B = C$), data A are coming from an input device to be added to data B, which are already stored in memory. The answer C is stored into memory. A copy of answer C is then sent to a printer and the program is finished. The computer goes through each memory location and performs the operation indicated by the op code. When the program is run on a computer, the resulting action will go something like this (refer to both figures together):

1. The first memory location, 0000, says to select an input device to obtain data A (02).
2. The second memory location, 0001, says to store data A (02) into memory location 0020.
3. The third memory location, 0002, says to clear the accumulator in the ALU and add the contents of memory location 0020, data A (02).
4. The fourth memory location, 0003, says to add memory location 0021, data B (03), to the accumulator.

Computer reads locations 000-0006	Tells computer what to do	Where to find data	Abbreviation of instructions
Memory location	Instruction (op) code	Operand address	Mnemonic code
0000	02	—	RIN
0001	50	020	STR A
0002	10	020	CLA A
0003	20	021	ADD B
0004	50	022	STR C
0005	70	022	PRT C
0006	99	—	HLT
.	.	.	.
.	.	.	.
.	.	.	.
0020	A(02)	—	—
0021	B(03)	—	—
0022	C(05)	—	—

Data { 0020, 0021, 0022

Figure 14-2 Sample program in machine language: A + B = C.

5. The fifth memory location, 0004, says to store the result of the add operation, data C (05) in the accumulator, into memory location 0022.

6. The sixth memory location, 0005, says to send a copy of the contents of memory location 0022, data C (05), to the output device to be printed.

7. The seventh memory location, 0006, says that the program is finished and tells the computer to stop.

14-2 FUNCTION OF KEY ELEMENTS

There are key elements that are required by a computer to interpret and execute the instructions given in a program (see Sections 13-18 and 13-27).

The *program counter* is used to maintain control of the program. It contains the address of the next instruction to be fetched from memory. Input to the program counter can be from a keyboard, memory, or microprocessor. The output of the program counter is to the address bus, which

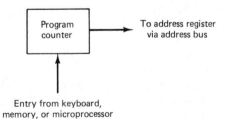

Program counter → To address register via address bus

Entry from keyboard, memory, or microprocessor

Figure 14-3 Program counter.

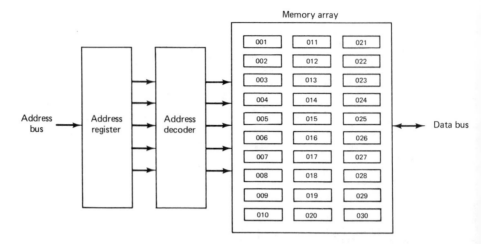

Figure 14-4 Address register and decoder.

goes to the address register in the memory section. During the running of a program pulses increment the contents of the program counter, so that in effect it points to sequential memory addresses as it steps through the program. The correct starting memory address (location) for a specific program must be placed into the program counter by the user for proper operation of that program.

The *address register* contains the location in the memory array where data can be stored or retrieved. The contents of the address register are decoded by the address decoder in order to select the desired location. When a "readout" is initiated, the contents stored at that location will be present on the data bus. When a "write in" is desired, the contents to be stored at the selected location must be placed on the data bus. The input to the address register is from the address bus, which might be connected to the program counter, instruction register, ALU, or I/O device. The output of the memory is on the data bus, which might be connected to the program counter, instruction register, ALU, or I/O device.

The *instruction register* is used to store the contents of the coded instruction being executed by the computer. Its input is from the memory via the data bus. The op-code part of the instruction is sent to the instruction decoder, where control circuits are set up to execute the instruction in various parts of the computer. The operand address part of the instruction is sent to the address register via the address bus to retrieve from memory the data that are to be operated by the instruction.

The *accumulator* is perhaps the most used register in a computer. It is usually involved in most operations and the results of an instruction are normally stored back in this register.

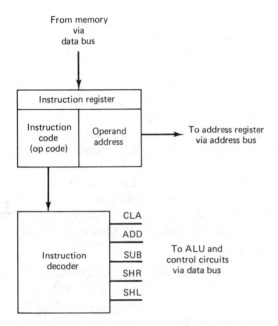

Figure 14-5 Instruction register and decoder.

4-3 THE INSTRUCTION CYCLE AND TIMING CONTROL

An instruction is performed by a computer with a series of controlled pulses. A master oscillator, called a *clock*, produces pulses that are applied to a timing generator (pulse distributor circuit). This circuit applies the input pulses on a single input to various outputs at different times.

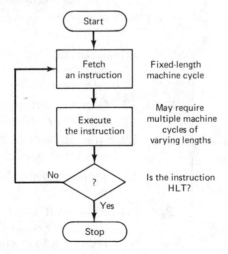

Figure 14-6 Instruction cycle.

The *instruction cycle* consists of two parts: a fetch operation (wher the coded instruction is retrieved from memory) and the execution opera tion (the performance of the instruction). A fetch operation usually require the same number of pulses or states and is referred to as a *fixed machin cycle*. The execution operation will vary, depending on the complexity o the instruction to be performed, and may require multiple machine cycles o varying pulses or states. The instruction cycle will be repeated and the pro gram continues to run until a halt (stop) instruction is encountered or jump instruction returns the computer to a monitor condition awaitin further user action.

As a concept of how *timing pulses* perform an instruction in a com puter, specific pulses can be listed as to their function. Using a single addres computer (read one instruction and perform one operation at a time) witl an instruction cycle of two machine cycles, the pulse nomenclature migh read:

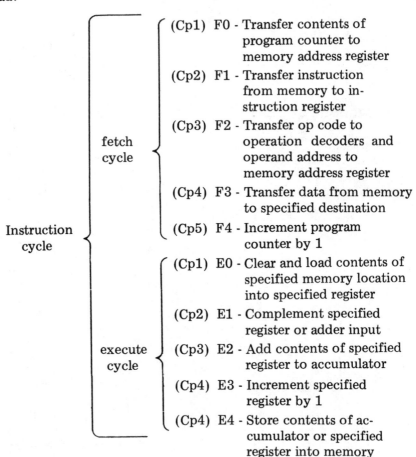

Instruction cycle

fetch cycle

(Cp1) F0 - Transfer contents of program counter to memory address register

(Cp2) F1 - Transfer instruction from memory to in- struction register

(Cp3) F2 - Transfer op code to operation decoders and operand address to memory address register

(Cp4) F3 - Transfer data from memory to specified destination

(Cp5) F4 - Increment program counter by 1

execute cycle

(Cp1) E0 - Clear and load contents of specified memory location into specified register

(Cp2) E1 - Complement specified register or adder input

(Cp3) E2 - Add contents of specified register to accumulator

(Cp4) E3 - Increment specified register by 1

(Cp4) E4 - Store contents of ac- cumulator or specified register into memory

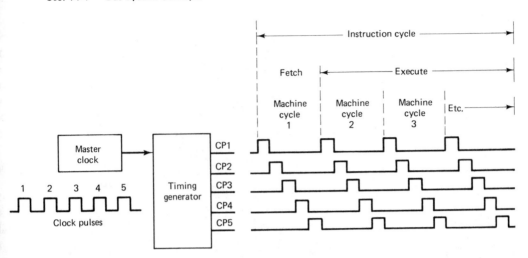

Figure 14-7 Timing-pulse distribution.

As an example of timing pulse control, let the instruction CLB mean to clear and load register B with data from memory. The fetch portion of the instruction cycle has been completed and the data are waiting to be loaded. One input to the gate is high, because of the decoded instruction CLB. At pulse time EO, the gate output goes low, which allows the data to be loaded into the register.

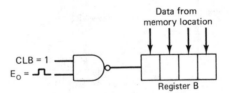

Figure 14-8 Controlling data entering a specified register.

14-4 BUS SYSTEM CONCEPTS

To provide a useful function, the various sections of a computer or microprocessor are connected together by a *bus system*, which is a switching network consisting of gated parallel connections that permits information to be transmitted between several locations. There are three major buses: the address bus, the data bus, and the control bus.

The *address bus* is unidirectional and allows information to travel to the

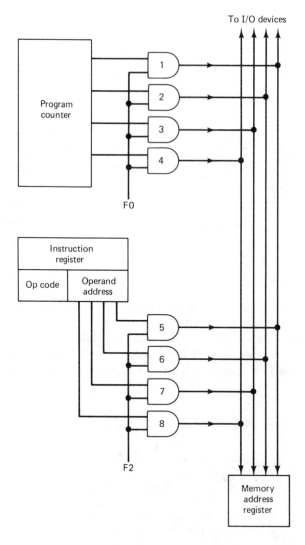

Figure 14-9 Address bus.

memory address register from such other units as the program counter, oper-
and address section of the instruction register, and I/O devices. Information
is allowed to travel from the program counter to the memory address register
when a timing control pulse F0 is present. At pulse F2 time, information
can travel from the instruction register to the memory address register.

    The *data bus* is bidirectional and allows information to travel from the
memory to the instruction register at pulse F1 time, or to the accumulator
at pulse E0 time, when the instruction is CLA (clear and add the contents of
the specified memory location to the accumulator). The output of the accumulator can be sent to the memory at pulse E4 time when the instruction

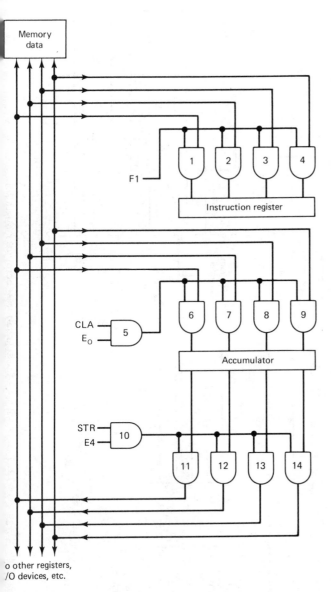

Figure 14-10 Data bus.

is STR (store the accumulator). The data bus is also connected to other registers, I/O devices, and so on.

The *control bus* operates in a similar manner to execute the instructions given to the computer.

Although AND gates are shown in the figures, three-state logic devices are used to prevent interaction of the binary information being transmitted (see Section 13-35).

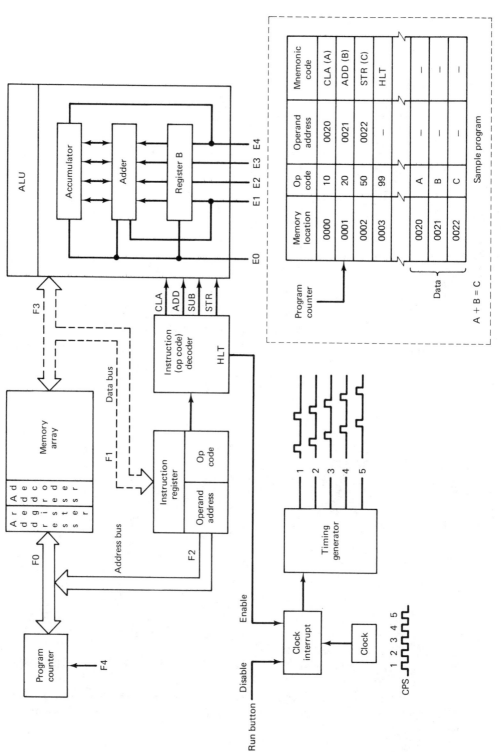

Figure 14-11 Basic concept of a single-address computer operation.

A concept of how a basic computer operates will enable a person to understand how instructions are interpreted and executed by the various elements and timing pulses in a computer. The timing pulses from the timing generator are sent to nearly all elements of the computer. These pulses initiate other timing sequences of varying lengths, which control the ALU, registers, buses, memory sections, and other special functions.

Assume that the program A + B = C has been loaded in memory and the program counter has been set to memory location 0000 by the user. When the run button is depressed, the clock is allowed to send pulses to the timing generator and the following sequences occur:

1. At pulse time F0, the contents of the program counter are transferred to the address register via the address bus.
2. At pulse time F1, the data (100020) of memory location 0000 are sent to the instruction register via the data bus.
3. At pulse time F2, the operand address (0020) is sent to the address register via the address bus and the op code (10) is sent to the instruction decoder to set up the necessary conditions for a CLA (clear and add to accumulator) operation.
4. At pulse time F3, the data (A) of memory location 0020 are present on the data bus.
5. At pulse time F4, the program counter is incremented by one and is pointing to memory location 0001, which contains the next instruction.

This completes the FETCH portion of the instruction cycle. The instruction decoder has set up the proper circuits in the ALU to execute the instruction.

The second machine cycle executes the instruction similarly:

1. At pulse time E0, the data (A) present on the data bus enter the accumulator.
2. At pulse time E1, nothing occurs because this pulse is not needed by the instruction CLA.
3. At pulse time E2, data (A) in the accumulator are sent to the adder and then placed back in the accumulator.
4. At pulse time E3 and E4, nothing occurs because they are not needed for this instruction.

In a similar manner, the instruction cycle is repeated for the ADD and STR operations. When the HLT instruction is fetched, a pulse on the enable line activates the clock interrupt and the computer stops.

The internal architecture of a microprocessor consists of the elements for performing arithmetic, control, and logical operations of a conventional computer. It occupies an area on an IC chip of less than 1 cm^2. It may also contain a working memory section. The microprocessor's capabilities are somewhat less than a conventional computer CPU, with a basic unit of infor-

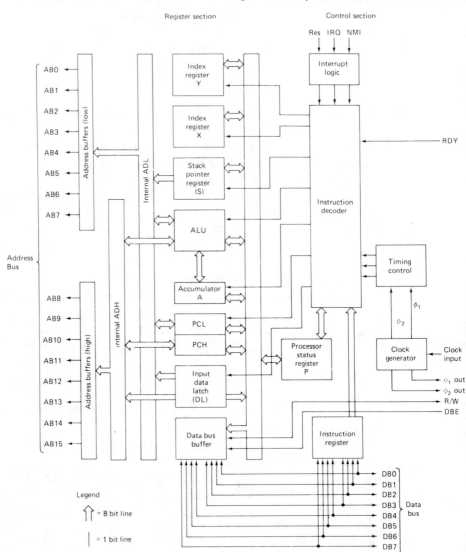

Figure 14-12 Microprocessor internal architecture. (Courtesy of Commodore Business Machines, Inc., 681 Moore Road, King of Prussia, PA 19406.)

mation being one byte (8 bits). However, considering its physical size, the versatility of the microprocessor is phenomenal. A typical microprocessor has the following elements:

1. *Clock generator*, which accepts pulses from an external oscillator and develops two or more clock pulse phases.

2. *Timing control*, which develops the pulses necessary to execute the instructions of the program.

3. *Interrupt logic*, which stops the normal processing of the microprocessor to answer the request from an internal condition, such as a flag, or from an external I/O device, such as a keyboard or printer. After the interrupt has been serviced, the normal program continues to be processed.

4. *Instruction register*, which temporarily holds the instruction code (op code) to be performed.

5. *Instruction decoder*, which decodes the op code into a series of control functions for operation on the data.

6. *Processor status register*, which is a group of independently controlled flip-flops called status bits or status flags that show conditions about the results of the previous instruction. Status flags may indicate a negative result, an overflow indication, a brake command (where the microprocessor stops and is externally restarted), decimal mode (operates with numbers 0 to 9), interrupt disable, zero result, and a carry result. The microprocessor examines the status register, which can result in a decision to modify a program by changing the sequence of the program counter with a jump or branch operation. The status register is also controlled by programming instructions, which are used for proper operation of a program.

7. *Index registers (X and Y)*, which are auxiliary registers used to perform complex instructions and special addressing mode operations.

8. *Stack pointer register*, which locates an area in memory for temporary storage. It is used in the handling of interrupts and subroutines. When the main program is interrupted, the status of the program is preserved in a stack (locations in memory) on a last-in, first-out (LIFO) basis. The contents of the status register, accumulator, and program counter are loaded into the stack. After the interrupt request is serviced, the contents of these registers are placed back into their respective registers and the main program continues operation.

9. *ALU (arithmetic/logic unit)*, which is the major element of the microprocessor, which performs arithmetic and logical functions on data.

10. *Accumulator*, which is the main operating register of the microprocessor, whose function is to act as the communications path between the data bus and the ALU. It will normally contain the result of an

arithmetic or logic operation; many instructions of a microprocessor instruction set are used to control this register.

11. *Program counter*, which is a 16-bit (usually) binary counter whose primary function is to address memory. Such a counter is able to address 65,536 (65K) memory locations. If the microprocessor uses only an 8-bit data bus, the program counter will have an 8-bit low-order section (PCL) and an 8-bit high-order section (PCH), which requires two memory fetch cycles. This counter has a preset capability and can be started from any count upon command, and it is incremented each time an instruction or data are fetched from memory.

12. *Input data latch*, which is a temporary storage register that traps the information, that is present on the data bus only for about 100 ns or less. The data can then be transferred to other elements of the microprocessor during the next control pulse phase.

13. *Data bus buffer*, which is a temporary storage register that holds data external to the microprocessor. It is used to compensate for the rate or flow of data between devices in a microprocessing system. (Note: I/O devices transmit data more slowly than the internal operation of a microprocessor.)

14. *Address buffers*, which are temporary storage registers, that store the addresses used in accessing other peripheral devices, such as ROM, RAM, and I/O. Its function is the same as that of the data bus buffer.

15. *Buses*, which are used to transfer digital information from one element to another in the microprocessor. Usually, there is an address bus, a data bus, and a control bus.

A typical microprocessor IC package such as the 6502 has 40 pins (20 pins per side). The pins can be grouped to show their relationship: the power supply is applied to pins 1, 8, and 21. An external master oscillator is applied to clock input $\phi0$ (pin 37) and a two-phase timing cycle input is present at pins 3 ($\phi1$) and 39 ($\phi2$).

The address bus consists of pins 9 to 22 and pins 22 to 25 with push/pull-type drivers capable of driving at least 130 pF and 1 standard TTL load. The data bus consists of pins 26 to 33, with the same features as the address bus.

The control pins consists of RDY (pin 2), which delays execution of a program fetch cycle until data are available from memory; SYNC (pin 7), used to identify those cycles in which the microprocessor is doing an opcode fetch (and can be used to control RDY to cause single-instruction execution); and R/W (pin 34), which controls the direction of data transfers between the microprocessor and peripheral support chips (ROM, RAM, PIA, etc.).

The interrupt pin $\overline{\text{IRQ}}$ (pin 4) is used to recognize an external interrupt

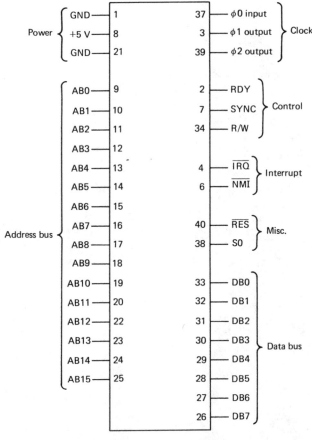

*Pins: 5, 35, 36 = N.C.; 21 = V_{ss}

Figure 14-13 Pin connections for the 6502 microprocessor (rearranged to show bus groupings).

provided that the I flag (interrupt inhibit flag of the processor status register) is a logic "0," and the $\overline{\text{NMI}}$ (pin 6), which is an uncontrolled interrupt, at which time the processor must stop to render service.

The $\overline{\text{RES}}$ (pin 40) is used to initialize the microprocessor from a power-down condition. The S0 (pin 38) is used to set the overflow and is designed to work with future I/O devices.

Pins 5, 35, and 36 have no connections.

14-7 MEMORY ORGANIZATION

The *memory system* of a microcomputer consists of RAM and ROM storage (see Sections 13-33 and 13-34). The working memory, including the stack, is RAM, where external data result from calculations and data movement, information to be printed, and so on, are stored. A ROM is used to store fixed

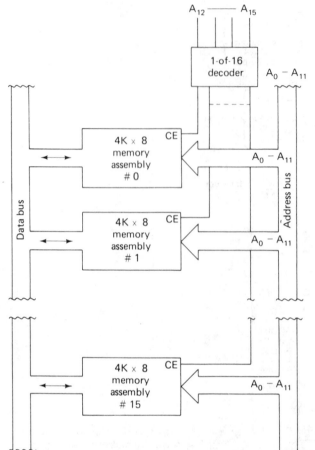

Figure 14-14 Microcomputer memory system.

subroutine programs, look-up tables, assemblers, compilers, routines for controlling I/O devices, and the program to initialize the computer.

A typical memory section in a microcomputer may contain 1024 (1K), 2K, 4K, 8K, 16K, 32K, or up to 65K memory locations. The memory ICs used may be organized into 4096 (4K) memory locations of 8 bits each (4K $\times$ 8). The address pins A_0 to A_{11} of each IC are connected to the address bus and the eight data pins of each IC are connected to the data bus. The information on the address bus and the data bus goes to all memory ICs, but only the desired IC is accessed by a 1-of-16 decoder that turns on the IC with a signal to the chip enable (CE) pin.

Memories are also organized into pages for addressing procedures. A page consists of all the locations that can be addressed by 8 bits (a total of 256 locations), starting at 0 and going through 255. The address within a page is determined by the lower 8 bits of the address and the page number

(0 through 255) is determined by the higher 8 bits of a 16-bit address. Therefore, only a single byte is needed to address a location in zero-page.

The total memory organization which allows specific areas for RAM and ROM can be called a *microcomputer memory map*. The first nine blocks (34K locations) contain RAM, where in an existing microcomputer only two blocks (8K) may be used with future expansion of an additional

Figure 14-15 Microcomputer memory map.

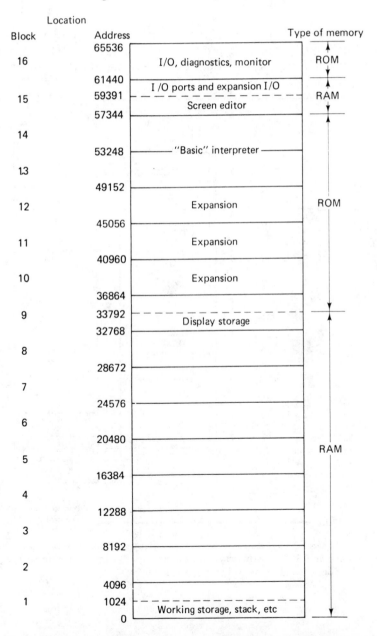

26K. Blocks 10, 11, and 12 may be allotted for ROM future expansion. A Basic interpreter may occupy blocks 13 and 14, which is used to compile and assemble Basic language into machine language. Block 15 is a RAM used for a screen editor to display characters on a CRT, for I/O ports and I/O expansion. The last block is a ROM used for I/O control, diagnostic programs, and the monitor program.

14-8 INTERFACING TECHNIQUES

Most I/O devices tend to be incompatible with microprocessors; therefore, *interface circuits* must be provided to convert the I/O electrical characteristics into electrical signals that are acceptable to the microprocessor. Inter-

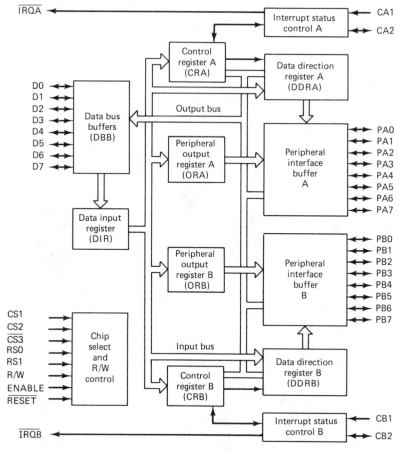

Figure 14-16 Peripheral interface adapter (PIA) internal architecture for a 6520. (Courtesy of Commodore Business Machines, Inc., 681 Moore Road, King of Prussia, PA 19406.)

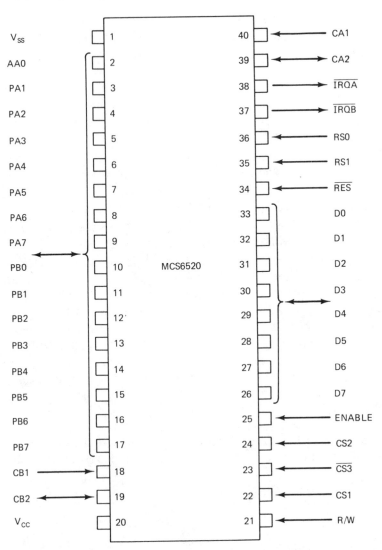

Figure 14-17 Pin connections for 6520 PIA. (Courtesy of Commodore Business Machines, Inc.)

face circuits are used to convert voltage/current levels, serial data to parallel data, and vice versa, encoding, decoding, and to provide temporary storage for matching the operating speeds of interconnected devices. Discrete logic devices, such as gates, flip-flops, counters, and registers, in conjunction with solid-state devices, such as transistors and thyristors, are used for interface circuits. Manufacturers have also produced interface devices known as peripheral interface adapter (PIA), programmable peripheral interface (PPI), or parallel I/O (PIO) to match their microprocessors to the outside world.

A typical PIA, the MCS6520, provides two 8-bit bidirectional peripheral data ports and an 8-bit bidirectional data bus that connects to the microprocessor. The elements and their functions of the PIA are:

1. *Data bus buffers (DBB)* provide the necessary voltage and current drive to assure proper microprocessor operation.
2. *Data input register (DIR)* temporarily holds data appearing on the data bus from the microprocessor.
3. *Control registers A and B (CRA and CRB)* control the respective interrupt status controls and the data direction registers.
4. *Peripheral output registers A and B (ORA and ORB)* store the output data which appear at the peripheral I/O parts.
5. *Interrupt status controls A and B* are used by I/O devices to request service from the microprocessor.
6. *Data direction registers A and B (DDRA and DDRB)* control the direction of the respective 8-bit peripheral I/O ports. Placing a "0" in the data direction register causes the corresponding peripheral I/O line to act as an input. A "1" causes it to act as an output.
7. *Peripheral interface buffers A and B* provide the necessary voltage and current drive to assure proper I/O control operation.
8. *Chip select and R/W control* is used to select the proper peripheral interface device and controls the data flowing to and from the microprocessor.

The selection as to the direction of the I/O ports is accomplished through programming procedures. As an example, port A (PA0 through PA7) could be used for an 8-bit input and port B (PB0 through PB7) could

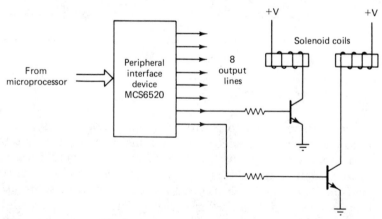

Figure 14-18 PIA application.   (Courtesy of Commodore Business Machines, Inc.)

be used or an 8-bit output. Single port lines can also be selected to provide output control to transistors that operate solenoids.

14-9 MICROPROCESSOR SOFTWARE

Each microprocessor has circuits designed into its instruction decoder that produce the desired operations in response to specific instructions. Every microprocessor has its own instruction set. The instructions are usually in hexadecimal form (machine language), but are referred to by their mnemonic code identification to facilitate writing and understanding programs. The instructions fall into five groups of operations, as shown in Table 14-1.

A single instruction may have one or several op codes that can be used, depending on the operation and addressing mode required. The various addressing modes extend many times the basic instruction set into a very powerful address-oriented operation system. Some of the types of addressing modes are:

ADC	Add Memory to Accumulator with Carry	**JSR**	Jump to New Location Saving Return Address
AND	AND Memory with Accumulator	**LDA**	Load Accumulator with Memory
ASL	Shift Left One Bit (Memory or Accumulator)	**LDX**	Load Index X with Memory
		LDY	Load Index Y with Memory
BCC	Branch on Carry Clear	**LSR**	Shift One Bit Right (Memory or Accumulator)
BCS	Branch on Carry Set		
BEQ	Branch on Result Zero	**NOP**	No Operation
BIT	Test Bits in Memory with Accumulator	**ORA**	OR Memory with Accumulator
BMI	Branch on Result Minus	**PHA**	Push Accumulator on Stack
BNE	Branch on Result Not Zero	**PHP**	Push Processor Status on Stack
BPL	Branch on Result Plus	**PLA**	Pull Accumulator from Stack
BRK	Force Break	**PLP**	Pull Processor Status from Stack
BVC	Branch on Overflow Clear	**ROL**	Rotate One Bit Left (Memory or Accumulator)
BVS	Branch on Overflow Set	**ROR**	Rotate One Bit Right (Memory or Accumulator)
CLC	Clear Carry Flag	**RTI**	Return From Interrupt
CLD	Clear Decimal Mode	**RTS**	Return From Subroutine
CLI	Clear Interrupt Disable Bit	**SBC**	Subtract Memory from Accumulator with Borrow
CLV	Clear Overflow Flag	**SEC**	Set Carry Flag
CMP	Compare Memory and Accumulator	**SED**	Set Decimal Mode
CPX	Compare Memory and Index X	**SEI**	Set Interrupt Disable Status
CPY	Compare Memory and Index Y	**STA**	Store Accumulator in Memory
DEC	Decrement Memory by One	**STX**	Store Index X in Memory
DEX	Decrement Index X by One	**STY**	Store Index Y in Memory
DEY	Decrement Index Y by One	**TAX**	Transfer Accumulator to Index X
EOR	EXCLUSIVE-OR Memory with Accumulator	**TAY**	Transfer Accumulator to Index Y
INC	Increment Memory by One	**TSX**	Transfer Stack Pointer to Index X
INX	Increment X by One	**TXA**	Transfer Index X to Accumulator
INY	Increment Y by One	**TXS**	Transfer Index X to Stack Pointer
JMP	Jump to New Location	**TYA**	Transfer Index Y to Accumulator

Figure 14-19 Typical microprocessor instruction set.

TABLE 14-1 MICROPROCESSOR INSTRUCTION SETS

Data transfer		Arithmetic		Shift and branch		Control and misc.		Logic
LDA	STX	ADC	DEY	ASL	BVC	BRK	NOP	AND
LDX	STY	BIT	INC	BCC	BVS	CLC	SEC	EOR
LDY	TAX	CMP	INX	BCS	JMP	CLD	SED	ORA
PHA	TAY	CPX	INY	BEQ	JSR	CLI	SEI	
PHP	TSX	CPY	ROL	BMI	LSR	CLV		
PLA	TXA	DEC	ROR	BNE	RTI			
PLP	TXA	DEX	SBC	BPL	RTS			
STA	TYA							

1. *Accumulator addressing*: A one-byte instruction requiring an operation on the accumulator, such as ASL, LSR, ROL, and ROR.
2. *Implied addressing*: A one-byte instruction requiring an operation on some element in the microprocessor, such as BRK, CLC, CLD, CLI CLV, DEX, DEY, INX, INY, NOP, PHA, PHP, PLA, PLP, RTI, RTS SEC, SED, SEI, TAY, TSX, TXA, TXS, and TYA.
3. *Immediate addressing*: A two-byte instruction with an 8-bit operand contained in the second byte, such as ADC, AND, CMP, CPX, CPY EOR, LDA, LDX, LDY, ORA, and SBC.
4. *Direct addressing*: A two-byte instruction with the first byte showing the op code and the second byte indicating the operand address (the memory location where the data are stored).

   4a. *Absolute addressing*: A three-byte instruction with a 16-bit operand address contained in the second byte (8 low-order bits) and third byte (8 high-order bits). An entire 65 kilobyte memory is accessible with this mode of addressing, with instructions such as ADC, AND, ASL, BIT, CMP, CPX, CPY, DEC, EOR, INC, JMP, LDA, LDX, LDY, LSR, ORA, ROL, ROR, SBC, STA, and STY.

   4b. *Zero-page addressing*: A two-byte instruction similar to absolute addressing, but utilizing only the first 256 memory locations. Careful use of this addressing mode can result in a significant increase in program efficiency. Its instructions include ADC, AND, ASL, CMP, DEC, EOR, INC, LDA, LDY, LSR, ORA, ROL, ROR, SBC, STA, and STY.
5. *Relative addressing*: A two-byte instruction used in branching, with the second byte establishing a destination for the branch. The second byte shows an "offset" which is added to the contents of the lower 8 bits of the program counter when the counter is set at the next instruction. The range of the offset can be from -128 to $+127$ bytes of the memory location of the next instruction. Its instructions include BCC, BCS, BEQ, BMI, BNE, BPL, BVC, and BVS.
6. *Indexed addressing*: This mode of addressing forms a memory address in which future data are stored by adding the current data included

with the instruction to the contents of some register or memory location. Indexed addressing is used in conjunction with zero page, absolute, and indirect addressing. Its use is more complex, but it greatly increases program efficiency.

7. *Indirect addressing*: This mode of addressing shows an instruction with an address that indicates a memory location or a register that, in turn, contains the actual address of an operand. The indirect address may be included with the instruction, contained in a register or a memory location.

The software of a microprocessor consists of the instruction set and any documentation explaining the procedures for programming and operation.

14-10 PROGRAMMING CONCEPTS

The microprocessor performs the instructions in a *program* according to the sequence in which they are written. *Flowcharts* are used as a guide to establish the proper sequence of operations (instructions) when writing a program and/or can be used in the understanding of complex programs. In a basic addition program, $A + B = C$, a flowchart may simply state: load A into accumulator (place the contents of memory location where A is stored into the accumulator), add B to the accumulator (add the contents of memory location B to the contents of the accumulator), store answer C in memory (store the sum of A + B in a location in memory), end (stop operation). Some programs may contain loops that will repeat a specific operation until a certain

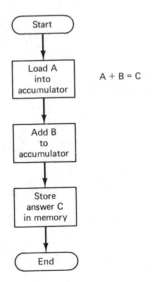

Figure 14-20 Flowchart for adding two numbers.

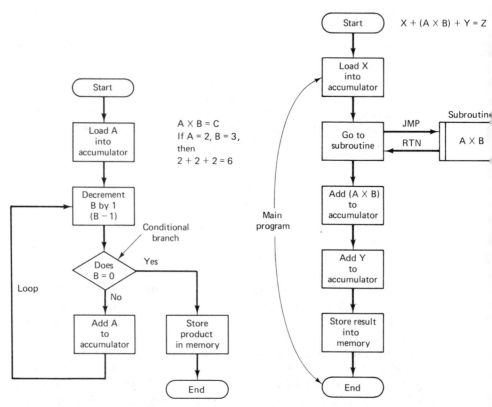

Figure 14-21 Flowchart for
multiplying two numbers.

Figure 14-22 Flowchart showing a
subroutine operation.

condition is met, and then branches to another instruction. Such may be the
case with a multiplication problem, A × B = C. Multiplicand A is added to
the accumulator B (multiplier) number of times to produce the product C.
The instruction, decrement B by 1 (B - 1), keeps track of the number of
times A is added to the accumulator. When B equals zero, the program
branches to the store instruction and then stops.

Many initial programs can be saved and used in more complex programs. These initial programs are then referred to as subroutines. For
example, in the problem X + (A × B) + Y = Z, X is loaded into the accumulator, and the computer then jumps (JMP) to the subroutine (A × B). When
the multiplication is finished, the computer returns (RTN) to the main program, adds the product (A × B) to the accumulator, adds Y to the accumulator, stores the answer Z in memory, and then stops.

A programmer will write a program in a high-level language, such as
FORTRAN, COBOL, or BASIC, using macroinstructions. The program is
then run on a computer using a compiler and/or assembler program which

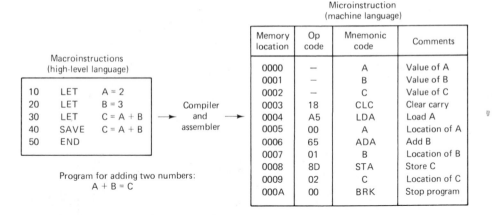

Figure 14-23 Comparison of high-level language to machine language.

translates it into machine language of instructions, known as *microinstructions*. Machine language is comprised of the coded instructions, in binary form, that are accepted and used by the computer. For example, in the microinstruction program, line 30 reads "let C = A+B," which will cause the computer to execute the instructions in memory locations 0003 to 0007 [clear the carry, load A, and add B (shown in the microinstruction)].

Microprocessors were not solely intended to be "number crunchers" used for data processing, but rather to be used in a variety of applications involving the control of systems. Anyone involved with microprocessors will have to know some programming techniques in order to operate, diagnose, test, and maintain systems effectively using microprocessors.

14-11 MICROCOMPUTER ORGANIZATION

The basic elements of a microcomputer system are the clock generator, microprocessor, memory (RAM and ROM), and a peripheral interface device (PIA) that connects to the I/O ports. The clock generator is the "heart" of the system, which produces pulses to control all signal transitions within the system. The microprocessor's inputs and outputs are used to control properly the other elements in the system. It fetches the first instruction in the program and executes the very simple task dictated by the specific pattern of bits (1's and O's) in the instruction. The next instruction is then fetched and executed. This simple operation is repeated over and over and in this way the program instructs the processor to bring about the desired system operation. The data memory (RAM) temporarily stores the current working program, input data, the results of arithmetic operations, and so on. The program memory (ROM) permanently stores programs for subroutines,

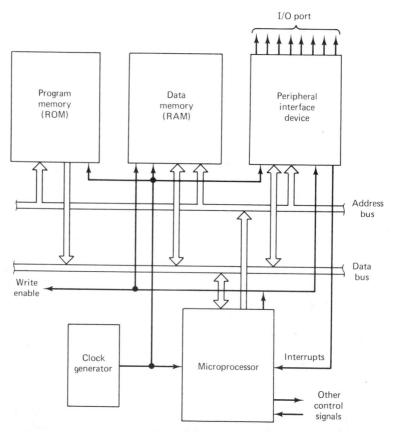

Figure 14-24 Organization of a microcomputer system. (Courtesy of Commodore Business Machines, Inc.)

compilers, assemblers, for control of I/O devices, and so on. The peripheral interface device (peripheral interface adapter PIA) controls the communication between the microprocessor and the I/O devices, such as printers, keyboards, displays, and so on.

These elements are tied together by three major buses: the address bus, the data bus, and the control bus. The microprocessor places an address on the address bus which can be sent to memory (RAM or ROM) or the PIA. The data bus is bidirectional and, depending on the operation, data may be coming from or going to the memory, the microprocessor, or the I/O ports. The control bus carries signals to all the elements to control data transfers and operations. Explanations of four commonly used data transfer operations can aid in understanding the flow of data on the buses:

1. *Move data from memory to processor*: The processor places the desired memory location address on the address bus. A signal on the control

bus allows the memory to be accessed. The data stored in the selected memory location are placed on the data bus. Another signal on the control bus allows the data to enter the processor.

2. *Move data from processor to memory*: The processor places the desired memory location address on the address bus. A signal on the control bus allows the memory to be accessed. The data in the processor are placed on the data bus. Another signal on the control bus allows the data to enter the specified memory location.

3. *Move data from PIA to microprocessor*: The processor places the address of the desired input device on the address bus. A signal on the control bus selects the desired device via the PIA. The PIA places incoming data on the data bus. Another signal on the control bus allows the data to enter the processor. The processor then executes the data transfer of operation 2 and places the data at a specified memory location for future use.

4. *Move data from processor to PIA*: The processor has completed data transfer operation 1 and contains the specified data. The processor places the address of the desired output device on the address bus. A signal on the control bus then selects the desired device via the PIA. The processor places the data on the data bus. Another signal on the control bus tells the processor that the data were accepted by the output device.

Transferring data between memory and the I/O ports via the microprocessor is time consuming. In systems using external high-speed tape or disk, the data transfer speeds are greater than the processor can handle. Therefore, a direct transfer from memory between I/O ports (called *direct memory access*, DMA) can be accomplished with the help of a special device called a DMA controller.

14-12 OVERVIEW OF MICROCOMPUTER TROUBLESHOOTING

14-12.1 Test Equipment

Standard electronic test instruments such as the voltmeter and oscilloscope can be used to test microprocessor-based systems. However, specialized digital test equipment is more efficient and often easier to analyze. Solid failures such as shorts or opens can be located with relatively simple and easy-to-use digital test instruments such as the logic pulser (see Section 13-39), logic probe (see Section 13-40) (also see Section 13-42) and logic monitor clip, which fits completely over the IC and gives the logic condition of all pins simultaneously. Some ICs mounted on printed circuit boards are

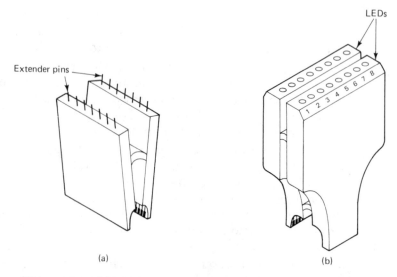

Figure 14-25 IC test devices: (a) IC test clip; (b) logic monitor clip.

difficult to reach with test probes, so an IC test clip with extender pins may be used to facilitate testing.

Intermittent failures and moving data in proper sequence are difficult to detect with conventional test equipment; therefore, a signature analyzer and/or a logic analyzer may be used. The signature analyzer resembles a digital voltmeter that mathematically converts the pattern of l's and O's to a four-digit pseudo-hexadecimal display that should appear at various points within a system. The unit is used in conjunction with looping test programs. A logic analyzer is a special multichannel oscilloscope that usually contains a memory section. The unit allows the display of parallel data from a working circuit on one-half of the screen to be compared to the parallel display from a nonfunctioning circuit on the other half of the screen. The memory section can show the pulses of a circuit before transition and at the same time show the pulses after the transition.

14-12.2 Techniques for Localizing Problems

Many microprocessor-based systems have diagnostic programs which the technician can run in order to localize and find a malfunction. Often these programs can be found in the ROM of the system. It may be necessary to write a simple loop program to determine if data are being properly moved between the elements of a system. An example might be to load the accumulator with the contents of a memory location and then store it into another location. The symbolic program might look something like that shown in Table 14-2.

TABLE 14-2 SYMBOLIC PROGRAM

Mnemonic code	Comment
LDA	Load accumulator with contents of specified memory location.
STA	Store contents of accumulator into specified memory location.
JMP	Jump to beginning of program.

(Loop brackets LDA, STA, JMP)

A program such as this could be used to check out the accumulator, address bus, data bus, memory, and partial control of the instuctions used.

14-12.3 Circuit Testing Procedures

A general troubleshooting procedure may be followed to locate faulty elements or components.

1. A simple test program can be run to determine if the desired results were obtained and the data were located in the proper elements. If this test proves negative, proceed to step 2.
2. Single-stepping (manual control of the clock generator) of the program through its instructions may indicate where the malfunction is occurring. When this test proves positive, proceed to steps 3 and 4. If it proves negative, proceed to step 5. (Note: this procedure is also useful in debugging programs.)
3. A specific element may be tested by injecting pulses at the inputs and observing the outputs for corresponding conditions.
4. If a particular element or component is suspected of malfunctioning, a unit known to be good can be substituted in its place and step 1 can be repeated.
5. A dynamic, in-circuit test can be performed using a loop program and observing the data flow with such test equipment as a signature analyzer or a logic analyzer.

14-13 GLOSSARY OF OTHER COMMONLY USED MICROPROCESSOR TERMS

Access The process of storing or retrieving data in a memory section.

Address A number that designates a specific memory or I/O location.

Algorithm A term used to describe the sequence of operations that defines the solution to a problem.

Alphanumeric A character set that contains letters, numerals, and special characters ($\sqrt{}$, ?, /, ;, :, etc.)

ASCII code The American Standard Code For Information Interchange. An 8-bit character code with the parity bit or a 7-bit character code without the parity bit.

Assembler A program that translates mnemonic code to machine language code. It translates source programs to object programs.

Base *See* radix.

Binary The base 2 number system, in which all numbers are expressed as powers of 2 and represented with 1's and 0's.

Binary-Coded Decimal (BCD) A code by which decimal numbers (0 through 9) are represented as binary values using 1's and 0's (i.e., $5_{10} = 0101_{BCD}$).

Bit A *binary* digit, that is, the smallest unit of information that can be represented. The binary digits are 1 and 0.

Branch (or jump) An instruction that causes a program jump to a specified address and begins processing the instruction at that address.

Byte A sequence of adjacent bits operated upon as a unit, usually containing 8 bits.

Carry The result of binary addition where the sum exceeds 1 (i.e., $1_2 + 1_2 = 10_2$). The leftmost bit is the carry to the next position.

Chip-enable (CE) A control input that when active allows operation of the IC and when inactive restricts the operation of the IC.

Clear To remove data from elements in a system to an initial condition, usually "0."

Compiler A program that translates high-level languages (FORTRAN, COBOL, BASIC) to machine language code.

Complement The 1's complement changes all 0's to 1's and all 1's to 0's. The 2's complement is the 1's complement with a 1 added to it.

Conditional branch (or jump) A program instruction that will cause the computer to branch from the normal program sequence to another specified location in the program when specified criteria are met.

Contact bounce The uncontrolled making and breaking of a contact when the switch or relay contacts are operated. The bounces can produce unwanted pulses that can upset a digital system.

Debounced A switch or relay that no longer exhibits contact bounce, usually because of added circuits such as a NAND gate latch.

Debug Detect, locate, and correct problems in a program (software) or digital system (hardware).

Decimal The base 10 number system, in which all numbers are expressed as powers of 10 and represented with the numbers 0 through 9.

Decoder A digital device with several parallel inputs which recognize bit patterns and produce a signal on the various outputs.

Decrement To decrease the value of a binary word, usually by a value of 1.

Firmware The contents of a ROM.

Flag A status flip-flop which indicates that a certain condition has occurred as a result of arithmetic, logical manipulations or data transmission between a pair of digital devices. Some flags can be tested and used for determining subsequent actions.

Flowchart A graphical representation of an algorithm required to solve a problem.

Handshake Interactive communication over the control bus between two system devices, often required to prevent loss of data.

Hardware Physical mechanical, electrical, and/or electronic devices in a digital system.

Hexadecimal The base 16 number system, in which all numbers are expressed as powers of 16 and represented by the characters 0 through 9, A, B, C, D, E, and F.

Increment To increase the value of a binary word, usually by the value of 1.

Input/Output (I/O) devices Computer hardware or external devices by which data are entered into a digital system or by which data are sent from a digital system to be recorded by immediate or future use.

Interrupt A break in the normal process of a digital system such that the process can be resumed from that point at a later time. An interrupt may be internal or external.

Leading edge The transition of a pulse that occurs first.

Load In programming, to enter data into a specific memory location or element.

LSB or LSD (least significant bit or digit) The last bit or digit of a number which has the smallest numerical value.

Machine language code A numerical code that a digital device or system is designed to recognize.

Masking A software process that uses a bit pattern to select bits from a data byte for use in a subsequent operation. Masking is used to enable specific I/O port lines.

Mnemonic code Alphabetic characters representing machine language-coded instructions, which allow easy identification of the functions represented.

Monitor Software or hardware that observes, supervises, controls, or verifies a digital system operation.

MSB or MSD (most significant bit or digit) The first bit or digit of a number, which has the largest numerical value.

Nibble A sequence of four adjacent bits of a computer word, or half a byte (i.e., 1000 1111, 8 F).

Octal The base 8 number system, in which all numbers are expressed as powers of 8 and represented by the numerals 0 through 7.

Operand Data that are, or will be, operated on by an arithmetic or logic instruction.

Overflow A condition where the results of an arithmetic operation is too large to be held in memory. This condition often sets a flag in the status register.

Parity A method of checking the accuracy of binary numbers, especially after transmission. A parity bit may or may not be added to the word. If even parity is used, the sum of all 1's in the word will always be even. If odd parity is used, the sum of all 1's will always be odd.

Peripheral A device or subsystem external to the microprocessor that provides additional system capabilities.

Polling Periodic interrogation of each of the peripheral devices that share a common communications line to determine whether any require servicing.

POP Retrieving data from a stack.

Port A device or network of lines through which data may be transferred, observed, or measured.

Processor A shorthand word for microprocessor.

PUSH Putting data into a stack.

Radix Also called the base of a number system. It is the total number of distinct marks or symbols used in a numbering system (i.e., binary has a radix of 2, decimal has a radix of 10, and hexadecimal has a radix of 16).

Read To retrieve data from memory and transmit them to other digital devices.

Reset *See* clear.

Set To condition a flip-flop or device to a 1.

Single step Execution of a program one instruction at a time from the control on a computer. A method to debug programs and malfunctioning hardware.

Software A computer instruction set, programs, procedures and any associated documentation concerned with the operation of a data processing system.

Stack A specified section of sequential memory locations used for the automatic storage of program data, subroutine return addresses, and contents of important registers.

Stack pointer A register that keeps track of the memory locations of a stack. It is automatically incremented or decremented as instructions perform operations with the stack.

Symbolic code A mnemonic code that uses symbolic names and addresses that do not depend on their hardware-determined locations.

Trailing edge The transition of a pulse that occurs last.

Unconditional branch (or jump) A program instruction that interrupts the normal process of obtaining the instructions in an ordered sequence and specifies the address from which the next instruction must be taken.

Word The maximum number of bits that can be stored in a single memory location of a specific digital system. A microcomputer usually uses an 8-bit (one-byte) word.

Write To store data into a memory device from another digital device.